THE ARCHITECTURE OF
NATURAL COOLING

Overheating in buildings is commonplace. This book describes how we can keep cool without conventional air-conditioning: improving comfort and productivity while reducing energy costs and carbon emissions. It provides architects, engineers and policy makers with a 'how-to' guide to the application of natural cooling in new and existing buildings. It demonstrates, through reference to numerous examples, that natural cooling is viable in most climates around the world.

This completely revised and expanded second edition includes:
- An overview of natural cooling past and present.
- Guidance on the principles and strategies that can be adopted.
- A review of the applicability of different strategies.
- Explanation of simplified tools for performance assessment.
- A review of components and controls.
- A detailed evaluation of case studies from the USA, Europe, India and China.

This book is not just for the technical specialist, as it also provides a general grounding in how to avoid or minimise air-conditioning. Importantly, it demonstrates that understanding our environment, rather than fighting it, will help us to live sustainably in our rapidly warming world.

Brian Ford *(RIBA FRSA)* is an architect, an environmental design consultant and Emeritus Professor at the University of Nottingham. He was in private practice for over 25 years, including Peake Short & Partners and Short Ford Associates, where he has worked on innovative low carbon projects in Europe, USA, India, Australia and China. He initiated a series of multi-partner EU funded research projects on natural ventilation and passive cooling and has served as a member of the UK Government's Zero Carbon Task Force for Schools, an advisor to Building Green Futures (Bologna, Italy), and was until recently Vice-President of the International PLEA organisation.

Rosa Schiano-Phan *(BSc DiplArch MSc PhD)* is Principal Lecturer in Architecture and Environmental Design at the University of Westminster and co-director of Natural Cooling Ltd. She worked as an architect and environmental design consultant internationally before gaining a PhD at the Architectural Association and becoming a Research Fellow at the University of Nottingham, working on EU research into thermal performance of buildings and passive cooling. She taught at the AA and subsequently moved to the University of Westminster where in 2014 she set up new BSc and MSc courses in Architecture and Environmental Design.

Juan A. Vallejo *(BSc DiplArch MSc PhD)* is Visiting Lecturer in Architecture and Environmental Design (MSc) at the School of Architecture + Cities of the University of Westminster in London (UK) and collaborates with the School of Sustainability (SOS) Team founded by Mario Cucinella Architects in Bologna (Italy). He has worked in Natural Cooling Ltd as an environmental design consultant on projects in the UK and abroad and he is associate of the International Passive Low Energy Architecture (PLEA) organisation. His expertise lies in the field of building environmental simulation software tools, natural ventilation and passive evaporative cooling.

"Natural cooling is one of the most important aspects to tackle while designing, and surely one of the trickiest to master. Through a well-calibrated collection of theoretical principles and practical recommendations, all supported by a set of thoroughly analysed and presented case studies, this volume provides a reference book for the many professionals who, in response to global trends, are just approaching the matter, as well as for those in search for a deeper understanding of natural cooling principles and their possible practical applications."

Mario Cucinella
Architect, Hon FAIA, Int. Fellow RIBA
(Founder MCA Bologna and New York)

"This book brings together a uniquely comprehensive body of history, architectural science and contemporary practice that will stand as a primary source in the foreseeable future. The understanding of the specificity of climate lies at the heart of the book, serving as one of the foundations for natural cooling design. At the other pole of environmental design, human comfort is an essential parameter. Between these poles the book presents a systematic and comprehensive methodology for design that embraces precedent, science and technology."

Dean Hawkes
Emeritus Professor, Welsh School of Architecture
& Darwin College, Cambridge

THE ARCHITECTURE OF
NATURAL COOLING

SECOND EDITION

BRIAN FORD, ROSA SCHIANO-PHAN AND JUAN A. VALLEJO

Routledge
Taylor & Francis Group

LONDON AND NEW YORK

Second edition published 2020
by Routledge
2 Park Square, Milton Park, Abingdon, Oxon, OX14 4RN

and by Routledge
52 Vanderbilt Avenue, New York, NY 10017

Routledge is an imprint of the Taylor & Francis Group, an informa business

© 2020 Brian Ford, Rosa Schiano-Phan and Juan A. Vallejo

The right of Brian Ford, Rosa Schiano-Phan and Juan A. Vallejo to be identified as authors of this work has been asserted by them in accordance with sections 77 and 78 of the Copyright, Designs and Patents Act 1988.

All rights reserved. No part of this book may be reprinted or reproduced or utilised in any form or by any electronic, mechanical, or other means, now known or hereafter invented, including photocopying and recording, or in any information storage or retrieval system, without permission in writing from the publishers.

Trademark notice: Product or corporate names may be trademarks or registered trademarks, and are used only for identification and explanation without intent to infringe.

First edition published by PHDC Press 2010

Publisher's Note
This book has been prepared from camera-ready copy provided by the authors.

British Library Cataloguing-in-Publication Data
A catalogue record for this book is available from the British Library.

Library of Congress Cataloging-in-Publication Data
Names: Ford, Brian, 1949- author. | Schiano-Phan, Rosa, author. | Vallejo, Juan A., author.
Title: The architecture of natural cooling / Brian Ford, Rosa Schiano-Phan, Juan A. Vallejo.
Other titles: Architecture & engineering of downdraught cooling.
Description: Second edition. | Abingdon, Oxon ; New York : Routledge, 2020. | First edition published by PHDC Press 2010. | Includes bibliographical references and index.
Identifiers: LCCN 2019024744 (print) | LCCN 2019024745 (ebook) | ISBN 9781138629059 (hardback) | ISBN 9781138629073 (paperback) | ISBN 9781315210551 (ebook)
Subjects: LCSH: Buildings–Environmental engineering. | Buildings–Thermal properties. | Architecture and climate. | Evaporative cooling. | Air-conditioning–Design and construction–Environmental aspects.
Classification: LCC TH6025 .F66 2020 (print) | LCC TH6025 (ebook) | DDC 697–dc23
LC record available at https://lccn.loc.gov/2019024744
LC ebook record available at https://lccn.loc.gov/2019024745

ISBN: 978-1-138-62905-9 (hbk)
ISBN: 978-1-138-62907-3 (pbk)
ISBN: 978-1-315-21055-1 (ebk)

Typeset in Futura
by Juan A. Vallejo

CONTENTS

PREFACE	x
ACKNOWLEDGEMENTS	xiv
COOLING WITHOUT AIR CONDITIONING	xvi
FOREWORD	xviii

PART 1

CHAPTER 1 | ORIGINS AND OPPORTUNITIES — 2
- 1.1. Origins — 3
- 1.2. Opportunities — 12

CHAPTER 2 | PRINCIPLES AND STRATEGIES — 24
- 2.1. Site & microclimate analysis — 25
  - Urban morphology — 25
  - Urban heat islands — 25
  - Effects of vegetation — 27
  - Traffic noise and air borne pollution — 28
  - Solar radiation effects — 29
  - Wind effects — 30
- 2.2. Natural cooling design strategies — 32
  - External air – natural convection — 32
  - Night-time ventilation — 33
  - Night sky radiation — 34
  - Ground cooling — 36
  - Evaporation — 38
- 2.3. Airflow strategies: updraught — 40
  - Single sided — 40
  - Cross ventilation — 40
  - Central atrium/concourse (open or enclosed) — 41
  - Perimeter stack (ventilated façade) — 43
- 2.4. Airflow strategies: downdraught — 43
  - Central atrium or concourse — 43
  - Central shaft or concourse — 45
  - Perimeter tower — 45
- 2.5. The cooling source for downdraught systems — 47
  - Porous media — 47
  - Shower towers — 49
  - Misting systems — 51
  - Active downdraught cooling — 53

CHAPTER 3 | CLIMATE APPLICABILITY MAPPING — 56
- 3.1. Interpreting weather data — 57
- 3.2. Applicability of passive cooling methods — 59
  - Natural convective cooling — 60
  - Passive evaporative cooling — 62
  - Active downdraught cooling — 63
- 3.3. Passive cooling applicability mapping — 65
  - Applicability maps for Europe — 66
  - Applicability maps for China — 68
  - Applicability maps for the USA — 70

CHAPTER 4 | INTEGRATED DESIGN — 76
- 4.1. Why integrated design? — 77
- 4.2. Refurbishment options for existing buildings — 82
  - Urban morphology and cooling options — 83
  - Stock analysis in southern Europe — 87
- 4.3. Integration in existing buildings — 88
- 4.4. Integration in new buildings: temperate climates — 94
  - Natural ventilation design — 94
  - Acoustic attenuation on airflow paths — 100
- 4.5. Integration in new buildings: humid tropics — 102
- 4.6. Integration in new buildings: hot dry climates — 108

CHAPTER 5 | TESTING THE STRATEGY — 116
- 5.1. Heat gains — 117
  - External heat gains — 117
  - Internal heat gains — 118
  - Total heat gains — 119
- 5.2. Defining the air volume flow rate required — 120
- 5.3. Sizing aperture areas for stack-driven cooling — 121
- 5.4. Sizing an evaporative cooling system — 124
  - Misting systems — 124
  - Porous materials — 128
  - Active downdraught cooling — 133

CHAPTER 6 | PERFORMANCE ANALYSIS	138
6.1. Performance criteria: thermal comfort	139
6.2. Steady-state methods	140
Solar control	140
Natural ventilation: Optivent tool	145
6.3. Dynamic methods	150
Modelling evaporative cooling	154
6.4. Computational fluid dynamics (CFD)	156
CHAPTER 7 | COMPONENT DESIGN	160
7.1. Effective integration	161
7.2. Ventilator design	162
Improvements in actuator design	165
7.3. Shafts, ducts and cool towers	166
Supply & exhaust air termination conditions	166
Wind baffles	168
7.4. Controls	169
Flow rates through ventilators	170
Combined user and automated control	170
7.5. Water systems	171
Water consumption in evaporative cooling	171
Water treatment	172
Low and high pressure misting systems	172
Maintenance of nozzles	174

PART 2

INTRODUCTION	176
CASE STUDY 1 | TORRENT RESEARCH CENTRE, AHMEDABAD, INDIA	180
CASE STUDY 2 | CENTRE FOR SUSTAINABLE ENERGY TECHNOLOGIES, NINGBO, CHINA	194
CASE STUDY 3 | FEDERAL COURTHOUSE, PHOENIX, ARIZONA, USA	212
CASE STUDY 4 | STOCK EXCHANGE, VALLETTA, MALTA	224
CASE STUDY 5 | ZION NATIONAL PARK VISITOR CENTRE, SPRINGDALE, UTAH, USA	238
CASE STUDY 6 | GLOBAL ECOLOGY RESEARCH CENTRE, PALO ALTO, CALIFORNIA, USA	252
CONCLUSIONS	268
IMAGE CREDITS	274
INDEX	278

PREFACE

Background

The construction industry is notoriously conservative and slow to change, and until recently, only the largest companies have engaged in research. However, concerns over global warming and the need for sustainable development have led to substantial investment (particularly in Europe) into the research and development of technologies and design approaches which will reduce our dependency on fossil fuels. The application of these techniques to building projects has spread around the world, in parallel with the research programmes, and is slowly becoming part of the mainstream, but there is a perceived knowledge and skills gap among construction professionals, which is seriously restricting further take-up. There is widespread understanding of the 'need', but a lack of widespread 'know how'. The transfer of knowledge from research into practice is therefore vital. But this is difficult in an industry which habitually dislikes change and which has traditionally undervalued research.

Over the last 30 years, developments in materials science, building energy technologies and design tools have been derived from a significant increase in research investment into energy efficiency in buildings. Major advances have been made in both our understanding and the application of bioclimatic design principles in buildings, and similar advances have been made in new and renewable energy devices, and their integration in buildings. Strategic thinking about these issues needs to be part of the early design process, and rigorous testing of options (using a wide range of analytic tools) is required in order to give clients the confidence to proceed with a radical solution. Not all projects can afford specialist testing and analysis, but simplified techniques exist, and this is where the *'know how'* really counts.

It is important for us to look at the scale of the task that faces us, in terms of reducing carbon emissions and their impact on global warming. Atmospheric carbon today constitutes approximately 410ppm (January 2019) compared with the pre-industrial concentration of 280ppm (pre 1850). Human activities are estimated to have already caused approximately 1.0°C of global warming above pre-industrial levels. It is clear that a radical reduction of global carbon emissions is necessary to avoid going above 1.5°C and the risks of catastrophic climate change. The IPCC's Special Report *(IPCC, 2018)* indicates that to limit global average rise to 1.5°C, global net anthropogenic CO_2 emissions must decline by about 45% from 2010 levels by 2030, reaching net zero around 2050.

The built environment has a huge impact on the level of emissions, and everyone has a part to play. For an energy strategy in buildings to be based on a combination of efficiency measures and the integration of renewables, both professional practice and education within the construction industry must change. The transfer of knowledge gained from recent research, into practice, must form part of this process. But in order for carbon emission targets to be achieved, gaps in knowledge and skills also need to be addressed among all who work in the industry. This book is intended as a contribution to this process.

The benefits of natural cooling

Global demand for cooling is increasing at a spectacular rate. In 2010–11 world sales of air-conditioning went up by 13% *(Cox, 2010)*. Data from 2016 indicates that in the USA 87% of all buildings are air-conditioned *(EIA, 2012)*, and that air-conditioning represents 42% of the peak demand for electrical power. Investment in renewables is increasing, but new fossil fuel power stations are still coming on stream every year, and demand side efficiencies have a huge contribution to make to reducing greenhouse gas emissions from air-conditioning.

Alternatives to conventional air-conditioning are needed urgently. The rise in demand for air-conditioning in the USA, and the current dependency on it, is unsustainable. While in Europe and the USA peak load demand for air-conditioning threatens to disrupt supply, in India, China, Africa and parts of the Middle-East supply disruption is a regular occurrence, forcing many large consumers to invest in expensive back-up generating equipment. Summer demand for power often outstrips supply, resulting in rationing and the closure of factories and offices. Additionally, the use of local generators impacts severely the air quality thus limiting the potential for natural ventilation and increasing the anthropogenic waste heat dumped in urbanised areas. Alternatives to conventional air-conditioning are therefore urgently required to reduce peak load demand for electricity, and thus reduce the risk of supply disruption. Since the alternatives also exploit ambient heat sinks and are inherently 'low carbon', they will also reduce the need for investment in supply infrastructure.

This book presents both the historical tradition and contemporary practice in the design and application of natural (passive) cooling solutions in different parts of the world. Historically, many different strategies and techniques have been used to promote passive cooling during extreme summer heat. Evaluation of the contemporary applicability of these cooling strategies in different climate zones has generated guidance for practitioners *(Givoni, 1994; Santamouris, 2007)*. Improved understanding of performance issues along with post occupancy evaluation of buildings has led to greater confidence in our ability to anticipate user needs, and to avoid some of the pitfalls. Technical developments have expanded the range of options and the range of building types in which passive cooling can be successfully applied.

It is now a viable alternative to conventional air-conditioning in many parts of the world and is beginning to enter the mainstream.

Concerns over global climate change, our dependency on fossil fuels, and the implications of 'peak oil' (Hirsch et al., 2005) are changing the way we view our use of mechanical air-conditioning. This change is no longer taking place at the margins. Policy makers and politicians are now joining academics, environmentalists and built-environment professionals, to promote improvements in building performance and alternatives to mechanical air-conditioning.

There is now a need to both improve building performance so that demand for cooling is reduced, and to develop cost competitive alternatives to conventional air-conditioning to meet the residual cooling need. In Europe, the EU Performance of Buildings Directive (EPBD) promotes the alignment (and improvement) of building performance regulations in member countries. This has led to a radical reduction in the need for heating and cooling in new buildings, although air-conditioning use is still increasing in new buildings. Also, demand for cooling continues to increase in relation to the existing building stock, and so viable alternatives to conventional air-conditioning have a major contribution to make. Elsewhere in the world, policies designed to reduce demand for cooling (through improved building performance) have become a priority, and in many situations natural cooling is providing the low-carbon design-based solutions required for our future.

REFERENCES

- Cox, S. (2010). *'Losing our Cool'*. The New Press, New York.
- EIA. (2012). 'Commercial Buildings Energy Consumption Survey (CBECS)' [Online]. https://www.eia.gov/consumption/commercial/. [Accessed 05/2017].
- Givoni, B. (1994). *'Passive and Low Energy Cooling of Buildings'*. Van Nostrand Reinhold, pp.139–143.
- Hirsch, R., Bezdek, R. & Wendling, R. (2005). *'Peaking of World Oil Production: Impacts, Mitigation, and Risk Management'*. US Department of Energy: 1–91.
- IPCC. (2018). *'Special Report: Global Warming of 1.5°C'* [Online]. https://www.ipcc.ch/sr15/chapter/summary-for-policy-makers/. [Accessed 02/2018].
- Santamouris, M. (2007). *'Advances in Passive Cooling'*. Earthscan.

ACKNOWLEDGEMENTS

Much of the content of this book has its origins in a series of projects which were funded under the European Commission's 5th & 6th Frameworks for Research & Development, and in this regard the authors would like to thank and acknowledge the assistance of: Professor Servando Alvarez, Dr Jose Molina and Dr Jose Salmeron of AICIA, University of Seville, Spain; Joanna Spiteri-Staines and Alberto Miceli-Ferrugia of AP architects, Malta; Jim Meikle and Paul Thomas of Davis Langdon Consultancy, London; Mario Cucinella, MC Architects, Bologna & New York, and Elizabeth Francis, formerly of MC Architects and now Atelier Francis, Bologna, Italy; Q. Xu of the Shanghai Research Institute for Building Science, Shanghai, China; Parul Zaveri and the late Nimish Patel of Abhikram Architects, Ahmedabad, India; Dr Evyatar Erell, Bengurion University of the Negev, Israel.

In writing the historical background (Chapter 1) we would like to acknowledge the work of: Dr Raha Ernest; Mark Hewitt, ICAX, London, UK; Dr Simos Yannas, AA Graduate School, London, UK; Prof Dean Hawkes, Cambridge; Benson Lau, University of Westminster, London, UK; Zhang Hongru and Dr Henrik Schoenfeldt, University of Kent, Canterbury, UK.

Chapters 2, 4 and 7 refer to the work of many architects and engineers including: Enrique Browne Architects, Santiago, Chile; Professor Edward Ng, HKUP, China; Estudio Carme Pinos, Spain; Bennetts Associates, London, UK; Renzo Piano Building Workshop, Genoa, Italy; ARUP engineers, London, UK; Atelier 10, London, UK; Nina Maritz Architects, W Namibia; Raphael Vinoly Architects, New York, USA; Pringle Richards Sharratt, London, UK; Richard Weston Architect, Cardiff, UK; Karan Grover Architects, Baroda, India; Design Inc, Melbourne, Australia; A.P. Architects, Valletta, Malta; Richard Meir Architects, USA; Transsolar Climate Engineers, Stuttgart, Germany; Foster & Partners, London, UK; Sauerbruch Hutton Architects, Berlin, Germany; Edward Cullinan Architects, London, UK; Mario Cucinella Architects, Bologna, Italy; Marsh Grochowski Architects, Nottingham, UK; WOHA architects, Singapore; and Ingeniatrics/Frialia, Seville, Spain.

In writing the chapters on climate mapping (3), testing the strategy (5) and performance analysis (6) we would like to thank Dr Jose Salmeron, AICIA, University of Seville, Spain; Dr Pablo Aparicio-Ruiz, University of Seville, Spain; Dr Huang Xuan, SJT University, China; Dr Camilo Diaz, formerly of BFA and now with WSP Group, London UK; Dr Maureen Trebilcock of University of

BioBio, Concepcion, Chile; and the Centre for Built Environment, University of California, Berkeley, USA.

Part 2 describes the detailed application of natural cooling principles and strategies in a series of case study buildings which also explore performance in practice. The compilation of these case studies would not have been possible without the assistance of both the designers and facilities managers. We therefore wish to thank:

Case Study 1: Dr Dutt, Director of the Torrent Research Centre, Nimish Patel and Parul Xaveri of Abhikram Architects; Ramesh Borad, Engineer and Systems Manager Torrent Research Centre; George Baird of University of Wellington; Lisa Thomas of University of Technology, Sydney and Adrian Leaman of Building Use Studies, UK.

Case Study 2: Mario Cucinella Architects; Professor Jo Darkwa, University of Nottingham, UK; Dr Huang Xuan and Dr Mingwei Sun, formerly of University of Nottingham.

Case Study 3: Mr Kevin Winschel, Project Manager, General Services Administration, Phoenix and Mr Mahadev Raman, Arup New York.

Case Study 4: Eileen Muscat of the Malta Stock Exchange; Joanna Spiteri-Staines, A.P. Architects and Steven De Bono MTS Ltd, Malta.

Case Study 5: Mr James Lutterman, Zion National Park, Utah and Mr Paul Torcellini, National Renewable Energy Laboratory, Colorado.

Case Study 6: Chris Field and Linda Longoria; Joe Berry, building manager and Ari Kornfeld research associate; EHDD architects and Rumsey Engineers.

We are very grateful to Robin Nicholson & Roddy Langmuir of Edward Cullinan Studios, Peter Fisher and Ben Hopkins of Bennetts Associates, Dr Nick Baker, Prof Sue Roaf and Prof Fergus Nicol for commenting on key parts of the text. Finally, we would like to thank Mario Cucinella for his overview and Professor Dean Hawkes for setting this work in context with his Foreword.

COOLING WITHOUT AIR-CONDITIONING

As the world around us is changing, so too should architecture. Indeed, today, professionals find themselves at a major crossroad. At stake is the question of how we should engage with the available resources, while also pursuing comfort, health, beauty and innovation. The global lack of resources and climate change are, in fact, the fundamental design problems of our time. With buildings accounting for some 30% of both total global energy use and CO_2 emissions (according to the UN Global Status Report, 2015) and Europe planning to achieve, despite the alarming population-growth and urbanisation trends, an incredibly ambitious target of an 80% reduction in carbon emissions already by 2050 (Roadmap 2050, commissioned by the European Climate Foundation), architects cannot afford to improvise any longer.

In this compelling transition towards a low-carbon future, we are now absolutely forced to develop a radically new approach to architecture, thus minimising the use of energy-intensive technologies and reviving some low-tech solutions such as passive heating, cooling and, of course, ventilation. Such solutions can only be inspired and supported by a deeper knowledge and connection with the climate, the culture and the natural environment of a place. It is in this context that vernacular architecture, with people developing clever ways to address their needs and achieve protection and comfort from little or no available resources, can offer a reference that is worth mentioning. If Marco Polo could, to his surprise, enjoy an ice-cream during one of his travels to the Far East in the 13th century, thanks to a local construction typology able to store ice in the middle of the desert, we will probably, with the same surprise, constantly discover a wide array of passive local solutions able to effectively support and strengthen each of our design proposals and lead them towards their net-zero-energy goal. In other words, we are now chasing a sustainable future whose roots might still somehow lie in the past.

It is against this background, and given the possibilities offered by a more empathic relationship with the context in which they are called to take action, that professionals need to commit themselves to a deep paradigm shift, thus leading to an actual revolution in building design and construction. And to achieve this fundamental and urgent change, architects must acquire a better understanding of the principles of environmental design, while also developing a clear and systematic methodology allowing for their proper application at all design stages, from the very conceptual ones to advanced component design.

Natural cooling is definitely one of the most important aspects to tackle while designing (especially in hot climates) and surely one of the trickiest to master. Through a well-calibrated collection of theoretical principles and practical recommendations, all supported by a set of thoroughly analysed and presented case studies, this volume can therefore represent a reference book for the many professionals that, driven by the aforementioned global trends, are just approaching the matter, as well as for those in search for a deeper understanding of natural cooling principles and their possible practical applications.

Mario Cucinella

Architect.
Founder and Principal Architect at
"Mario Cucinella Architects"
(based in Bologna and New York).
Hon FAIA, Int. Fellow RIBA.

FOREWORD

01. Allegorical engraving of the Vitruvian primitive hut. Frontispiece of Marc-Antoine Laugier: Essai sur l'Architecture 2nd ed. 1755 by Charles Eisen (1720-1778).

The image of the *'Primitive Hut'*, the frontispiece of the Abbé Laugier's Essai sur l'architecture (1755) (fig. 01), beautifully represents one of the most fundamental purposes of the art of building, that is to provide shelter from the rigours and unpredictability of the natural environment. The transition from the unselfconscious building practices of the primitive to the self-conscious art of architecture was marked by the codification of the knowledge implicit in such buildings to allow it to be consciously applied to the design of buildings. In the western world, one of the first such codifications was Vitruvius' De architectura (The Ten Books on Architecture) *(Vitruvius, 1960)*. Written in the first century BC, this provided guidelines for the design of buildings for the wide range of climates embraced by the geographical span of the Roman Empire, from the chill, northerly latitudes of the British Isles to the warmth of the Mediterranean. In Book VI, Vitruvius directly addressed the relation of climate and building.

In the north houses should be entirely roofed over and sheltered as much as possible, not in the open, though having a warm exposure. But on the other hand, where the force of the sun is great in the southern countries that suffer from heat, houses must be put more in the open and with a northern or north-western exposure. Thus, we may amend by 'art' what nature, if left to herself, would mar.

Some thirty years ago I derived from De architectura a 'Vitruvian tri-partite model of environmental design' in which the environmental function of architecture is, as with the Primitive Hut, to mediate between, on the one hand, climate and, on the other, a notion of comfort *(Hawkes, 1996)*. In setting out such clear and well-founded principles for design, Vitruvius' 'art' is the distant and distinguished antecedent of the methods of modern architectural science. In the twenty-first century the relationship between climate and architecture has acquired new and significant relevance as, in response to global environmental challenges, we seek to make buildings that achieve a new and necessarily more harmonious partnership with the natural environment. The present book stands as an important contribution to this growing and significant line of environmental theory and practice.

To establish the background, let's paraphrase the history of environmental design in architecture. For almost two millennia, from Vitruvius up to the end of the eighteenth century, the environmental, sheltering function of architecture was achieved, in all climates, by the organisation of the form and material of the building enclosure in relation to the conditions of climate. In The Architecture of the Well-tempered Environment *(Banham, 1969)*, Rayner Banham proposed a taxonomy of historical modes of *'environmental management'*, the *'conservative'* and the *'selective'*, that were defined by their relations of material and form to climate. In defining a third *'mode'*, the *'regenerative'*, Banham recognised the transformation that occurred at the end of the eighteenth century, with the innovations of new technologies for heating and ventilating and then of artificial lighting that came with the industrial revolution in Europe. Now buildings became less intimately connected to the natural environment as they could be warmed on cold days and lit after dark. In the twentieth century, after mechanical cooling systems were developed, it became possible to make the interiors of buildings cool in the hottest of climates. In a Darwinian process of architectural evolution, the new technologies were, for the whole of the nineteenth century and for the first decades of the twentieth, applied to buildings that were, in their fundamental conception, *'pre-mechanical'*, with their form still influenced by the requirements of natural illumination and, often, of natural ventilation. In other words, these buildings remained connected to the natural environment in which they were placed. But, at some point in the middle years of the twentieth century, the balance shifted, and buildings became deeper in plan and their enclosing envelopes, often diaphanous glass skins, became sealed as mechanical cooling and ventilation and artificial lighting provided all the elements of their internal environment. Technology had transcended nature. This strategy soon became universal and cities on all continents, with their clusters of glass skyscrapers in diverse climates, became almost indistinguishable one from another.

Since the middle of the twentieth century an alternative line of environmental thought has emerged in architecture. One of the first and most comprehensive expressions of this was Victor Olgyay's book, Design with Climate (Olgyay, 1963). The title, with its emphasis on designing with not against climate, stands in clear distinction from the predominant, mechanically dominated approach of conventional practice. In addition, the book was one of the earliest expressions of the idea of regionalism in architectural discourse, predating by twenty years Kenneth Frampton's seminal essay on the subject (Frampton, 1983). Here was a view of architecture working in harmony with nature and responsive to the challenges and opportunities of place, of each unique climate, to make buildings fit for the complex needs of modern society. In the half century since the publication of Olgyay's book, the idea of passive environmental design has gained wide support in both theory and practice, and there is now an extensive literature in the field that embraces a vast body of new research and practice that is applied to buildings of all types and in all climates.

It is against this background that the present book takes its place. In the array of passive possibilities in architecture the potential of passive cooling is particularly important. As the book shows, this is one of the oldest means of achieving comfort in buildings in hot places, with precedents from at least the thirteenth century in the Middle East, and there is a continuous history of its application in buildings of diverse function down to the present day. The authors have a distinguished record in the field, both as scholars and practitioners. In this book, they bring together a uniquely comprehensive body of history, architectural science and contemporary practice that will stand as a primary source in the foreseeable future. As with both Vitruvius and Olgyay, the understanding of the specificity of climate lies at the heart of the book, with climates as diverse as those of Europe, China and the USA, and the immense variations within these territories, serving as one of the objective foundations for natural cooling design. At the other pole of environmental design, human comfort is once again an essential parameter. Between these poles the book presents a systematic and comprehensive methodology for design that embraces precedent, science and technology. All this is reinforced by the six detailed Case Studies, gathered from around the globe, that illustrate the process by which theory becomes form, the final and most persuasive test.

Dean Hawkes

Emeritus Professor.
Welsh School of Architecture &
Darwin College, Cambridge.

REFERENCES

- Banham, R. (1969). 'The Architecture of the Well-tempered Environment'. The Architectural Press, London.
- Frampton, K. (1983). 'Towards a Critical Regionalism: Six Points for an Architecture of Resistance'. in Hal Foster (ed.), 'Postmodern Culture', Pluto Press.
- Hawkes, D. (1996). 'The Environmental Tradition'. E & FN Spon, London.
- Olgyay, V. (1963). 'Design with Climate: Bioclimatic Approach to Architectural Regionalism'. Princeton University Press, Princeton.
- Vitruvius, M. (1960). 'De Architectura (The Ten Books on Architecture)', M.H. Morgan (ed.), Dover Publications, New York.

PART 1

CHAPTER 1
ORIGINS AND OPPORTUNITIES

The natural cooling of buildings developed empirically over many centuries, based on an understanding of seasonal and diurnal changes in the local climate, and through a process of trial and error, to provide relief from the extreme heat of summer. Different techniques developed in response to local conditions, often reflecting a profound understanding of the local environment. Anecdotal evidence of how these buildings worked has been available for many years, but it is only recently that significant research has contributed to a more detailed understanding worldwide. This increased interest in the origins of climate-responsive architecture is reflected in the many journal papers on this tradition, in the annual PLEA Conferences, and in recent books *(see Weber & Yannas, 2014; and Hawkes, 2012)*. This research is not just of academic interest, as the move to a low carbon future is now a major driver among design professionals globally. Many buildings now demonstrate that a 'climate aware' approach to design is both practical and cost effective. At the very least it can reduce dependency on mechanical air-conditioning, and in many locations can be completely avoided.

1.1. ORIGINS

The tradition of 'Natural Cooling', which incorporates a range of design responses to climate, has its origins in Egypt, where frescoes from the 13th century BC (fig. 01) depict buildings with a *'malqaf'* (traditional windcatcher) used to help ventilate and cool the interior (figs 02–03) *(Fathy, 1986)*. This approach subsequently spread eastwards as part of the Islamic tradition through the Middle East and Iran to north India (with the Moghul empire), and westwards across North Africa to southern Spain with the Moors.

This tradition, which has been largely overlooked, is characterised by dramatic achievements. Travelling through the Iranian desert on his journey to China in the 13th century AD, Marco Polo commented on his being offered fruit flavoured water-ices to relieve the summer heat. The creation and storage of ice in the desert was made possible at the time through the construction of huge natural refrigerators. These typically consisted of a shallow pool, protected and shaded on its south side by a huge earth brick wall, and connected to a domed ice storage pit (figs 04–05). These shallow pools were provided with water from man-made underground water conduits *'qanat'* bringing water sometimes hundreds of miles from surrounding mountains to the desert *'caravanserai'* along the silk-route to China *(Beazley & Harverson, 1986)*.

Under the clear night sky of the desert, long wave radiation from ground to sky causes surface temperatures to drop low enough to

01. Wall painting from the tomb of Nebamun. Egyptian Dynasty 18, about 1475 BC.

02. Traditional windcatcher *malqaf*, Dubai.

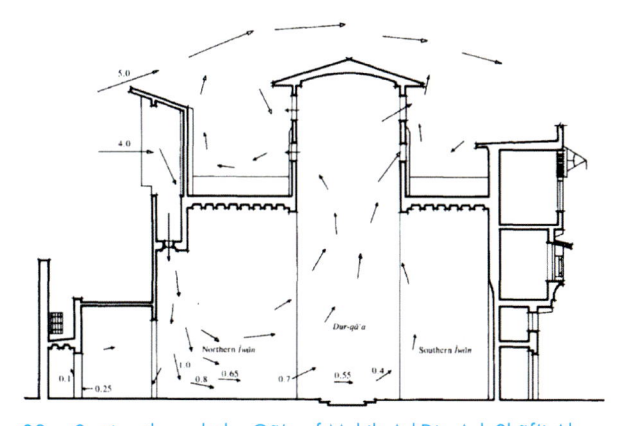

03. Section through the Qā'a of Muhib Ad-Din Ash-Shāf'i Al-Muwaqqi, showing how the malqaf and wind-escape produce internal air movement. All wind and airspeeds are given in metres per second.

CHAPTER 1 | ORIGINS AND OPPORTUNITIES   3

freeze a thin layer of water, introduced to the shallow pool from the *qanat*. On successive nights the depth of ice built up until it was about 300mm thick. The ice was then cut up and stored below ground level in a massive domed ice house. The temperature of the surrounding walls of the ice pit, al-

> UNDER THE CLEAR NIGHT SKY OF THE DESERT, LONG WAVE RADIATION FROM GROUND TO SKY CAUSES SURFACE TEMPERATURES TO DROP

though several degrees above freezing, were low enough and sufficiently stable to store the ice for many months. In this way, the extremes of the environment were turned to the advantage of the people living there. This reflects an attitude of respect and symbiosis between people and their environment, working with it rather than against it, to enhance the quality of their lives.

This understanding is elaborated further in a long tradition in the Middle East of using various techniques to encourage air movement and natural cooling both within and between buildings. Water jars mounted in specially designed window openings in Muscat, Oman (fig. 06), cool the air passing over them into the room by the process of evaporation, at the same time keeping the water at a stable temperature. These windows have a sophisticated design which simultaneously allowed for shading, through the external shading, control of the evaporative cooling effect, through the internal shutters and stack ventilation, through the low and high level openings. In a similar manner, woven '*khus*' matting suspended over window and door

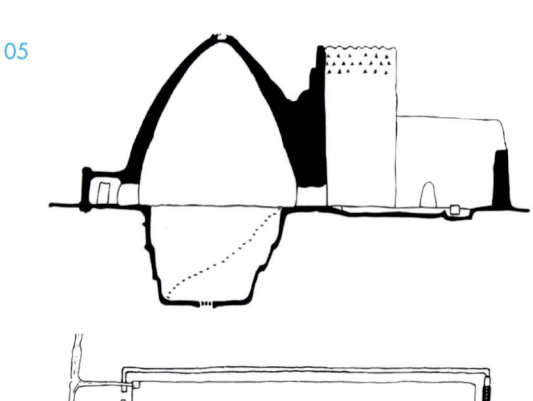

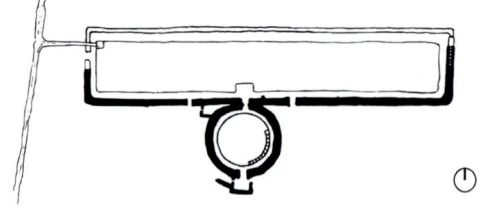

04. General view of the Ice House in Yazd, Iran.
05. Section and plan of the Ice House in Yazd, Iran.

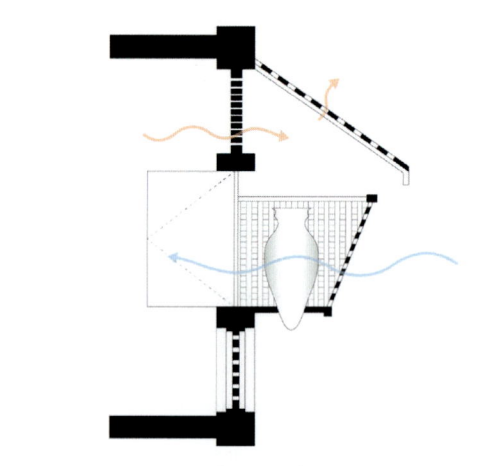

06. Porous ceramic cooling window in Muscat, Oman.

4 **CHAPTER 1** | ORIGINS AND OPPORTUNITIES

openings are still commonly used in parts of northern India. *Khus* mats are made from the root of a plant from the Jasmin family, and add a delightful fragrance to the air as it passes into the interior.

Passive cooling and ventilation of buildings in Iran, incorporating wind catchers, porous water pots and *'salsabil'* (figs 07–08), have been widely applied and very effective for several centuries. In this tradition, wind-catchers guide outside air over water-filled porous pots, inducing evaporation and bringing about a significant drop in temperature before the air enters the interior. Hassan Fathy was also aware of this tradition in Egypt, adapting, developing and re-applying these techniques to cool and ventilate schools and housing projects *(Fathy, 1986)*. In a field study which Fathy describes, on measurements of temperature and air velocity within a house with '*malqafs*', the pattern of airflow and the benefits of enhancing air movement within the building are illustrated.

The movement of cool air between adjacent courtyards is remarked upon by Fathy, who refers to the tradition of promoting natural convection between adjacent courtyards having different hygro-thermal properties. This strategy was so effective that a pavilion placed between the two courtyards became a highly desirable location during the heat of the summer day. This space was known as '*taktabus*' in Arabic and as '*tablinum*' in the Roman tradition. This natural cooling strategy is embodied in many residences across the southern Mediterranean and north

07. Sloping *salsabil* in Red Fort, Delhi, India.
08. Salsabil and lotus pool in Red Fort, Delhi, India.

CHAPTER 1 | ORIGINS AND OPPORTUNITIES

Africa from the 5th century BC to the present day.

In Seville in southern Spain, a sixteenth century house, the Casa de Pilatos, incorporates this strategy within its design (figs 09–10). Seville (latitude 37°N) experiences mild winters and hot dry summers (mean max air temperature of 37°C in July, with relative humidity of 35%). The Casa de Pilatos is a mixture of Italian Renaissance and the Spanish 'Mudéjar' tradition (Irwin, 2004). Mudéjar was the term given to the architecture which (after the Moors had been ejected) had inherited Islamic features.

> A PAVILION PLACED BETWEEN THE TWO COURTYARDS BECAME A HIGHLY DESIRABLE LOCATION DURING THE HEAT OF THE SUMMER DAY

From the outer court, the visitor passes through a protected arrival gateway into the central formal paved courtyard, around which the building is organised. The whole courtyard is paved, and the visitor naturally keeps to the shade of the cloisters, lined with beautiful tiles 'azulelos'. Glimpses of green are obtained through grilled openings 'rejas' in the narrow rooms surrounding the courtyard, and on entering, the cool of the interior provides relief and where window seats invite a moment of rest to view the adjacent courtyard, full of an almost tropical greenery: palms, citrus trees, flowers and a central fountain (figs 11–12). The seated visitor feels a gentle cooling breeze across face and arms, as air moves from the cool green courtyard to the central paved courtyard. The movement of air and the cooling it induces is not just a happy accident. The building has been designed to exploit this phenomenon.

The central courtyard, being entirely paved, and with no greenery at all, was heating up under the burning sun (the surface temperature of the sunlit paving could reach 50°C+). This caused a plume of warm air to rise from the central courtyard, pulling fresh air from the significantly cooler green courtyards on either side, through the openings in the narrow ground floor rooms, and providing relief to the occupants in the process. Due to the thermal capacity of the paving in the courtyard, this effect would continue through the night as well. The variable influence of the wind would be minimised due to the central urban location of this house, so it is highly likely that these localised thermal differences in temperature would dominate in driving air through the building.

It is evident from the layout of the house that the ground floor rooms were occupied during the summer, and the family moved up to the first floor in winter, to occupy south facing rooms overlooking the gardens. This 'nomadic' adaptation to the changing seasons has been common in many cultures, to cope with the extremes of climate.

The Casa de Pilatos raises many questions: was this natural cooling strategy reliable? Did it result in significant cooling of the building interior? How did it respond to changing conditions of wind and sun? Research has demonstrated that the strategy was reliable in providing thermal comfort for the building's occupants, and in achieving significant cooling *(Ernest, 2011)*. The research has shown that the airflow pattern and the convective cooling achieved are driven by the thermal differences between the courtyards, and there is little correlation with wind speed or direction, demonstrating that the strategy is robust under varying environmental conditions *(Ernest & Ford, 2012)*. The controlled grilled openings' size, the scale and proportion of the spaces and the careful integration of different techniques such as shading, evapotranspiration and thermal mass all contribute to the successful cooling of this building in summer.

In north India the Mughal palaces and gardens exploited evaporative cooling to provide thermal relief and to delight the senses. Thin water chutes '*salsabil*' and other evaporative cooling techniques were features of Mughal architecture from the 13th to the 17th centuries. The intense dry heat and dust of the summer in north India calls for the creation of an internal 'refuge' or haven from the extremes of the external world. The diurnal swing in temperature is '*dampened*' by the mass of stone and earth, and the air is further cooled by the evaporation of water in the ventilation airflow path. The coolness of their interiors and the use of evaporative cooling has been widely commented on, but it was only relatively recently that measurement of conditions within these buildings was undertaken *(Ford & Hewitt, 1996)*. Results of measurements in a number of step wells, well-houses '*baoli*' and bath houses '*hammams*' reveal that they were very suc-

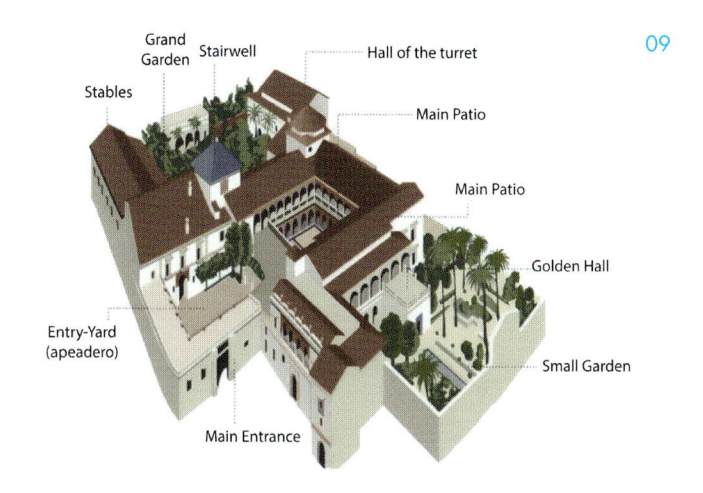

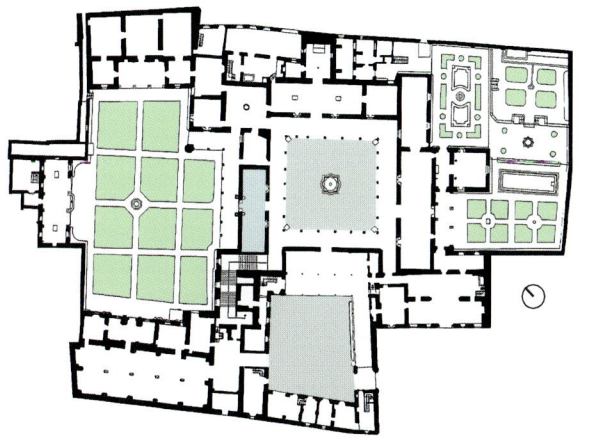

09. 3D view of Casa de Pilatos, Seville, Spain.
10. Floor plan of Casa de Pilatos.
11. Green courtyard in Casa Pilatos.
12. Window seat next to open grille in Casa Pilatos.

CHAPTER 1 | ORIGINS AND OPPORTUNITIES

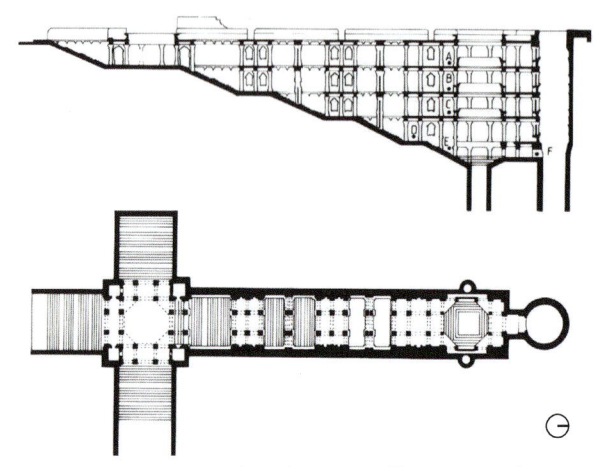

13. Plan and section of Adalaj Step Well, Gujarat, India.

14.

15.

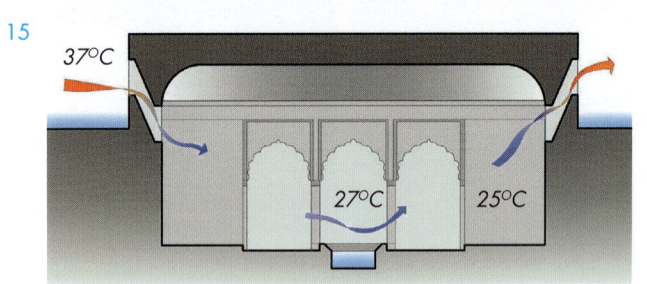

14. Rai Praveen Mahal, Summer Room.
15. Section of Summer Room, Rai Praveen Mahal.

cessful in providing shelter from the extremes of summer heat.

The '*step well*', which can still be found very widely in Gujarat and Rajasthan, was constructed as a source of water, a shrine, and a meeting place and retreat in the hot season. Usually strictly aligned on a north-south axis, the structure consists of a series of broad flights of steps and covered platforms leading down to the water. In contrast, the '*well-house*' consists of a group of retreat rooms arranged around the well shaft, and sometimes on a number of levels. The '*bath house*', imported by the Moghuls from Persia and comprising a series of interlinked chambers, is usually set above ground but constructed from massive stone or earth brick walls supporting vaulted '*double*' roofs.

The main staircase of the Adalaj step well is over 70 metres long, falling in five flights to a depth of 20 metres below ground (fig. 13). At each level a pavilion structure provides bracing to the retaining walls of the staircase, as well as cool retreats and meeting places, becoming progressively cooler as you descend. Measurements made in April 1995 showed a progressive lowering of the air temperature from ground level (35°C) to the lowest level (23°C), and progressive increase in absolute humidity (2.8gr/kg above external). Within thermally massive structures coupled to the earth (which also have low air change rates), one can expect stable air temperatures close to the annual mean external air temperature. However, in this case the very low air temperature (4 degrees below the annual mean) reflects

CHAPTER 1 | ORIGINS AND OPPORTUNITIES

the additional effect of evaporative cooling. The porous stone walls of the well act like a wick in drawing up water which then evaporates, absorbing the latent heat in the water and thus reducing the air temperature while raising the internal absolute humidity. The Adalaj step well has been providing a cool retreat for locals and travellers for over 500 years, and continues to do so today.

The design strategy of combining high thermal capacity with evaporatively cooled air was adopted in many Mughal era buildings, and is exemplified perfectly in the beautifully atmospheric summer room in the Rai Praveen Mahal in Orchha, Madhya Pradesh *(Ford & Hewitt, 1996)*. Orchha (latitude 25°N) experiences a composite climate and summers (from March to June) are extremely hot (daily max 45°C+) and dry (below 20% relative humidity).

> WITHIN THERMALLY MASSIVE STRUCTURES COUPLED TO THE EARTH ONE CAN EXPECT STABLE AIR TEMPERATURES CLOSE TO THE ANNUAL MEAN EXTERNAL AIR TEMPERATURE

Built towards the end of the 16th century, the main rooms in the house look north over an orchard and garden. On the north side, underneath a raised terrace, is a room partially sunk into the ground and flanked by two large open water tanks. As the visitor descends into this elegant room, the relief from the intense summer heat and the peace and tranquillity of the space leave a deep impression. Although the pools are empty now, the original intention of the architect is still clear: small openings on each side of the space with deep reveals to prevent solar gains encourage the movement of evaporatively cooled air across the space (figs 14–15). Even without water, spot measurements at mid-day in August revealed a 10°C difference between inside air temperature and outside in the shade *(Ford & Hewitt, 1996)*. The presence of the water pools would have induced evaporative cooling of the airstream and made this room comfortable in the most extreme heat of summer. The form and architectural expression of these buildings derived from the need to respond to the cultural and climatic context while creating a habitable and comfortable environment in what were often extreme conditions.

In comparison with the hot dry summer conditions of north-west India, the warm humid summer conditions of south-west China prompted another response to promote convective cooling. The fishing village of Zhouzhuang (latitude 31°N) is located at the junction of two large lakes (Baixian & Beibai) about 30km south-east of Souzhou, and west of Shanghai. The region experiences cold winters (mean minimum January air temperatures of 1°C), while summers are warm and humid (mean max 32°C in July with absolute humidity of 20g/kg). The settlement grew around a network of natural and man-made waterways, and many of the buildings within the old town date back to the Ming Dynasty *(Lau et al., 2014)*.

The entrance to Zhang's house faces almost due west, adjacent to a landing stage from the canal. It is the house of a wealthy

CHAPTER 1 | ORIGINS AND OPPORTUNITIES

merchant and follows the traditional layout of public reception and ceremonial halls in the front with more private living spaces for the family behind (figs 16–18). Frontage to the river/canal would have been very valuable, so the houses have a relatively narrow frontage and are stretched out as a series of pavilions and courts, the geometry of which is manipulated to provide light and air while maintaining privacy and security.

The contemporary visitor proceeds directly on axis via a very narrow (2m wide) court into a second reception hall, from where one looks through a very imposing doorway within a blank masonry wall, towards a slightly raised and much bigger ceremonial hall which is reached across a larger courtyard. In the past the central screen doors between the reception pavilions may have been closed and the visitor forced to go either side of the first narrow court, not gaining a glimpse of the ceremonial hall until arriving at the imposing central doorway. The door here is heavy and solid and sits within a completely opaque masonry wall, not allowing light or air to pass through, and providing a line of security and privacy. Standing just inside this doorway one has a clear view of the imposing ceremonial hall opposite and looking up one sees the roof of the hall which completely screens the sky (figs 19–20).

The main ceremonial hall is reached up three steps to a platform under a magnificent roof of dark beams and tiles. This is where the family will have received and entertained their guests, in a space diffusely

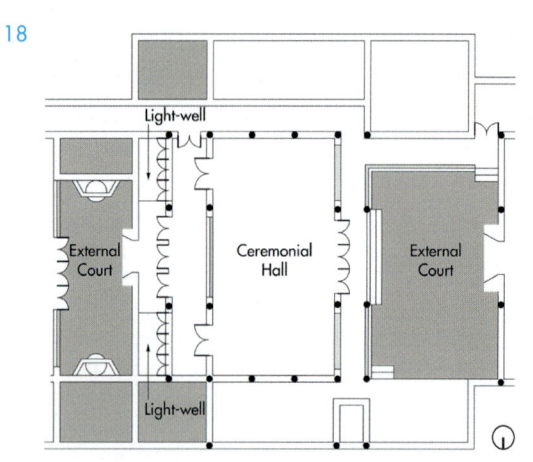

16. Entry from canal to Zhang's House, Zhouzhuang, China.
17. Natural ventilation strategy of the Ceremonial Hall in Zhang's House.
18. Plan of Ceremonial Hall and related courts in Zhang's House.

CHAPTER 1 | ORIGINS AND OPPORTUNITIES

lit via paper covered screens which fill both the longer sides of the space. Gable walls are rendered brick incorporating columns to support the roof trusses. The central screen wall opposite the entrance into this Hall will normally have been closed, so that members of the family passing through to the more private domain beyond go via the doors either side of the screen. One is then within a corridor space. But it is more than a corridor space. Two 'crab eye' lightwells provide reflected light to this space, while also providing the backdrop for a bamboo plant which appears in silhouette like a traditional painting, delighting the eye and softening the space. The geometry of this lightwell prevents direct sunlight penetration, while providing reflected light off the white rendered masonry wall. This diffuse light will have had a psychological as well as physical effect in achieving 'light without heat', compared with the harsh bright light of the outside world.

> DIFFUSE LIGHT WILL HAVE HAD A PSYCHOLOGICAL AS WELL AS PHYSICAL EFFECT IN ACHIEVING 'LIGHT WITHOUT HEAT'

Closer observation reveals a further function for these 'lightwells'. Openable screens rise from about 500mm from the floor to the ceiling, and they would have been opened in the summer to promote air movement. The screens sit above a low bench, the so called 'beauty seat', where the ladies of the house apparently sat in the summer. Why sit here in this 'corridor'? The answer involves one of the most subtle aspects of this design. Air from the front of the Hall (which has a very large open area) passes through the doors (a much smaller area) and accelerates as

19. Entry courtyard of Zhang's House, Zhouzhuang, China.
20. Interior of Ceremonial Hall in Zhang's House.

CHAPTER 1 | ORIGINS AND OPPORTUNITIES

it moves across the bench seat and up the *'lightwell'*. Spot measurements on a recent summertime visit revealed air velocities of 0.1 to 0.3m/s through the screens at the front of the Hall compared with 0.7 to 1.2m/s through the doors adjacent to the *'beauty seats' (ibid)*. On a hot and humid summer day this would therefore have been an ideal place to sit; a calm and tranquil place, promoting not just thermal comfort but an experience which combines air movement, the rustle of bamboo leaves, diffuse light from above, and oblique views through the Hall and into the reception courtyard from where guests would approach. One can imagine the pleasure and excitement of the occupants of the house on formal occasions, when the house would have provided an exquisite setting for ceremony and entertainment.

> AIR FROM THE FRONT OF THE HALL (LARGE OPEN AREA) PASSES THROUGH THE DOORS (SMALLER AREA) AND ACCELERATES AS IT MOVES ACROSS THE BENCH SEAT AND UP THE 'LIGHTWELL'

The multiple functions and the subtle interactions of the different elements promote admiration for the designers of this house. But this is not the work of a single designer (and certainly not a single period). The different forms and elements of this house are representative of the traditional house found in many parts of China, which has developed over millennia. It is interesting to speculate about the process of gradual, empirically based improvement which has occurred to achieve such a synthesis.

1.2. OPPORTUNITIES

Nearly 50 years ago, Reyner Banham was beginning to piece together a very novel form of architectural history, which was to become a classic of its time. *'The Architecture of the Well-Tempered Environment'*, described the impact of the development of building services engineering on architectural design, and the nature of the architect's role in the design team.

The invention of mechanical air-conditioning by Willis Carrier in 1906 has had a profound influence on the design of buildings and on the thermal comfort of their occupants. It has also had an influence on the role of the architect within the design team. Banham was the first to reveal that one of the bi-products of the air-conditioned interior was that the form and fabric of the building envelope had ceased to function as the primary moderator of the external climate on the internal environment. In the process, many architects lost their knowledge of, and interest in, the environmental control functions of the building itself. Indeed, in many contemporary buildings the external envelope only serves to exaggerate dependency on mechanical conditioning to achieve a satisfactory environment. This of course has been a gradual process, and many of the great architects of the twentieth century proved to be exceptions to this rule *(Hawkes, 2008)*. However, by the end of the 20th century, the control of the internal environment had come to be regarded as almost entirely the province of the engineer, and reliant on energy derived from fossil fuels.

But the last 20 years have seen rapid change. Many (architects and engineers alike) believe that it is no longer acceptable for engineers to have to make good by brute force what others have failed to achieve by design. There is renewed interest in the rich, varied and subtle vocabulary of an architecture which successfully moderates the impact of the external climate on the internal environment to achieve thermal as well as visual comfort and delight, without total dependence on mechanical intervention. Alternatives are being sought for conventional air-conditioning, which is now recognised as a significant contributing factor to global warming and climate change.

This search has led to a re-discovery of the principles of environmental control through the manipulation of the building form and fabric: design principles which enable cooling without mechanical air-conditioning. Prior to the development of mechanical and electrical devices to control the environment within buildings, architects had developed an empirically based understanding of the relationship between climate, and building form and fabric. This formed the basis of a book describing an environmental history of British Architecture 1600–2000 under the title *Architecture & Climate* (Hawkes, 2012).

An example is provided by Christopher Wren's Sheldonian Theatre in Oxford (completed 1669). Its primary purpose was to provide space for the graduation ceremony of Oxford University students (the Encaenia) and was designed to be daylit and naturally ventilated (figs 21–22). Much of the glazing

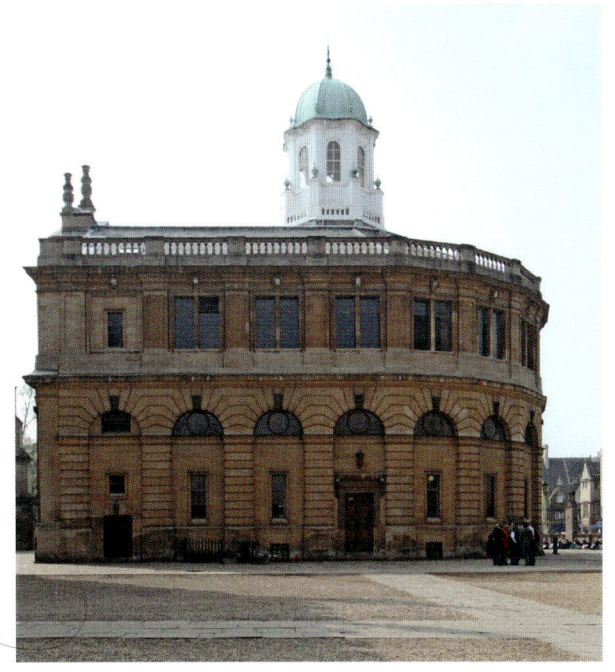

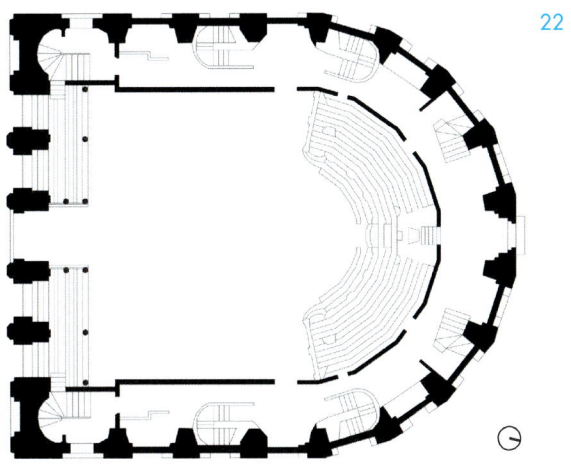

21. Exterior of the Sheldonian Theatre, Oxford University, UK [Christopher Wren].
22. Plan of the Sheldonian Theatre.

CHAPTER 1 | ORIGINS AND OPPORTUNITIES

has a northerly aspect, providing an even light and reducing solar heat gain, and was fitted with opening ventilators to promote a combination of cross and stack ventilation. The graduation ceremonies regularly took place in June and July, two of the warmest months in the UK, when the graduating students would also have been wearing heavy robes. The theatre has a seating capacity of 950 people, and to avoid overheating substantial airflow rates would be required to remove internal heat gains. Hawkes refers to *'the ingenuity of the design of the windows and their opening mechanisms that provide copious ventilation but prevent the ingress of rain'* (Hawkes, 2014). It is a testament to the robustness of the original design that the building is still used for its original purpose.

Christopher Wren was a fellow of the UK Royal Society and was both an architect and a scientist. Today, combining these disciplines would be regarded with suspicion, but in the 17th and 18th centuries the arts and sciences were perceived to have a natural synergy, and their understanding was the hallmark of an educated person. In the 19th century the rationalism of science and engineering began to create a split with those in the arts. This split can be exemplified by the unhappy relationship between David Boswell Reid (a physician before becoming a specialist in building ventilation) and Charles Barry (the architect of the 'new' Houses of Parliament, Westminster, London) in their professional collaboration (Schoenefeldt, 2016).

However, the value of painstaking practical research for subsequent application in practice (as exemplified by Reid) was also understood by Joseph Paxton, whose empirical investigations into the control of temperature and ventilation within greenhouses led to their ultimate application in the Great Exhibition Building of 1851, and the Crystal Palace at Sydenham *(Schoenefeldt, 2008; 2011)*.

Paxton's design for the Great Exhibition Building incorporated shading devices, provision for natural ventilation and evaporative cooling, intended to maintain comfortable temperatures on hot days within the huge glazed pavilion. On being challenged whether his design would actually work, he claimed that his previous experience through small-scale experiments in conservatory design at Chatsworth demonstrated the effectiveness of his ventilation and cooling strategy. Extensive monitoring of conditions within the building during the Great Exhibition provided an insight into the strengths and weaknesses of the environmental design strategy, and enabled Paxton to make significant improvements to the subsequent Sydenham building (the Crystal Palace).

The building for the Great Exhibition was naturally ventilated by means of manually operable ventilators at low and high levels. The opening mechanism was operated by means of a *'cord and wheel'*, while calico shades to the roof reduced solar gain and diffused the light. Aware that his strategy was not capable of lowering indoor temperatures below the potential high summer outdoor temperatures, Paxton considered additional passive cooling options, including hanging coarse canvas sheets (dampened with water) in front of the ventilators, to cool the incoming air stream. Temperatures were manually recorded at 2 hour intervals within the building from May to August 1851. The results indicate that, while occasional extremes (>30°C) were recorded, the general response to the ventilation system was very favourable, particularly when compared with other large public buildings of the time.

The development of successful environmental design strategies, based partly on the testing of prototypes, can be described as part of an 'empirical tradition' which goes back many centuries and is continuing today. Monitoring and performance analysis, along with occupant feedback, is central to determining how buildings perform in use. However, for most of the 20th century this tradition was gradually ignored in favour of the new technologies of the mechanical and electrical conditioning of buildings (as described by Banham and Hawkes). This became completely dominant by the end of the 20th century, although there is gathering evidence of renewed interest in the empirical tradition.

> MONITORING AND PERFORMANCE ANALYSIS, ALONG WITH OCCUPANT FEEDBACK, ARE CENTRAL TO DETERMINING HOW BUILDINGS PERFORM IN USE

The testing of ideas at a small scale in advance by Paxton and Reid (and Wren before them) prior to their application in practice raised confidence that their innovative ideas

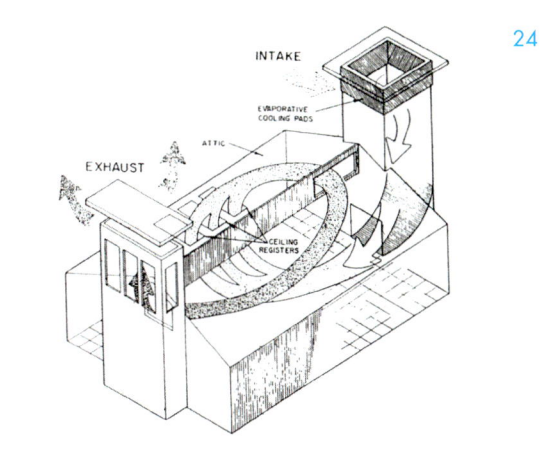

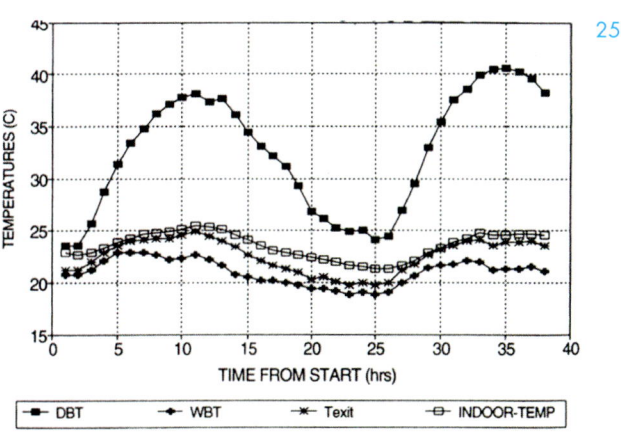

23. General view of Experimental Building, University of Arizona, Phoenix, USA.
24. 3D diagram of Experimental Building.
25. Temperature data from the Experimental Building.

CHAPTER 1 | ORIGINS AND OPPORTUNITIES   15

26. Cellulose mats and spray nozzles mounted in test rig, Seville, Spain 1991.

were worth pursuing. Full scale *'mock-ups'* can be costly but underline the value of research and is as relevant today as it was in the past.

In an experimental building in Tucson, Arizona in 1986 a downdraught tower incorporating wetted cellulose pads demonstrated the effectiveness of direct evaporative cooling in driving a substantial airflow through the building. In the extreme dry climate of Arizona, the potential for effective cooling by direct evaporation is very significant (figs 23–24). *Givoni (1994)*, in analysing the test data, confirmed the effectiveness of this strategy. From a sample of the published results the exit air temperature from the cooling tower was measured at 24°C when the outdoor dry bulb temperature (DBT) reached 41°C and the wet bulb temperature (WBT) was 22°C (fig. 25). The large drop in temperature through the cooling tower, coupled with high air change rates (induced by buoyancy alone), indicates not only the significant cooling potential of this technique, but also that it drives the ventilation air through the building. Since this experimental work, several buildings in south-west USA have been designed incorporating cool towers with wetted cellulose pads, to achieve internal thermal comfort.

Cunningham & Thompson's experiments demonstrated that buoyancy forces alone can achieve very high air change rates (30 air changes per hour were recorded in their experimental building). This is significant if applied to a building, because fans normally required to drive air around the building can

CHAPTER 1 | ORIGINS AND OPPORTUNITIES

be avoided. Fan power can represent 25% to 35% of the electrical energy required by a conventionally air-conditioned building. The innovation in relation to contemporary buildings is therefore not the use of evaporative cooling per se, but rather the rediscovery that it can drive the airflow through the building. Typically, in a new air-conditioned office building in southern Europe this might represent a saving of 70–80kWh/m² per year (35–45kg CO_2/m² per year).

BUOYANCY FORCES ALONE CAN ACHIEVE VERY HIGH AIR CHANGE RATES

Many aspects of the historic tradition outlined above were also reviewed by the designers of the Expo site in Seville for the 1992 World Fair. This included the 30 metre high 'cool towers' of the Avenue of Europe (figs 27–29), which employed high pressure water misting nozzles (micronisers) to induce a downdraught of cool air on the heads of passers-by (Alvarez et al., 1991).

The application of misting nozzles within a tower serving a prototype office building was tested as part of the PDEC research project at the laboratories of Conphoebus in Catania, Italy 1996–1998 (figs 30–31).

In parallel with this research and experimentation, many buildings have applied direct evaporative cooling in its various forms. Case studies of some of these buildings are presented in Part 2 of this book, and while some of these buildings are not without their

27. Avenue of Europe in Seville Expo '92.
28. Misting towers in Seville Expo '92.
29. Interior of misting tower in Seville Expo '92.

CHAPTER 1 | ORIGINS AND OPPORTUNITIES

faults, they demonstrate the benefits of natural cooling in different parts of the world, where it can displace the need for conventional air-conditioning.

In parts of the world where external night-time temperatures drop below desirable internal daytime temperatures, cool night air can be used to remove heat gains from the previous day, and pre-cool the building for the following day. Night-time convective cooling is now used very widely in offices and other commercial buildings in Europe and North America.

One of the first commercial buildings to successfully demonstrate this was the Process building for Farsons Brewery in Malta *(Ford & Short, 1988)*. This building maintains internal temperatures 10-15°C below summer-time peaks, by promoting buoyancy driven night-time convective cooling coupled to a high thermal capacitance interior (figs 32–34). In this relatively deep plan building, towers are constructed to encourage air movement even in the absence of wind. Control of ventilation is achieved by opening and closing vents linked to temperature and daylight sensors. The building responds to daily and seasonal changes in climate, exploiting the thermal characteristics of the building form and fabric to moderate the internal environment.

> NIGHT-TIME CONVECTIVE COOLING IS NOW USED VERY WIDELY IN OFFICES AND OTHER COMMERCIAL BUILDINGS IN EUROPE AND NORTH AMERICA

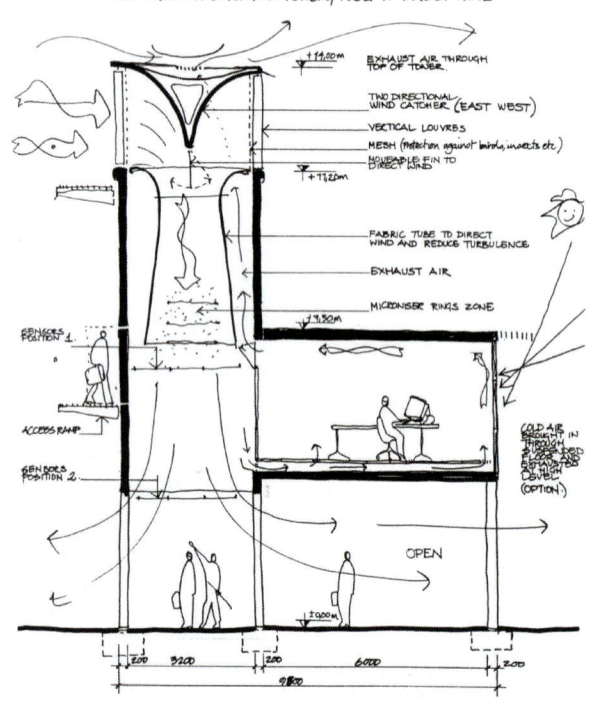

30. Experimental Building, Conphoebus, Catania, Italy.
31. Section and airflow strategy, Experimental Building.

18 **CHAPTER 1** | ORIGINS AND OPPORTUNITIES

The design of the building was based on a four-part environmental strategy:
- Insulation Effects and Thermal Buffer-in, to minimise daytime solar heat gain by preventing direct solar radiation entering the Process Hall and providing a *'buffer zone'* around the Process Hall.
- Capacitance Effects, to exploit the thermal capacitance of the floors, walls and roof to even out diurnal fluctuations.
- Convective Cooling, to encourage ventilation at night, exploiting lower night-time temperatures to remove heat build up during the day within the roof and walls.
- Daylighting, to provide good daylight to the Process Hall without attendant solar gains by providing reflected light around the perimeter plus diffuse north light.

The Process Hall is surrounded by a *'jacket'* of circulation and services, providing a thermal buffer between the interior and the outside world. High level glazing allows diffuse and reflected light to enter the Process Hall without permitting direct solar gain. The process area is ventilated via high level openings in the *'towers'* above the jacket. The towers are designed to enhance both wind and stack induced ventilation. Daytime ventilation of the Process Hall is kept to a minimum by keeping the vents closed between approximately 8.00am and 7.00pm (in fact, the vents close on a daylight signal in the morning, and open when the outside temperature drops below the inside temperature). The jacket is continuously ventilated, experiencing a larger swing in temperature than would be acceptable in the Process Hall, but allowing economies in vent

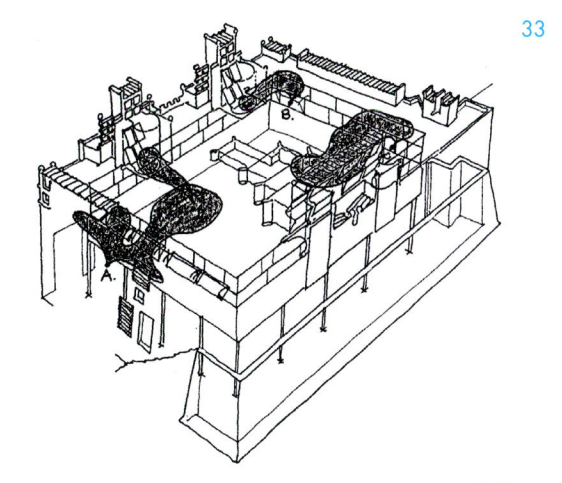

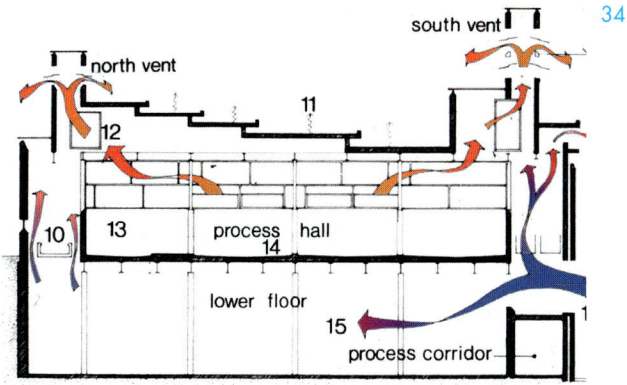

32. South elevation of the SFC Brewery Malta [Peake Short], ©Peter Cook.
33. Axonometric of the SFC Brewery Malta showing smoke test.
34. Section to show night ventilation of the SFC Brewery Malta.

CHAPTER 1 | ORIGINS AND OPPORTUNITIES

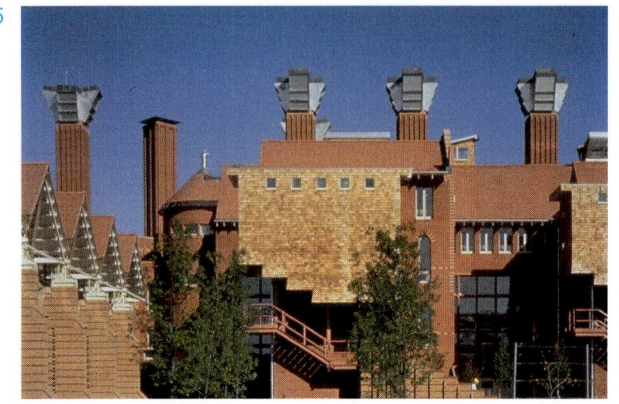

opening gear without causing a significant loss of performance.

It was established by subsequent field measurements *(Ford, 1993)*, that the design of the building produces thermally stable conditions internally, and that temperatures remain below 27°C for all but extreme days in August. The project was an early demonstration of the adoption of a successful natural cooling strategy as part of a non-domestic building in southern Europe.

The success of the passive strategy for the Brewery building in Malta gave the architects confidence that a similar strategy could work in the UK as well.

The main environmental design strategy of the engineering laboratory building for De Montfort University (The Queens Building), was to provide a highly insulated, thermally massive envelope around a plan which responds to the nature of the activities within, and for the entire building to be daylit and naturally ventilated *(BRE, 1997)*. Air is supplied, under the control of the occupants, at the perimeter, and exhausted at high level via vertical shafts and ridge vents which are automatically controlled (figs 35–36). The main part of the building, which includes classrooms, lecture theatres, laboratories and offices, is over 30 metres deep and zoned into separate fire compartments, but still provides daylight and natural ventilation to all spaces. A full height central concourse links all these spaces, acting as a lightwell and thermal zones for adjoining spaces.

35. South façade of the Queens Building, Leicester, UK [Short Ford Architects], ©Peter Cook.
36. Section of the Queens Building.
37. 'Salt bath' test for natural ventilation airflow evaluation of the Queens Building.

20 **CHAPTER 1** | ORIGINS AND OPPORTUNITIES

To reduce noise levels at a nearby terrace of private houses, the naturally ventilated machine hall is flanked on the western façade by a two storey block of specialist laboratories. This also provides a secondary function of resisting the lateral forces of the travelling gantry crane. These forces are opposed on the east elevation by a series of buttresses. Each buttress is hollow, providing an attenuated fresh air inlet duct, with similarly lined voids over and between ground floor offices supplying air from the west façade. Gable glazing provides very even daylight across the plan, while being protected by deep roof overhangs.

By contrast the electrical laboratories are housed in two shallow plan, four storey wings, on either side of the entrance courtyard. These labs exploit simple cross ventilation to remove the high internal heat gains from computers and equipment. The narrow section also promotes good daylighting which is moderated by light shelves, designed to minimise glare.

> THE FIRST APPLICATION OF THE 'SALT BATH' TECHNIQUE WAS USED TO PREDICT BOTH AIRFLOW RATES AND TEMPERATURES WITHIN THE BUILDING, ENABLING THE COMPARISON OF DIFFERENT DESIGN OPTIONS AND THE REFINEMENT OF THE DESIGN

During the development of the design of the different elements of this building, the thermal, ventilation and daylighting performance was tested using both computer based and physical modelling tools. This modelling included the first application of the '*salt bath*' technique for representing airflow in buildings (fig. 37). It was used to predict both airflow rates and temperatures within the building, in conjunction with the dynamic thermal model ESPr, enabling the comparison of different design options and the refinement of the design. It also gave both the design team and the client the confidence to proceed with the design solution. This was one of the first of a new generation of naturally ventilated buildings in the UK to be tested in this way, and one of the first to demonstrate that large non-domestic buildings with high internal heat gains can be successfully naturally ventilated and avoid the need for mechanical cooling.

The origins and opportunities for natural cooling described above form part of the background to the current move towards low-carbon architecture, which is influencing policy and practice in many parts of the world today. The physical principles on which natural cooling is based, and the strategic options available to designers, are discussed in the next chapter.

CHAPTER 1 | ORIGINS AND OPPORTUNITIES

REFERENCES

- Alvarez, S., Rodriguez, E. & Moina J.L. (1991). *'The Avenue of Europe at Expo 92: Application of Cool Towers'*. Proc. of 9th PLEA Conference, Seville Spain. Pergamon Press.
- Beazley, E. & Harverson, M. (1986). *'Living with the Desert: Working Buildings of the Iranian Plateau'*. Aris & Phillips Ltd.
- BRE (1997). *'The Queens Building, De Montfort University: Feedback for Designers and Clients'*. New Practice final report 102, published by BRECSU on behalf of the DoE's Energy Efficiency Best Practice Programme, TSO N25837 07/97.
- Ernest, R. (2011). *'The role of multiple courtyards in the promotion of convective cooling'*. PhD Thesis. Dept. of Architecture, University of Nottingham, UK.
- Ernest, R. & Ford, B. (2012). *'The Role of Multiple Courtyards in the Promotion of Convective Cooling'*. Architectural Science Review 55(4): 241–249.
- Fathy, H. (1986). *'Natural Energy and Vernacular Architecture: Principles and Examples with Reference to Hot Arid Climates'*. University of Chicago Press.
- Ford, B. (1993). *'Passive Night Vent Cooling: Results from Field Measurements in Malta'*. 3rd European Conference on Architecture, Florence, Italy. 17–21 May.
- Ford, B. & Hewitt, M. (1996). *'Cooling without Air-conditioning: Lessons from India'*. Architectural Research Quarterly 1(4): 60–69.
- Ford, B. & Short, A. (1988). *'Passive Cooling of a New Brewery Process Building in Malta'*. Proc. Plea Conference, Porto, Portugal.
- Givoni, B. (1994). *'Passive and Low Energy Cooling of Buildings'*. Van Nostrand Reinhold.
- Hawkes, D. (2008). *'The Environmental Imagination: Technics and Poetics of the Architectural Environment'*. Routledge.
- Hawkes, D. (2012). *'Architecture & Climate: an Environmental History of British Architecture 1600–2000'*. Routledge.
- Hawkes D. (2014). *'The Origins of Building Science in the Architecture of Renaissance'*. Ed. In: Wolkenkucksheim, Internationale Zeitschrift zur Theorie der Architektur 19(33).
- Irwin, R. (2004). *'Alhambra, Wonders of the World'*. London: Profile Books Ltd.
- Lau, B., Ford, B. & Hongru, Z. (2014). Chapter 6: *'Traditional Courtyard Housing in China: Zhang's House, Zhouzhuang, Jiangsu Province'*. In *'Lessons from Vernacular Architecture'*. Weber & Yannas (Editors), Routledge.
- Schoenefeldt, H. (2008). *'The Crystal Palace, Environmentally Considered'*. Architectural Research Quarterly, 12 (3–4): 283–294.
- Schoenfeldt, H. (2011). *'Adapting Glasshouses for Human Use: Environmental Experimentation in Paxton's Designs for the 1851 Great Exhibition Building and the Crystal Palace, Sydenham'*. Architectural History, 54: 233–273.
- Schoenefeldt, H. (2016). *'Architectural and Scientific Principles in the Design of the Palace of Westminster'*. In: Brittain-Catlin, Timothy, Bressani, Martin and De maeyer, Jan, eds. Gothic Revival Worldwide A.W.N. Pugin's Global Influence. KADOC Art Series. Leuven University Press, Leuven, Belgium, pp. 175–199.
- Weber, W. & Yannas, S. (2014). *'Lessons from Vernacular Architecture'*. Routledge.

CHAPTER 2
PRINCIPLES AND STRATEGIES

In clarifying the environmental design choices available for a particular project, it can be useful to structure the iterative design process to allow a considered review of the options available. Such a structure may include: understanding/defining the client's programme/brief/user's requirements; understanding the site location (physical and cultural context), its climate and microclimate (seasonal & diurnal variations), applicable strategies (solar control, ventilation etc.), before defining a proposition (strategy), testing & refining it (does it work?), and proceeding through to detailed design and construction. In this way, environmental design ideas embedded in early design stage thinking help to 'clarify the choices available', to make design propositions in tune with the rhythm of the natural environment.

This chapter presents an outline of the environmental design principles and strategies which can moderate the impact of climate on the internal environment to create comfortable conditions without recourse to significant mechanical assistance. It is intended as an introduction and is certainly not an exhaustive review of environmental design principles and strategies. An exploration of environmental design strategies for northern Europe and the USA can be found in *Baker & Steemers (2000)*, for the Middle East & India in *Krishan et al. (2004)*, and for Australia/South-East Asia in *Hyde (2003)*.

2.1. SITE & MICROCLIMATE ANALYSIS

The location and the physical characteristics of any site (topography, urban morphology, vegetation, orientation etc.) and its setting within its wider physical context often have a profound influence on design response. This is apparent in the work of many architects, as diverse as Geoffrey Bawa in Sri Lanka, Peter Zumthor in Switzerland, Kengo Kuma in Japan or Francis Kere in Burkino Faso.

The physical and cultural context of their work and the individual design responses of these architects are extremely diverse, but they all respond to the particularities of site and microclimate. There are a range of issues which need to be addressed by all architects and engineers when responding to a particular location. Thermal comfort is of central concern, and the different environmental design strategies available to achieve comfort are equally important to both architects and engineers in developing sustainable design solutions at the scale of the individual building and the city. The general principles of adaptive thermal comfort are described in *Nicol et al., (2012)* and are discussed in more detail in Chapter 6 on Performance Analysis.

URBAN MORPHOLOGY

Urban microclimatology has become a field of significant research over the last fifteen years, and much has been written on the subject *(Erell et al., 2011; Santamouris & Kolokotsa, 2016)*. As part of these studies, urban morphology has been identified as a significant factor in shaping the urban microclimate and the internal environment of buildings. In addition, the 'urban grain' – the pattern and hierarchy of streets, their orientation, and the relative width, height and density of surrounding buildings – can potentially have a significant impact on the local microclimate. The orientation of streets will define when sunlight falls on different elevations and whether wind is obstructed or channelled through a particular urban area *(Littlefair et al., 2000)*. The characteristics of urban form have often evolved partly in response to the climate, and an understanding of this evolution can help to frame an appropriate environmental design response. This is considered in more detail in Chapter 4 on Integrated Design.

URBAN HEAT ISLANDS

Urban heat islands are characterised by significantly elevated temperatures relative to the surrounding urban fringe or rural areas. Temperature differences may be particularly marked at night, although daytime variation can also be significant. Areas of generally higher average temperatures are associated with denser urban areas with dark surface colours and lacking green areas. Urban morphology and street pattern can also have a significant influence. A summary of urban heat island studies in Europe and North America is provided by *Santamouris (2001)*.

In the low rise sprawl of Los Angeles, the wide boulevards, dark roof surfaces and omnipresent air-conditioning contribute to a significant heat island in the city. A study in the 1990s *(Gartland, 2008)* revealed the urban heat island effect in Los Angeles, and this has now led to new building codes which promote lighter coloured roofs and roadways, and increased vegetation. Parks, gardens and green areas contribute significantly to lower average temperatures, and studies have shown that there can be significant variation in temperature within a small area *(ibid)*. However, the ubiquitous use of air-conditioning massively aggravates the problem.

> URBAN MORPHOLOGY HAS BEEN IDENTIFIED AS A SIGNIFICANT FACTOR IN SHAPING THE URBAN MICROCLIMATE

The sprawl of Los Angeles would appear to be the antithesis of high-rise high-density Hong Kong, but in fact Hong Kong also generates heat islands. While the high rise buildings shade the city streets during the day, they also block the path of prevailing winds and trap heat during the night due to the low sky view factor within the urban layout. In fact, UHI intensity in Hong Kong reaches 2.5 degrees above the urban fringe during the night! (fig. 38). These local urban heat island variations are the result of many different factors depending on the particular characteristics of both morphology and climate. However, one common result is a significant increase in cooling demand, which, if met by traditional air-conditioning, will exacerbate the situation and is clearly unsustainable *(Ng, 2010)*.

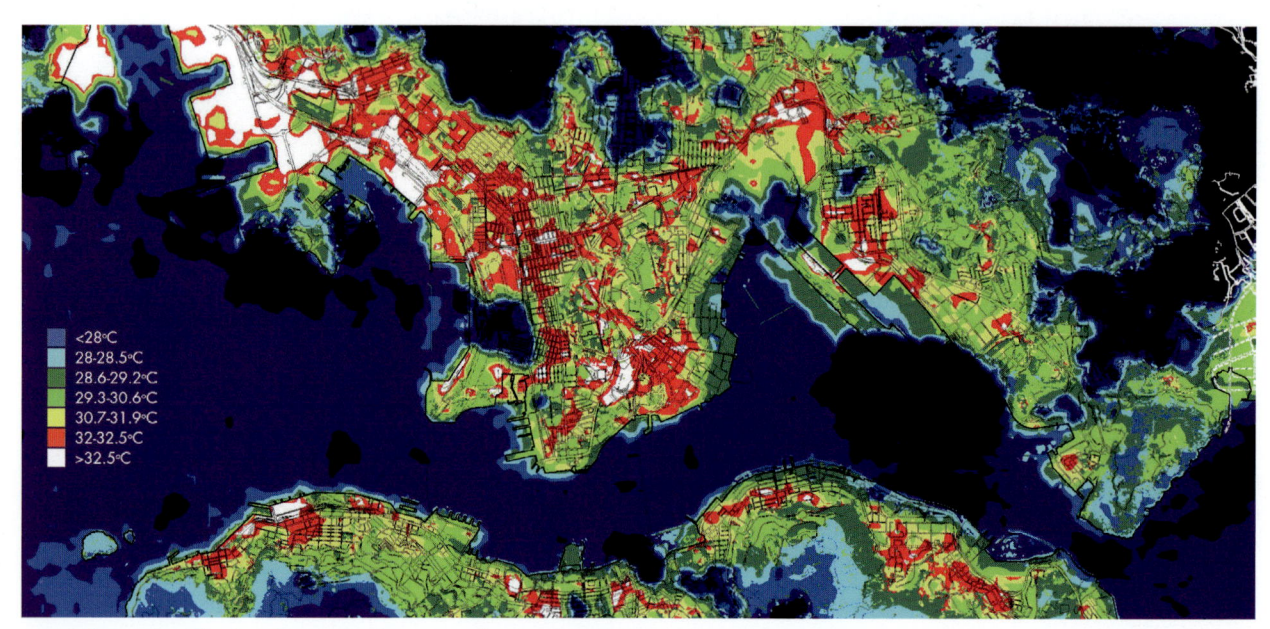

38. Urban Heat Island: night-time satellite image showing urban temperatures in Hong Kong.

CHAPTER 2 | PRINCIPLES AND STRATEGIES

EFFECTS OF VEGETATION

Vegetation can have a moderating influence on the local microclimate, both reducing heat gains (and temperature) during the day, and reducing heat loss by radiation at night. Shade from vegetation can reduce radiant temperatures and also increase water retention in the soil, while evapo-transpiration (the absorption of CO_2 and the release of water vapour by trees and other vegetation) can reduce air temperatures locally.

The use of vegetation to both shade and cool façades has been used in many different parts of the world. The Consorcio building in Santiago, Chile (architects Enrique Browne and Borja Huidobro) incorporates a vegetative screen in front of west facing glazing, and studies have revealed a significant contribution to reducing the cooling load and glare to offices behind this façade (*Browne, 2008*). The vegetative screen was found to contribute a reduction of 48% in energy consumption compared to a data base of ten typical office buildings in Santiago, and a reduction of 25% compared to the top floor that has no vegetative façade (figs 39–40). These effects have also been found to be significant where the vegetation is contained between buildings or within courtyards (as referred to in Chapter 1).

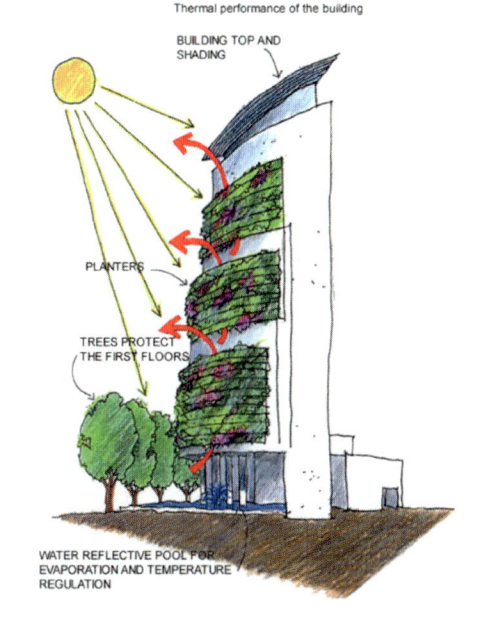

> SHADE FROM VEGETATION CAN REDUCE RADIANT TEMPERATURES AND INCREASE WATER RETENTION IN THE SOIL, WHILE EVAPO-TRANSPIRATION CAN REDUCE AIR TEMPERATURES LOCALLY

39. Concept design of Consorcio Building, Santiago, Chile [Enrique Browne Arquitectos].
40. Vegetation to west wall of Consorcio Building.

CHAPTER 2 | PRINCIPLES AND STRATEGIES

TRAFFIC NOISE AND AIR BORNE POLLUTION

Traffic noise and air borne pollution may be a challenge, but not necessarily an impediment to natural ventilation and passive cooling. The occurrence of traffic noise and air borne pollution around a particular site can be mapped, as it may impact on the environmental design strategy. Analysis of noise and pollution for a particular site will reveal where sound baffling may be required and also where the cleanest fresh air can be obtained to supply the building. The intensity of noise and pollution is greatest in proximity with busy streets and reduces with distance from the source. However, concentrations of nitrogen dioxide (NO2) and particulate matter (PM10 & 2.5) in many European cities regularly contravene the European Standard (Dir 2008/50/EC) which specifies benchmarks for air pollutants based on annual averages or daily/hourly averages (NO2 should not exceed annual average of 40μg/m3 and 50μg/m3 over 24hrs).

The health implications of exceeding these benchmarks have become very significant. Pollution levels in cities in India, South America and China are known to be even worse than those in Europe and North America. Action by politicians, pressure groups, industrialists and consumers will be required to bring down pollution to acceptable levels, and building designers must play their part by ensuring that supply air is taken from the least polluted and quietest side of the building.

41. Former Post-Office building renovation site plan, Milan, Italy [Mario Cucinella Architects].
42. Glazed screen wall for Former Post-Office building renovation, Milan, Italy.

28 **CHAPTER 2** | PRINCIPLES AND STRATEGIES

By mapping the noise and pollution characteristics of a site, passive design options can be evaluated and can potentially be as viable in city centre locations as they are in suburban and rural areas, depending on the particular characteristics of the site. The addition of secondary glazed screens to provide acoustic buffering to central urban office buildings for example has become commonplace in Europe & North America. The recent refurbishment of an office complex adjacent to a busy roundabout in Milan, Italy by Mario Cucinella Architects (figs 41–43) incorporates a secondary glazed skin to reduce noise ingress and solar heat gain. (These issues are discussed further in section 4.3 of Chapter 4).

SOLAR RADIATION EFFECTS

The need for cooling will depend on the characteristics of the site and microclimate (as discussed above), and the nature of the building (activities and density of occupation). In much of Europe, the historical tradition of masonry buildings with small windows meant that the need for cooling was rare, and when it arose could be dealt with by natural convection whenever the external air temperature was below the desired internal air temperature. Shutters prevented solar gain during the day in summer, while continuing to allow ventilation, and the high thermal capacitance interior helped to stabilise internal temperatures.

43. Before and after Former Post-Office building renovation [Mario Cucinella Architects].

CHAPTER 2 | PRINCIPLES AND STRATEGIES

The contemporary context and expectations are completely different. The thermal performance of the external 'skin' of buildings has improved significantly, and internal heat gains from people and equipment (coupled with solar heat gains) increase the risk of overheating and therefore the need for cooling. The need for cooling has often been exacerbated by the design of the building itself (often incorporating large unprotected glazed surfaces). Excessive solar heat gains can be avoided if the implications of solar geometry, façade orientation and shading opportunities are fully understood.

> THE PRIMARY CONSIDERATION IS THEREFORE TO PROVIDE EFFECTIVE SHADING FOR VULNERABLE SURFACES OF THE BUILDING

The primary consideration is therefore to provide effective shading for vulnerable surfaces of the building (including the roof!). Wherever we are in the world, the design of appropriate sun-shading devices for buildings requires the designer to have a basic understanding of solar geometry, climatic variation and occupant needs. A number of design tools are available to assist the architect in quantifying solar radiation falling on a vertical or horizontal surface (fig. 44), and these are discussed in more detail in Chapter 6.

WIND EFFECTS

The interaction between urban morphology and the direction and strength of the prevailing wind can either promote air movement (as in the warm and humid tropical

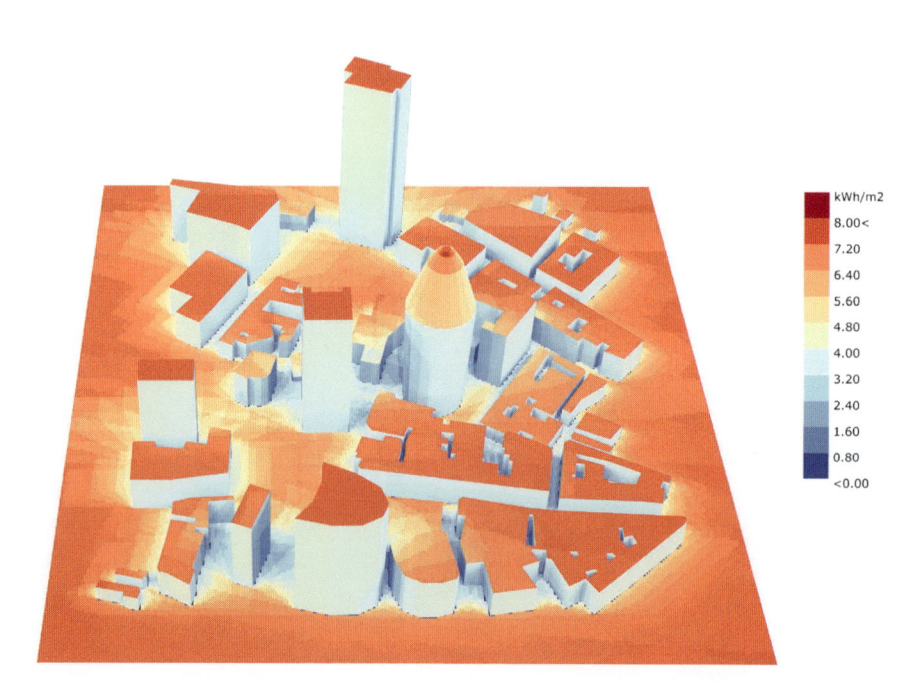

44. Incident solar radiation on a sunny day (July) in London, UK.

30   CHAPTER 2 | PRINCIPLES AND STRATEGIES

cities of Colombo or Singapore), or it can inhibit it, resulting in unhealthy and uncomfortable stagnant areas (as in Hong Kong). The wind environment in cities is often complex and highly variable, but can inform city development and planning guidance documents (figs 45–46). Analysis of the wind environment in Hong Kong has resulted in planning regulations to promote greater air movement and create a more 'porous' arrangement of city blocks to reduce the concentration of pollutants *(Ng, 2010)*.

The wind can also be exploited to promote natural ventilation and cooling of building interiors, but such strategies must be implemented with care to create robust solutions and avoid localised high air velocities internally. Prevailing wind speeds and direction need to be determined over the year to identify opportunities and constraints, depending on the nature of the building type and other characteristics of the site. Suburban and rural sites will generally be exposed to higher average wind speeds than city centres in the same regional location. Design strategies need to account for variations in wind pressure around the building, depending on wind speed and direction and the surrounding topography. Examples of spatial variation of wind pressure coefficients for rectangular buildings are given in numerical data for air infiltration and natural ventilation calculations *(CIBSE, 2005)*.

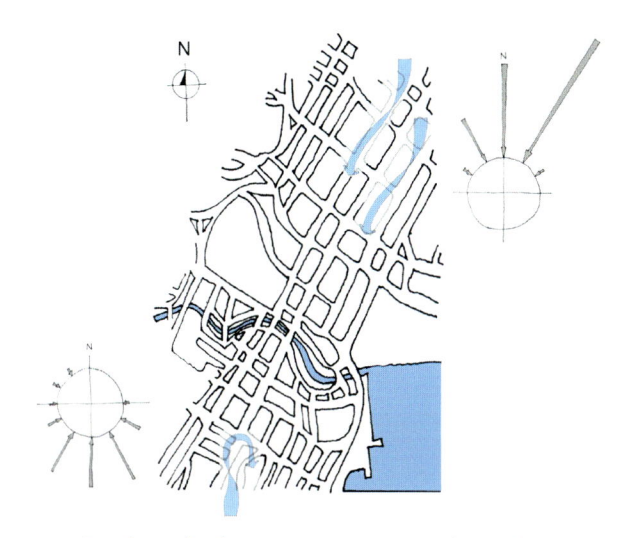

45. The relationship between street pattern and prevailing winds for central Singapore.

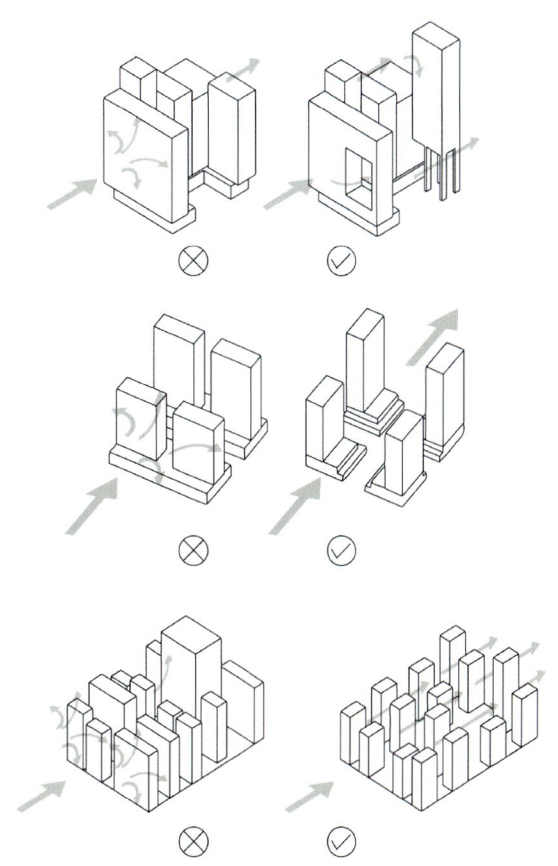

46. Proposed changes to Hong Kong Planning Standards and Guidelines, ©Edward Ng, HKPU, China.

CHAPTER 2 | PRINCIPLES AND STRATEGIES

2.2. NATURAL COOLING DESIGN STRATEGIES

'Natural' cooling, as described above, refers to the exploitation of an environmental heat sink to achieve cooling (as opposed to 'active' cooling which relies on mechanical intervention). Environmental heat sinks include external air (when below the desired internal air temperature), the night sky (often clear in hot dry climates), the ground (3+metres below the surface) and the evaporation of water. Heat from within a building can be removed through the movement of air or water between the building and the heat sink.

> ENVIRONMENTAL HEAT SINKS INCLUDE EXTERNAL AIR, THE NIGHT SKY, THE GROUND AND THE EVAPORATION OF WATER

The opportunities of developing design strategies to exploit environmental heat sinks passively are discussed below.

EXTERNAL AIR – NATURAL CONVECTION

All occupied buildings require a supply of fresh air, and this can be achieved by natural convection in most parts of the world, depending on diurnal and seasonal variations in climate. The use of natural convection to remove internal heat gains is dependent on the external (supply) air temperature being lower than that required internally. In northern Europe the external air temperature is generally below the desired internal temperature, so natural convection can be used during the day to remove unwanted heat through much of the year. In southern Europe the external air temperature in summer may be low enough to bring into the building during the day and at night, depending on the building's use and construction characteristics. However, when the external air temperature is above the desired internal air temperature, this implies that (in non-domestic buildings) the air supplied must be pre-cooled either naturally or mechanically, and that ventilators must be controlled appropriately. While generally applicable, the perception of thermal comfort will be influenced by internal surface temperatures and the thermal capacitance of the interior.

The first step is to understand the diurnal and seasonal variation for the location of the project. Having identified the variation in ambient air temperature, it is possible to identify times of the year (and day) during which natural convection can be used to remove heat from within the building. The thermal capacitance of air is low (volumetric specific heat of air = 1,200j/kg) which means that high volume flow rates are required. This also means that significant fan power is required if this is to be achieved mechanically (requiring 25–35% of the electrical load in many non-domestic buildings). However, in most circumstances, the air volume flow rates required can be achieved by natural convection, provided ventilation supply and exhaust openings have been sized and located appropriately (Simple calculations are described in Chapters 5 & 6).

Daytime natural ventilation is now being applied year round in a wide range of building types. Perhaps surprisingly, many tall office buildings are being naturally ventilated to achieve significant capital and running cost savings while maintaining comfort for occupants *(Wood & Salib, 2012)*. A striking example is the 17 storey Torre Cube in Guadalajara, Mexico by Carme Pinos architects (figs 47–48). This office tower promotes a combination of cross and stack ventilation by clustering three open plan office elements around a hollow core.

In temperate climates many non-domestic buildings can (potentially) be naturally ventilated throughout the year. However, as we have seen above, the site and microclimate analysis may influence the ventilation and cooling strategy significantly, and when external air temperatures exceed the desirable internal condition, other forms of passive or hybrid cooling may be required.

NIGHT-TIME VENTILATION

Wherever a relatively large diurnal swing in air temperature is experienced in summer, with night-time temperatures dropping below the desired threshold comfort temperature, this can be exploited to cool the building. Cool night air can be drawn into the building to flush out any residual heat from the day and to pre-cool the internal fabric for the following day (as described in Chapter 1). The coupling of the airflow path with internal components with high thermal diffusivity promotes convective heat transfer from internal surfaces to the air. This air is

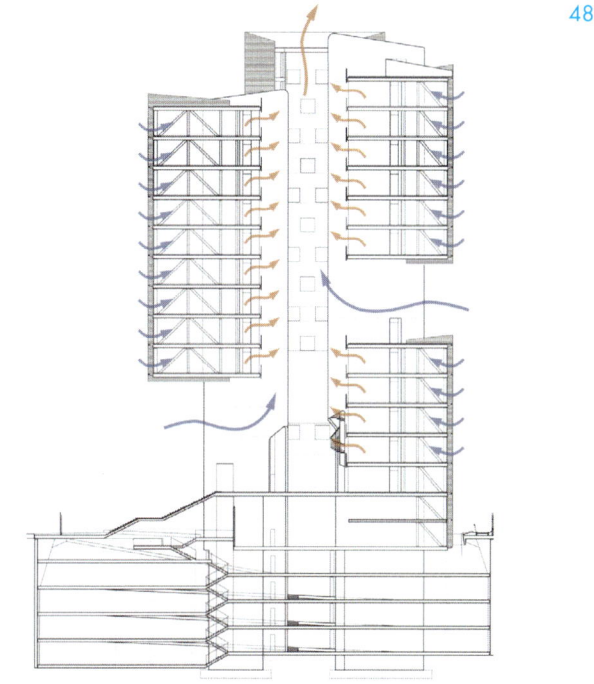

47. Torre Cube, Guadalajara, Mexico [Estudio Carme Pinós].
48. Building section: typical cross + stack ventilation in Torre Cube.

CHAPTER 2 | PRINCIPLES AND STRATEGIES   33

then exhausted via automatically controlled vent openings to regulate both the airflow rate and the temperature inside the building, promoting pre-cooling prior to the arrival of the buildings' occupants the next day. This technique is now widely used in both domestic and non-domestic buildings, wherever there is sufficient temperature difference between outside air and internal surface temperatures *(CIBSE, 2005)*. In Chapter 1 this strategy is exemplified by the SFC Brewery Process building in Malta, and is also applied in a number of the examples given in Chapter 4. Figs 49–51 show the ground breaking PowerGen (now EON) HQ offices by Bennetts Associates.

NIGHT SKY RADIATION

The Iranian ice houses described in Chapter 1 exploited the clear night skies to create ice in the desert. Hot dry regions of the world often will have clear skies at night, which can provide a potential heat sink by radiation from the relatively warm surface of the roof of a building, to the cold upper atmosphere. However, urban morphology may have a significant influence on the proportion of the sky available to a building (the 'sky view factor'). Atmospheric pollution can also have a significant effect in reducing potential radiant heat loss at night. Other factors to consider are the colour and emissivity of roof surfaces, and the thermal conductivity of the roof (since heat loss at night could also imply heat gains during the day). These factors influenced the development of the 'SkyTherm' roof ponds developed and applied in the south-west USA by Harold

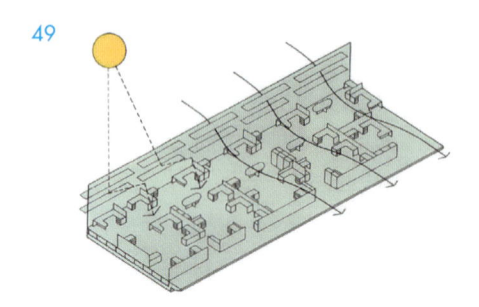

49. Cross ventilation strategy for PowerGen (now EON) HQ, Coventry, UK [Bennetts Associates].
50. Naturally ventilated offices for PowerGen (now EON) HQ, ©Peter Cook.
51. Atrium of PowerGen (now EON) HQ, ©Peter Cook.

34   **CHAPTER 2**| PRINCIPLES AND STRATEGIES

Hay in the 1970s *(Yannas et al., 2006)*. This technique has been applied in housing projects in the US, but it has not been widely applied in non-domestic buildings as it is only applicable to a single floor.

Most practical contemporary applications of radiative cooling are achieved by indirect systems. A simple and commonly applied technique is to circulate water over the surface of the roof, inducing cooling by radiation to the night sky, and by evaporation during the day. This strategy is exemplified by the Lowara office building (figs 52–54) in Italy by Renzo Piano Building Workshop (*Buchanan, 2003*), and the Global Ecology Centre at Stanford University (included as a case study in Part 2).

52. External view of Lowara Office Building, Montecchio Maggiore (Vicenza), Italy [Renzo Piano Building Workshop], ©Gianni Berengo Gardin.
53. Irrigated roof of Lowara Office Building, Montecchio Maggiore (Vicenza), Italy, ©Gianni Berengo Gardin.

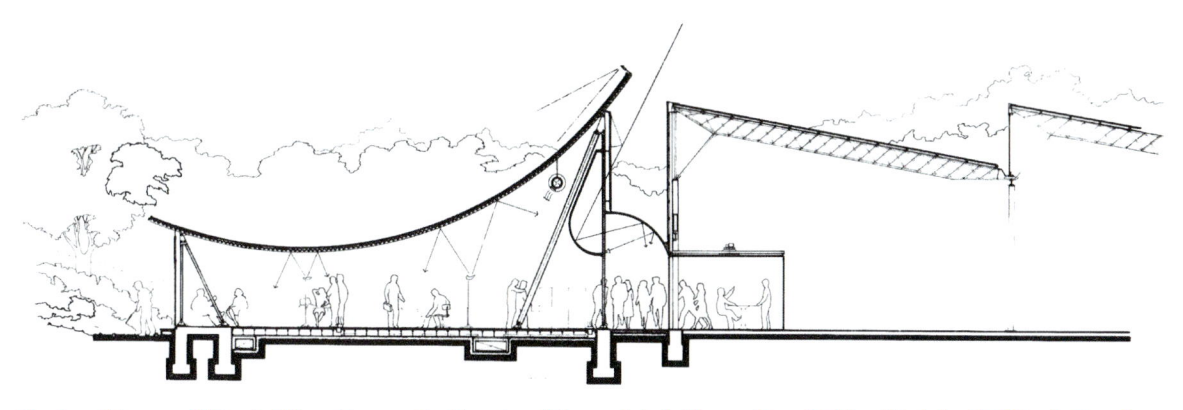

54. Section of Lowara Office Building, Montecchio Maggiore (Vicenza), Italy [Renzo Piano Building Workshop], ©Fondazione Renzo Piano.

CHAPTER 2 | PRINCIPLES AND STRATEGIES

55

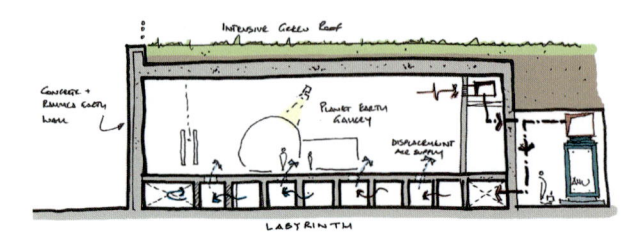

56

55. Typical room section for the Earth Centre Galleries in Doncaster, UK [Atelier Ten].
56. Ground cooling labyrinth for the Earth Centre Galleries.

BY INTRODUCING AIR TO THE BUILDING THROUGH A NETWORK OF BURIED PIPES, OR VIA AN UNDERGROUND LABYRINTH, IT CAN BE PRE-COOLED IN SUMMER AND PRE-HEATED IN WINTER

GROUND COOLING

The temperature of the Earth ten metres below ground level is generally very stable, and has been found to be equal to the annual mean air temperature for the location (anywhere in the world), varying perhaps by ±2°C according to the season (*Oke, 1987; Mulligan, 1983*). The earth is therefore a huge source of low grade heat, which can be used for either heating or cooling. For example, the annual average air temperature for central England is about 12°C. By introducing air to the building through a network of buried pipes, or via an underground labyrinth, it can be pre-cooled in summer and pre-heated in winter. This technique has been applied successfully in a number of educational buildings in northern Europe to passively pre-cool supply air to classrooms and to avoid the need for mechanical cooling. The height of the water table and the nature and thermal characteristics of the subsoil can be significant factors affecting feasibility of this technique.

The Digital Lab, Warwick University (Architect: Cullinan Studios, Engineer: Hoare Lea), which was completed in 2008, incorporates a labyrinth for pre-heating / pre-cooling of supply air. In order to assess the labyrinth's performance, temperature sensors were installed at two locations at the outlet of the labyrinth, before air is introduced to air handling units (AHUs) for mechanical heating or cooling, and monitoring was carried out for a ten-month period between January and November 2009. It was found that in winter, the temperature at the outlet of the labyrinth

CHAPTER 2 | PRINCIPLES AND STRATEGIES

varied between 12 and 17°C (higher than the outside air temperature). In summer, the temperature at the outlet of the labyrinth varied between 14 and 20°C (lower than the outside air temperature). This analysis indicates that the labyrinth has an important stabilising effect on the supply air, pre-heating it in winter and pre-cooling it in summer and therefore reducing energy consumption for space heating and mechanical cooling. The building's low heating and cooling demand would also support the idea that the labyrinth is useful at reducing energy consumption.

A similar strategy was also applied in the Earth Centre Galleries in Doncaster, UK, designed by Atelier Ten. One of the galleries buried in the earth incorporated a huge underground air distribution and thermal storage system integrated into the foundations actively cooling the supply air in the summer and warming it in the winter (figs 55–56).

At Federation Square in Melbourne, Australia, engineers Atelier 10 also provided a labyrinth located below a public plaza, from where it provided pre-cooled air to the large atrium space at the heart of the complex (figs 57–58). Measured data during the hottest month (January) has revealed that supply air from the labyrinth rarely goes above 23°C, even when external air temperatures are above 35°C.

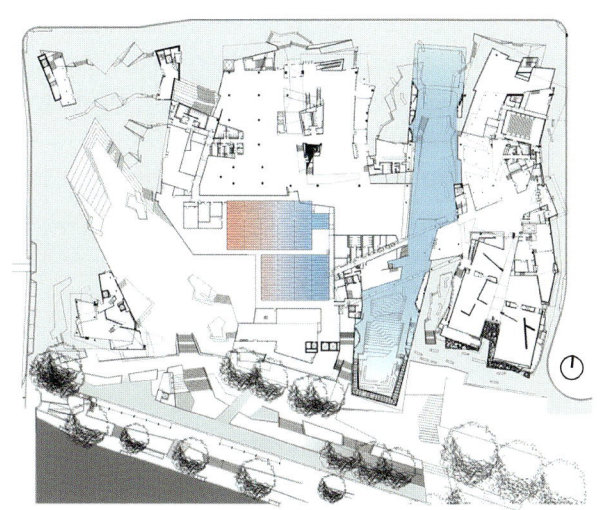

> A GROUND COOLING LABYRINTH HAS AN STABILISING EFFECT ON THE SUPPLY AIR, PRE-HEATING IT IN WINTER AND PRE-COOLING IT IN SUMMER

57. Federation Square site plan, Melbourne, Australia [Atelier Ten].
58. Ground cooling labyrinth for the Federation Square in Melbourne, Australia.

CHAPTER 2 | PRINCIPLES AND STRATEGIES

This source of heating and cooling can also be exploited by circulating water though pipes in the ground and then exchange heat through a heat pump to either increase or decrease the temperature according to the season, before delivering heat or 'coolth' via intermediate floor slabs within the building *(Mumovic & Santamouris, 2009)*.

EVAPORATION

As described in Chapter 1, the cooling potential of evaporating water has been exploited throughout southern Europe, the Middle East and northern India for many centuries. A contemporary application of evaporative cooling combined with a down-draught ventilation strategy is exemplified in the Habitat Resource and Development Centre in Katutura, Namibia designed by Nina Maritz Architects. In this project, several cooling towers induce evaporation through wet mats and distribute the air on the occupied zones (figs 59–61).

As a general rule, a reduction in supply air temperature of 70–80% of the dry bulb wet bulb depression can be achieved *(Givoni, 1994)*. This of course varies from only a few degrees when the air is relatively humid, to as much as 10 to 12°C or more when the air is relatively dry. The upper limit for the application of direct evaporative cooling is considered to be a wet bulb temperature of

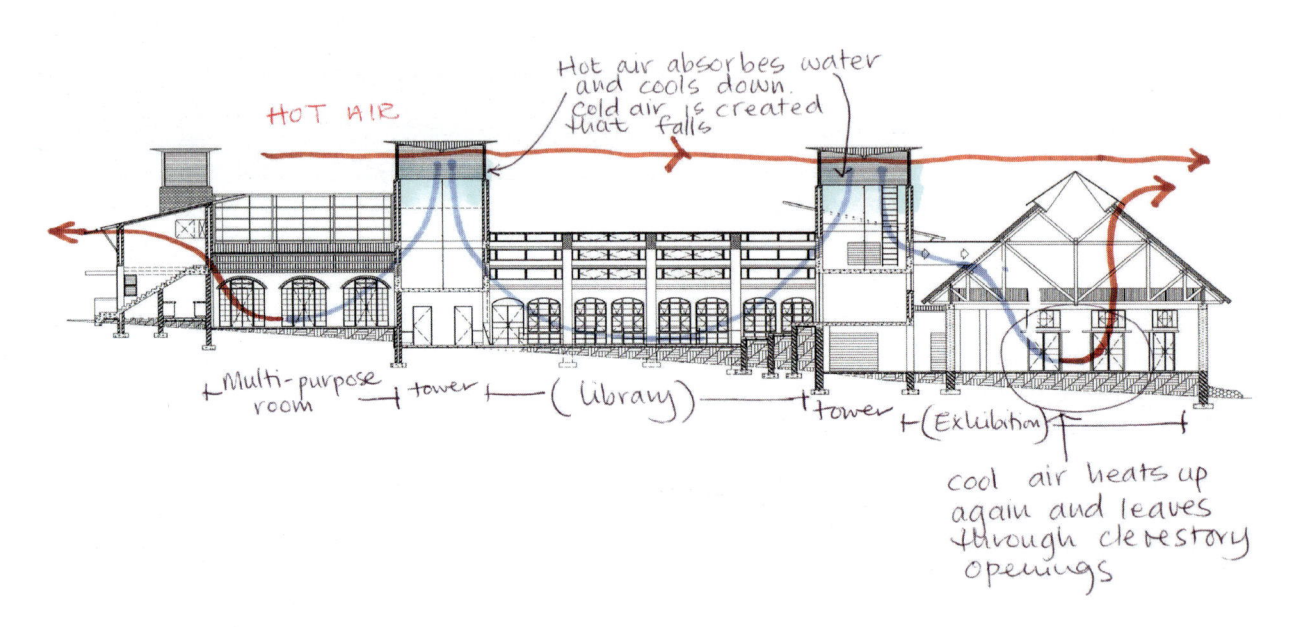

59. Passive downdraught evaporative cooling strategy for the Habitat Resource and Development Centre in Katutura, Namibia [Nina Maritz Architects].

38   **CHAPTER 2** | PRINCIPLES AND STRATEGIES

22°C, or an air moisture content of 12g/kg *(Szockolay, 2008)*. As with convective cooling, control of the rate of evaporation and the airflow through ventilation openings is vital to optimise performance and to avoid over-humidification.

Evaporative cooling occurs as a result of the latent heat of vaporisation – the amount of energy that is required to convert unit mass of water from its liquid to its gaseous/vapour phase (~2,260kJ/kg). This is appreciably greater than the specific heat required to raise the temperature of unit mass of water (4.18kJ/kg). In evaporative cooling, the energy is supplied by ambient air whose heat content and capacity to hold vapour are indicated by its dry bulb temperature and relative humidity. The transfer of heat energy from ambient air to water supports the phase change process and so reduces the temperature of the ambient air *(Bowman et al., 2000)*. For a given pressure, the wet bulb temperature depends on the air temperature and the humidity content of the air. The relationship between air temperature and humidity is best understood by reference to the psychrometric chart. The interpretation of weather data is discussed in more detail in Chapter 3.

> A REDUCTION IN SUPPLY AIR TEMPERATURE OF 70–80% OF THE DRY BULB-WET BULB DEPRESSION CAN BE ACHIEVED BY EVAPORATION

60. Exterior view of the Habitat Resource and Development Centre [Nina Maritz Architects].
61. Evaporative cooling tower in the Habitat Resource and Development Centre.

CHAPTER 2 | PRINCIPLES AND STRATEGIES

2.3. AIRFLOW STRATEGIES: UPDRAUGHT

It is apparent that for a natural cooling strategy to be successful, a viable airflow path must be embedded within the architectural proposition. For naturally driven convective cooling the airflow path may be essentially either an 'updraught' or a 'downdraught' strategy. An 'updraught' airflow strategy is one that, assuming still air conditions, makes provision for low level inlet openings and high level outlet openings. Sometimes referred to as the 'stack' effect, this relies on the fact that warm air rises and is replaced by relatively cooler (denser) air. This flow of air up through the building will continue as long as the low and high level vents remain open, and as long as there is a positive temperature difference between outside and inside the building. This pattern of airflow, in the absence of wind, is also referred to as a buoyancy driven airflow. When wind forces combine with thermal forces to drive airflow through a building, outlet openings at the top of the building are often desirable to exploit the suction (updraught) arising from the wind forces over the building, and also as a strategy to exhaust smoke in the event of a fire. The following section explores a number of different updraught airflow strategies.

> **UNDER BUOYANCY CONDITIONS THE AIR VOLUME FLOW RATE IS DIRECTLY PROPORTIONAL TO THE 'EFFECTIVE' AREA OF THE VENTILATION OPENINGS**

SINGLE SIDED

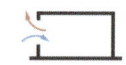

Possibly the simplest 'updraught' airflow strategy is to provide openings on one side of a single room, to allow cooler outside air to enter at low level and for the internal warmer air to rise and leave the space from the upper part of the opening. The airflow rate induced by this strategy can be enhanced by providing a low level 'supply' opening and a separate high level 'exhaust' opening. This strategy has been found to be effective for room depths of about twice the room height, although this depends on occupant density. While this may be satisfactory for a residential or low density office space, it is unlikely to be satisfactory for the much higher occupant density of a school classroom for example, so the strategy must of course account for internal heat gains from occupants, equipment etc. (see Chapter 5).

CROSS VENTILATION

Cross ventilation will generally be a more effective strategy than single sided ventilation, because cross ventilation takes advantage of pressure differences across the building. Pressure differences can be generated by wind forces or by buoyancy (due to relative difference in density and temperature of the air). Generally, cross ventilation has been found to be effective for building depths up to five times the room height, although the achievable airflow rate will be significantly influenced by the area of inlets and outlets and the height difference between the mid-point of these openings. Under buoyancy conditions (thermal forces only) the air volume flow rate is directly proportional to the 'effective' area of the ventilation openings (double the area and the flow rate is doubled), and that airflow rate increases in proportion to the square root of the 'stack' height. Preliminary calculations of airflow rate related to the effective area of the vent openings and the stack height are described in Chapter 5. A combination of natural stack and cross ventilation strategies are applied (during exhibition put-up and take-down modes) to the large daylit exhibition spaces in the Hanover Exhibition Building (Hall 26) in Germany by Thomas Herzog & Partner (1996), and in the David Lawrence Convention Center in Pittsburgh (fig. 62) by Raphael Vinoly Architects *(Ji & Plainiotis, 2006)*.

CENTRAL ATRIUM/CONCOURSE (OPEN OR ENCLOSED)

Single-sided and cross ventilation are limited by the floor to ceiling height of the space considered, and so for many non-domestic buildings airflow strategies which link multiple floors within a building can have significant advantages. The design of a central atrium or concourse can significantly enhance airflow rates through multi-floor buildings. In the case of an open office arrangement (where the total volume will be limited by fire compartmentation), the space can be treated as a single cell, while recognising the role of the openings at each floor level in achieving an even distribution of air. It is also important to ensure that the final exhaust is significantly above roof level (discussed in more detail in Chapter 5).

62. David Lawrence Convention Centre, Pittsburgh, USA [Raphael Vinoly Architects], ©Brad Feinknopf.

Alternatively, the atrium/concourse and the adjacent floors may be enclosed, in which case the linked spaces are treated as a multi-cell arrangement. This type of arrangement was adopted in the Powergen HQ office building in Coventry UK by Bennetts Associates (figs 49–51), one of the earliest of the new generation of naturally ventilated office buildings in Europe (*Baird, 2001*), and has since been applied in many large office and public buildings. The competition entry for the Manchester Civil Justice Centre, UK by Pringle Richards Sharratt architects (2003) adopted a central atrium strategy, but additionally integrated thermal and acoustic buffering and attenuation to protect courts and consulting rooms from the risk of either external or internal noise disruption (figs 63–64).

63. Building section for competition proposal for naturally ventilated Civil Justice Centre [Architects Pringle Richards Sharratt].
64. Central atrium model view' for competition proposal for naturally ventilated Civil Justice Centre.

CHAPTER 2 | PRINCIPLES AND STRATEGIES

PERIMETER STACK (VENTILATED FAÇADE)

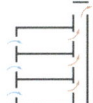

A further option for promoting airflow through multi-floor buildings is to provide a perimeter stack or ventilated façade to enhance both buoyancy and wind driven ventilation. Under buoyancy conditions, it is assumed that the air temperature within the stack is higher than in the occupied floors. In such cases care must be taken to ensure that the neutral pressure plane lies above the top floor to achieve adequate ventilation of upper floors. Airflow rate can be enhanced by encouraging solar gain within the stack, depending on orientation of the façade and latitude of the site. However, care must be taken if overheating of adjacent occupied spaces is to be avoided. The environmental design implications of double skin façades are reviewed in *Oesterle (2001)*. This approach is illustrated in Chapter 4 with the GSW Headquarters Tower in Berlin, Germany by Sauerbruch & Hutton in 1999. It exemplifies one of a range of strategies that can be applied to naturally ventilate tall buildings *(Wood & Salib, 2012)*.

2.4. AIRFLOW STRATEGIES: DOWNDRAUGHT

A downdraught airflow strategy relies on the effect of gravity on a body of (relatively) cold air, to create a draught directed downwards, and thus circulate air from the source of cooling to the occupied zone within the building. The source of the cool air may be either 'passive' (e.g. through the evaporation of water) or 'active' (e.g. from chilled-water cooling coils or panels). The downdraught may be augmented (or inhibited) by wind effects but is often designed to rely on gravity alone.

Supply air can either be delivered at the centre of the plan or from the perimeter. We have identified three generic cases: a central atrium or concourse (open), a central shaft (enclosed) and a perimeter tower.

All three options assume buoyancy driven airflow, and that the impact of the wind is inhibited by baffles. Opportunities for exploiting the wind to increase the air volume flow rate are discussed separately below. It should also be noted that in all cases the airflow pattern can be reversed from downdraught to updraught at night to promote night-time convective cooling.

CENTRAL ATRIUM OR CONCOURSE

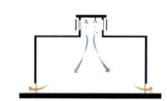

An open central atrium or concourse can be considered as a single cell, where the supply zone and the delivery zone are the same, and surrounding spaces are not served by the downdraught cooling system. Case study buildings of this type include the Phoenix Courthouse, the Malta Stock Exchange (both described in Part 2), and the competition entry for the Johannesburg Constitutional Court by architect Richard Weston (figs 65–66). For initial calculations, the temperature in this single zone can be assumed to be uniform. However, in practice, temperatures can vary significantly within an atrium served by downdraught cooling. This may not be a problem, since atria are often used as transition spaces, and can be allowed to have a wider range of temperatures and humidity than surrounding work spaces. However, where the atrium serves as work or recreation space, then potential variations need to be minimised.

The open geometry of an atrium or concourse also supports the incorporation of chilled water cooling coils (mounted at high level) as a back-up cooling source for times when the relative humidity is too high for direct evaporative cooling to operate effectively. This is one of the modes of operation in the Malta Stock Exchange (see case study in Part 2). The upper threshold for the internal relative humidity is normally set at 65%, and in many locations a hybrid strategy will be appropriate (this is discussed more in Chapters 6 and 7 on thermal comfort and controls).

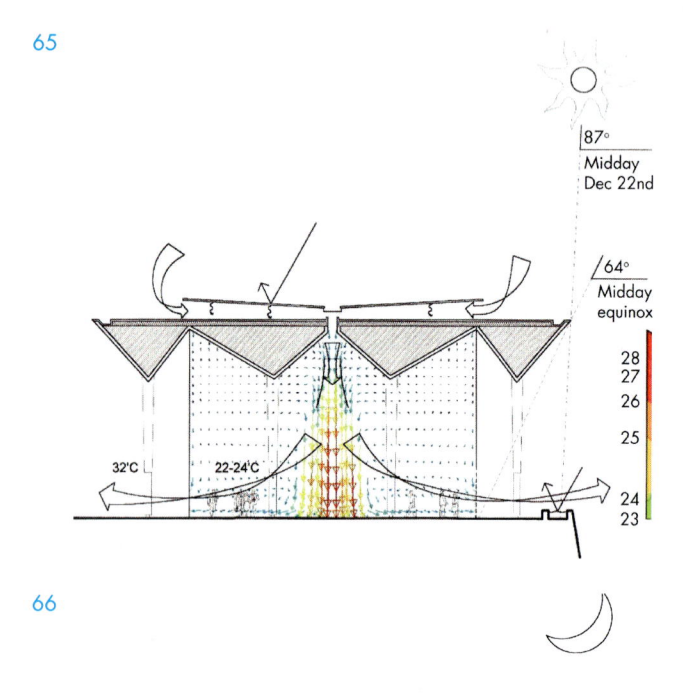

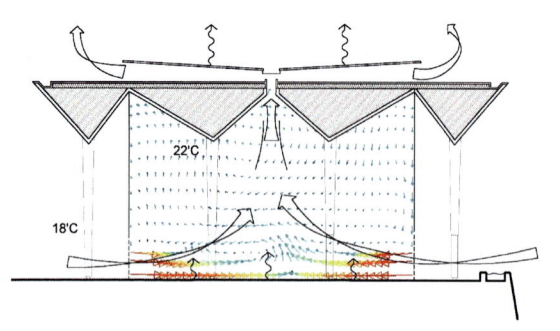

65. Daytime airflow strategy for competition proposal for Constitutional Court, Johannesburg [Richard Weston with Brian Ford Associates].
66. Night ventilation strategy for competition proposal for Constitutional Court.

44 **CHAPTER 2** | PRINCIPLES AND STRATEGIES

CENTRAL SHAFT OR CONCOURSE

An enclosed central shaft or concourse can be considered as the supply zone which serves a number of surrounding zones, and is therefore a 'multi-cell' building when considering airflow and thermal performance simulations. Case study buildings of this type include the Torrent Research Centre (figs 67–68), Ahmedabad, which is described in Part 2.

The temperature in the supply zone is likely to be different from that in other zones. Also, as the height between inlet and outlets within the central shaft or concourse will vary, so too will the vent opening areas required at each level vary, if the same volume flow rate is required at each floor level. The calculation of shaft sizes and vent area opening sizes is discussed in Chapter 5.

PERIMETER TOWER

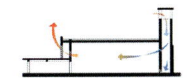

A perimeter tower attached directly to the served space or spaces can be considered a multi-cell case (discussed in more detail in Chapter 6). The Stanford Global Ecology Centre (described in detail in Part 2) and the auditorium in the CII Centre Bangalore, are case study buildings of this type. While a central shaft has the potential to deliver air to all the spaces around it, a perimeter shaft can only deliver cool air to one side.

In order to provide a sufficiently large opening from the tower into the adjacent

67. Central shaft of Torrent Research Centre, Ahmedabad, India [Abhikram Architects].
68. Exterior view of Torrent Research Centre.

CHAPTER 2 | PRINCIPLES AND STRATEGIES

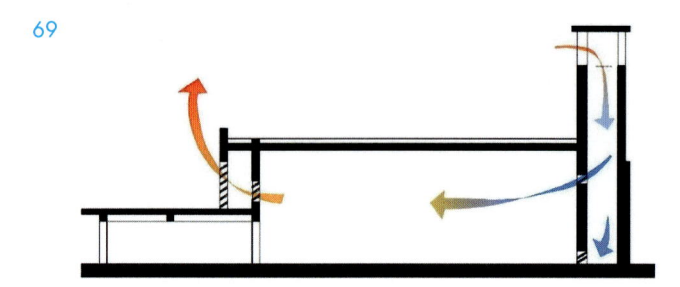

space (to achieve the required airflow rate), it may be necessary to create a plenum along one side of the served space. Such a plenum is provided along one side of the auditorium space at the CII Centre for Excellence, Bangalore. A continuous perimeter delivery zone could form an entry/foyer transition space, where conditions are allowed to vary more widely than in the attached 'working' spaces. Alternatively, a continuous narrow buffer space could be provided in the form of a perimeter double façade. The perimeter plenum and three towers serving the auditorium of the CII Centre in Bangalore are shown in figs 69–70 and section with CFD temperature plots overlaid in fig. 71. Delivery from the perimeter can be successful for open plan attached spaces, but where secondary cellular spaces are included, care must be taken to ensure a satisfactory airflow path is achieved.

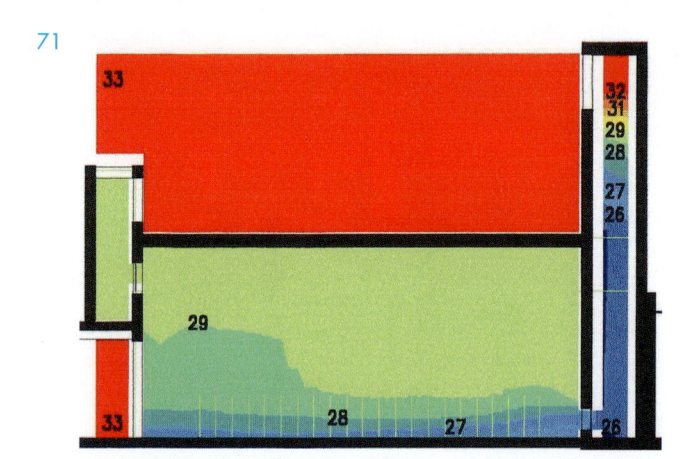

69. Section through auditorium, CII Centre, Bangalore, India [Karan Grover & Associates].
70. View of auditorium misting towers, CII Centre.
71. CFD temperature plot for a sunny day (external air temperature 33°C, still air), CII Centre.

> THE CHOICE OF DOWNDRAUGHT SYSTEM AND THE AIRFLOW STRATEGY CAN HAVE A SIGNIFICANT IMPACT ON BUILDING GEOMETRY, AND THEREFORE NEEDS TO BE CONSIDERED EARLY IN THE DESIGN PROCESS

CHAPTER 2 | PRINCIPLES AND STRATEGIES

2.5. THE COOLING SOURCE FOR DOWNDRAUGHT SYSTEMS

The downdraught systems available vary according to the technology applied, from 'low tech' porous media to relatively 'high tech' misting systems as the source of cooling. The options vary according to the climatic opportunity provided by the project location (see Chapter 3) and the source of the cool air, which may be generated by either:

• Porous Media: the irrigation of a cellulose matrix in the path of the air stream (e.g. wet pads in a 'Cool Tower'), or wet porous surfaces located within the air stream (e.g. porous ceramic in an 'Air Shaft').

• Misting Systems: a mist of water sprayed into the air stream (e.g. by misting nozzles within a delivery zone), or large droplets of water sprayed into the air stream (e.g. by a 'Shower Tower').

• Active Downdraught Cooling: chilled water coils or de-humidifiers creating a downdraught of cool air. The choice of downdraught system and the airflow strategy can have a significant impact on building geometry, and therefore needs to be considered early in the design process. There are a number of different geometric options which define the relationship between the source of cool supply air and the served spaces. The different options result in different airflow paths, areas of influence, and cooling effectiveness and comfort. For each delivery mode there are also different geometric options which will influence the pattern of airflow.

POROUS MEDIA

Cool towers

The evaporation of water through a cellulose matrix located at the top of a tower will create a downdraught which can potentially drive the airflow through a building. The towers can be located either at the perimeter (as at the Zion National Park Visitor Centre, fig. 72) or within the depth of the building. Most of the buildings constructed with cool towers in the USA are single storey, probably due to the large surface area required relative to the air volume flow rate achieved. While this represents the potential, cellulose pads within cool towers provide a significant resistance to airflow in still air conditions, and therefore large surface areas of pads are required in relation to the volume flow rate. Field studies have found that in windy conditions the airflow can carry water droplets into the building. Water consumption can also be high as there are substantial losses to ambient, and water is supplied to the pads for most of the day.

Ceramics

The historic use of porous ceramic jars to promote evaporative cooling can be reinterpreted today to provide cooling to individual (cellular) spaces. The production of a cool downdraught by the wetting of porous ceramic bricks within a tower has been shown to be a viable design approach in Greece *(Papagiannopoulos & Ford, 2003)*. The use of porous ceramic panels within a wall element was also tested as part of the EC funded EVAPCOOL research project which enabled the performance characteristics to be defined. The applicability of this technique to housing in southern Spain has also been studied *(Schiano-Phan & Ford, 2003)*, but has so far not been applied widely in practice. Porous ceramic products have been developed for the cooling of transitional spaces between inside and outside. The 'E-Cooler' *(StudioKahn, 2010)* has been designed to form a screen between inside and outside (figs 73–74), and the influence of ceramic form and airflow rate on cooling performance has been evaluated in recent doctoral research by *Vallejo et al. (2017)*.

When integrated within a building the air supply path containing the ceramic panels must be directly connected to the outside, which implies either a 'double' envelope or integration within a shaft or duct within the external wall, or the roof or within a standalone shaft within the building. Porous ceramic panels can be applied to cellular spaces and the shaft height will normally be limited to a single storey height. An external louvred grille at high level will deliver air into the shaft, and exhaust cool air into the room at low level. An equivalent area will also have to be provided to exhaust air from the room. Laboratory tests have established that the velocities achieved by the shaft are low (less than 0.3m/s) and may be either enhanced or negated by external wind pressures *(Schiano-Phan & Ford, 2003)*.

72. Zion National Park Visitor Centre [NPS Denver Service Centre Architects].
73. Ecooler design concept by StudioKahn, ©StudioKahn.
74. Ecooler screen by StudioKahn, ©StudioKahn.

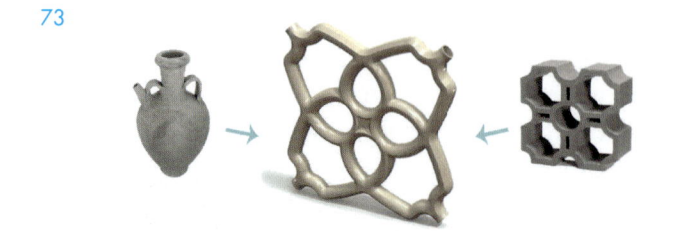

CHAPTER 2 | PRINCIPLES AND STRATEGIES

The 'wicking' effect of porous ceramic components was exploited in the Spanish Pavilion at the Zaragoza Expo 2008 (figs 75–76), and significant research and development has taken place in the application of porous ceramic components to moderate the external urban environment. Porous ceramic evaporators have also been used to provide both shading and evaporative cooling to the outside façade of the Sony Osaki office building in Tokyo, Japan by Nikken Sekkei. The façade design is said to be based on the principles behind 'sudare', or traditional Japanese screens. Rainwater collected from the roof area is fed through special porous ceramic pipes, and as the water evaporates, it reduces the surface temperature of the pipes and the adjacent air. The architects stated that ceramic surface temperature could be lowered by 10°C on the hottest day in the summer and claim it could help reduce the temperature in the surrounding walkways and the entrance hall by 2°C. Generally a large surface area is required if either porous ceramic or cellulose pads are to achieve significant cooling.

SHOWER TOWERS

To overcome some of the disadvantages of using cellulose pads, the direct evaporation of large droplets of water within the air-stream has been adopted as a strategy in a number of buildings. Hassan Fathy had experimented with running water over charcoal within an Egyptian 'Malqaf' *(Fathy, 1986)*, collecting the water in a pool at the base of the tower. The use of simple shower heads at the top of the tower within a small courtyard was also tested experimentally, first by Givoni and Alhemiddi in southern California *(Givoni, 1994)*, and then by

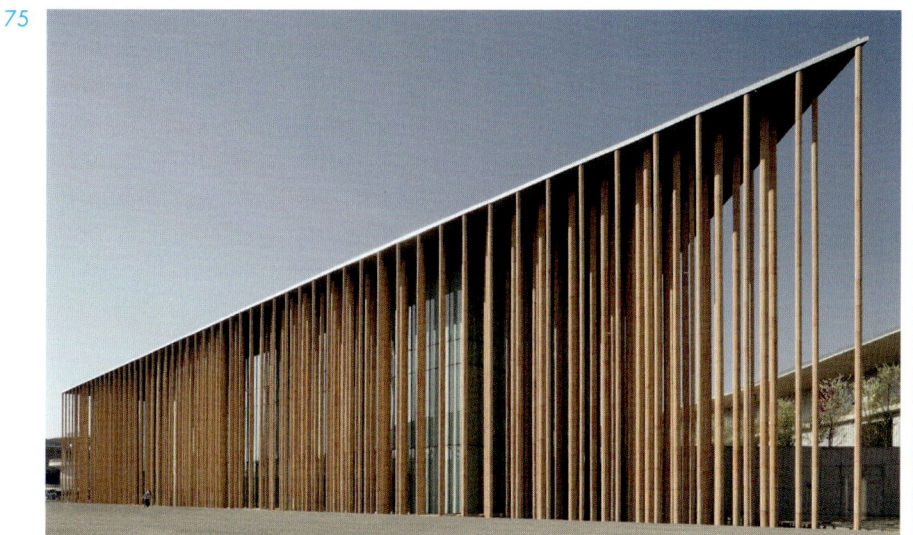

75. Spanish Pavilion in Expo Zaragoza 2008, Spain [Francisco Mangado], ©Pedro Pegenaute.
76. Porous ceramic columns in Spanish Pavilion, Expo Zaragoza 2008, Spain, ©Pedro Pegenaute.

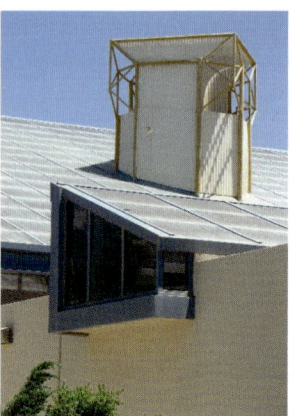

Etzion et al. (1997) in southern Israel. In the latter case, the shower tower serves an enclosed (glazed) courtyard, adjacent to an office building and library (figs 77–79). The use of large droplet shower heads means that much of the water does not evaporate, and has to be collected in a pool at the base of the tower. The influence of wind catcher design on air volume flow rate was compared with fan induced air movement. The performance of the Israeli project has been recorded over a number of years and reported (Erell, 2007).

Airflow strategies for shower towers and cool towers are essentially the same as for misting towers, except that with shower towers and cool towers cool supply air is generally delivered at the base of the tower, whereas with misting towers cool air can be delivered throughout the height of the building. Shower towers are generally enclosed, and require a collection pool at the base of the tower. The shower tower at the Blaustein Institute in Sede Boqer, Israel, is of the 'open' atrium type, while the Interactive Learning Centre at the Dubbo Campus in Australia incorporates four shower towers of the central enclosed type. Perimeter attached towers were incorporated as part of the cooling strategy for Council House 2 (CH2) in Melbourne, Australia (fig. 80).

Airflow obstructions have been identified as a recurring problem in the operation of 'cool tower' buildings. Yoklic & Thompson have reported that on occasion towers have been connected to a space "with inadequate or improperly located outlets to exhaust the

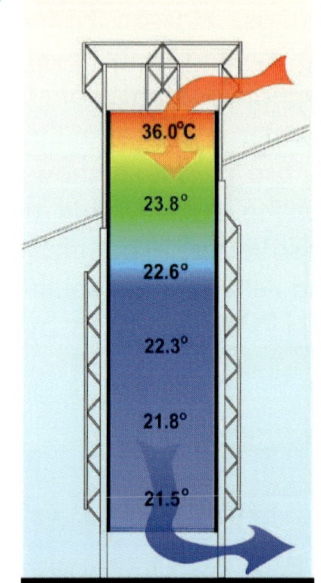

77. Interior view of the Blaustein International Center for Desert Studies Building, Israel, showing shower tower in atrium [Wolfgang Motzafi-Haller].
78. Exterior view of the Blaustein International Center for Desert Studies Building, Israel, showing shower tower wind catcher [Evyatar Erell].
79. Vertical temperature profile inside the Blaustein Institute tower, Israel, on a typical hot summer day. Figure based on temperature measurements [David Pearlmutter].

CHAPTER 2 | PRINCIPLES AND STRATEGIES

air" *(Yoklic & Thompson, 2004)*. It is clear that in some buildings, while inlet and outlet to the cool tower itself has been considered, the airflow path through the building has not been adequately considered. Exhaust from the building will only be successfully achieved if the outlet free area is at least equal to the inlet free area, and if the outlets are baffled from potential positive wind pressure (this was found to be a particular problem at the Dubbo campus building). Designers must ensure that inlet or outlet openings are not obstructed and that decorative grilles, fly screens, etc., are accounted for in the calculation of effective free area.

> AIRFLOW OBSTRUCTIONS HAVE BEEN IDENTIFIED AS A RECURRING PROBLEM IN THE OPERATION OF 'COOL TOWER' BUILDINGS

MISTING SYSTEMS

Many aspects of the tradition outlined above were reviewed by the designers of the Expo site in Seville for the 1992 World Fair *(Alvarez et al., 1991)*. This included the 30 metre high 'cool towers' of the Avenue of Europe, which employed high pressure water misting nozzles to induce downdraught cooling (illustrated in Chapter 1).

The first large scale application within a building was in the Torrent Research Centre at Ahmedabad in India, where misting towers have successfully provided cooling to pharmaceutical research laboratories and offices since 1998. Since then, a number of other buildings in India and the USA

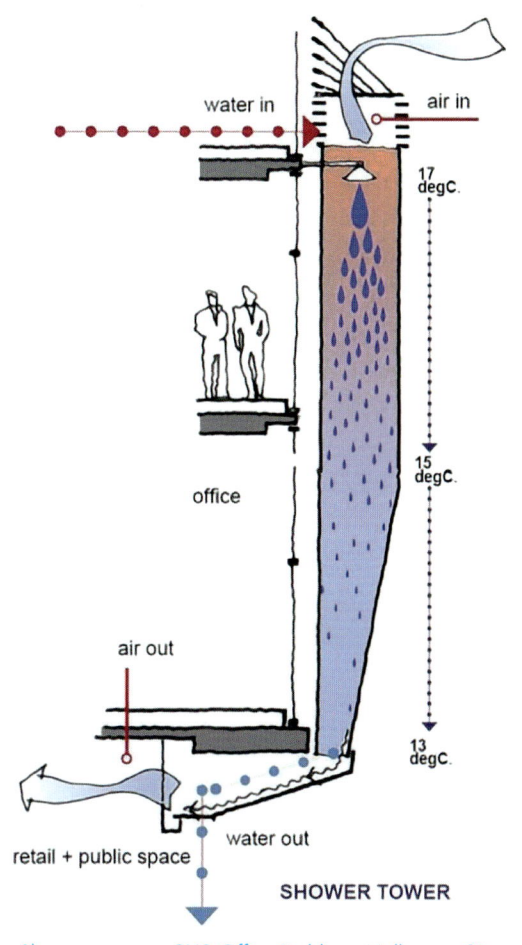

80. Shower tower at CH2 Office Building, Melbourne [Design-Inc].

CHAPTER 2 | PRINCIPLES AND STRATEGIES   51

have applied this approach. Experience in the design of these buildings suggests that the successful application of misting towers requires a 'void-to-floor' ratio of 0.05 (i.e. where 5% of the building footprint would be void). These voids normally take the form of vertical shafts, lightwells or atria, incorporating misting nozzles located above the top floor served. Automatic control of inlet and outlet vents is required at each level. These vents must also be sized to balance the airflow between floors, and baffles or exhaust stacks may be required to prevent counter-flow.

At the Phoenix Courthouse in Arizona, fully air-conditioned cellular offices and courtrooms are located on the south side, with open balconies on the north side within the atrium space (fig. 81). Linear diffusers provide additional cooling to the balcony corridors. The evaporative misting system creates a 'curtain' of cool air in front of the balconies, buffering the offices from the relatively higher temperatures of the atrium. (For more detail on the Phoenix Courthouse see case study in Part 2).

In an enclosed atrium or concourse arrangement, exhaust from the separate floors and individual spaces takes place via either dedicated perimeter shafts (as in the case of the Seville office building and the Torrent Research Laboratories), or via a 'ventilated'

81. Atrium of the Phoenix Courthouse in Arizona, ISA [Richard Meier Architects], ©Scott Francis.

or 'double' façade. This may be necessary in order to inhibit the impact of wind and to guarantee exhaust from the building under all conditions. It should be noted that under most conditions when evaporative cooling is operating, the air will exit the building from the base of the exhaust towers, due to the fact that exhaust air is at a much lower temperature than ambient. When the wind is blowing, air from the top floor may exit from the top of the shafts/double façade. This can be accommodated by providing a raised parapet to the top floor.

ACTIVE DOWNDRAUGHT COOLING

While evaporative cooling may be applicable in much of southern Europe, it is rarely applicable for 100% of the time (simply because at times the wet bulb temperature is too high and further evaporation could lead to saturation). Therefore, some form of back-up cooling may be required, depending on location, building type and internal heat gains. In other climatic regions (warm humid or cool temperate), evaporative cooling may never be appropriate due to continually high humidity. In seeking alternatives to conventional a/c systems, a back-up or stand alone system that mimics the downdraught operation of PDEC has considerable advantages. Chilled water cooling coils located high in an atrium or lightwell have been used successfully in a number of buildings to perform this role (figs 82–83). Chilled water cooling coils located at the top of a column of air can induce a downdraught of cool air into a space, displacing warmer air which rises to the height of the coils before

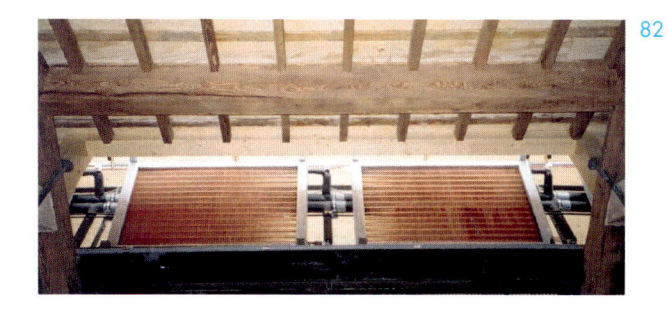

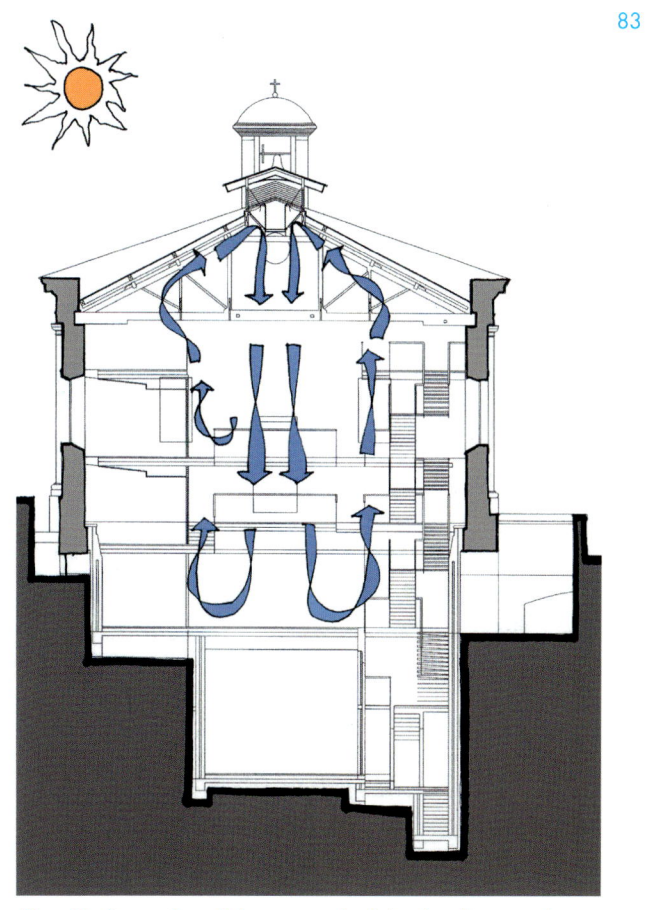

82. Cooling coils and drip tray on high level walkway within atrium of Stock Exchange, Malta.
83. Summer day cooling strategy using cooling coils (minimum supply air), Stock Exchange, Malta [Architecture Project].

CHAPTER 2 | PRINCIPLES AND STRATEGIES   53

being cooled again, thus setting up a natural buoyancy driven circulation of air. This can also be induced within a room (single floor height) or within a large volume (many floors in height). By avoiding the need for a full mechanical cooling system (including air handling units, ductwork, suspended ceilings etc.), significant capital and energy savings can be achieved. While requiring chilling equipment, the avoidance of fans, ductwork etc. also reduces maintenance and replacement costs. Room based systems such as 'Gravivent'® are available commercially (see TTC Timmler Technology GmbH).

Such an approach has been applied to a wide range of building types in different parts of the world, but rarely in conjunction with passive downdraught evaporative cooling. Both techniques exploit gravity to drive the airflow, avoiding the need for fans. They can be combined successfully to meet cooling needs where ambient conditions vary from hot and dry to hot and humid. Gravity driven mechanical cooling (using conventional cooling coils) has been applied in the Malta Stock Exchange and the School of Slavonic and Eastern European Studies, for UCL London. Buoyancy driven cooling has also been adopted in the CSET building at the University of Nottingham campus in Ningbo, China, as part of a hybrid cooling strategy, but in this instance the cool air is introduced into the top of the shaft via a fan driven de-humidifier. These buildings are described and discussed in detail in Part 2.

This chapter has presented an outline of the environmental design principles and strategies which can moderate the impact of the external climate on the internal environment, without recourse to significant mechanical assistance. It initially describes the different aspects of site and microclimate analysis which forms a crucial first step in defining which principles and strategies may be appropriate for a particular location. It goes on to explore the principles of natural cooling based on exploiting environmental heat sinks through natural convection, night sky radiation, ground cooling and the evaporation of water. The integration within the design process of appropriate strategies to achieve both 'updraught' and 'downdraught' convective cooling was then discussed. The interpretation of weather data, and the climatic applicability of different strategies in Europe, China and the USA is presented in the next chapter, while performance analysis and testing of these strategies is discussed in Chapters 5 & 6 on performance assessment.

REFERENCES

- Alvarez, S. et al. (1991). *'The Avenue of Europe at Expo 92: Application of Cool Towers in Architecture and Urban Space'*. 9th Passive and Low Energy Architecture, Seville, Spain. Pergamon, pp. 195–202.
- Baird, G. (2001). *'The Architectural Expression of Environmental Control Systems'*. Spon Press, pp. 44–46.
- Baker, N. & Steemers, K. (2000). *'Energy and Environment in Architecture – A Technical Design Guide'*. E&FN Spon.
- Bowman, N. T., et al. (2000). *'Passive Downdraught Evaporative Cooling: I. Concept and Precedents'*. Indoor and Built Environment 9 (5), pp: 284–290.
- Browne, E. (2008). *'Edificio Consorcio Santiago: Catorce Años Después'*. Revista CA N° 133.
- Buchanan, P (2003). *'Renzo Piano Building Workshop Complete Works – Vol 1'*. Phaidon, pp. 134–139.
- CIBSE (2005). *'Applications Manual 10: Natural Ventilation in Non-Domestic Buildings'*. London: Chartered Institution of Building Services Engineers.
- Erell, E. (2007) in Santamouris, M. (2007) *'Advances in Passive Cooling'*. Routledge/Earthscan.
- Erell, E., Pearlmutter, D. & Williamson (2011). *'Urban Microclimate: Designing the Spaces between Buildings'*. Earthscan.
- Etzion, Y., Pearlmutter, D., Erell, E. & Meir, I. A. (1997) *'Adaptive Architecture: Integrating Low-Energy Technologies for Climate Control in the Desert'*. Automation in Construction 6, September, pp 417–425.
- Fathy, H. (1986). *'Natural Energy and Vernacular Architecture: Principles and Examples with Reference to Hot Arid Climates'*. Chicago, Published for the United Nations University by the University of Chicago Press.
- Gartland, L. (2008). *'Heat Islands: Understanding and Mitigating Heat in Urban Areas'*. Earthscan.
- Givoni, B. (1994). *'Passive and Low Energy Cooling of Buildings'*. Van Nostrand Reinhold, pp.139–143.
- Hyde, R. (2003). *'Climate Responsive Design – A Study of Buildings in Moderate and Hot Humid Climates'*. Spon Press.
- Ji, Y. & Plainiotis, S. (2006). *'Design for Sustainability'*. China Architecture & Building Press, pp. 79–85.
- Krishan, A., et al. (2004). *'Climate Responsive Architecture – A Design Handbook for Energy Efficient Buildings'*. McGraw Hill.
- Littlefair, P. J., et al. (2000). *'Environmental Site Layout Planning: Solar Access, Microclimate and Passive Cooling in Urban Areas'*. Building Research Establishment.
- Mulligan, H. (1983). *'Environmental Characteristics of the Vernacular Underground Dwelling'*. Proc. Plea International Conference. Crete, Greece, 1983.
- Mumovic, D. & Santamouris, M. (2009). *'Handbook of Sustainable Building Design & Engineering'*. Earthscan/Routledge.
- Ng, E. (2010). *'Designing High Density Cities for Social and Environmental Sustainability'*. Routledge.
- Nicol, F., et al. (2012). *'Adaptive Thermal Comfort: Principles and Practice'*. London, New York. Routledge.
- Oesterle, E. (2001). *'Double-Skin Facades : Integrated Planning: Building Physics, Construction, Aerophysics, Air-conditioning, Economic Viability'*. Munich; London, Prestel.
- Oke, T. M. (1987). *'Boundary Layer Climates'*. Methuen, 2nd ed. London.
- Papagiannopoulos, G. & Ford, B. (2003). *'Evaporative Cooling Using Porous Ceramic Bricks: Experimental Results from Greece'* Proc. 20th Plea International Conference. Santiago, Chile. 9–12 November.
- Santamouris, M. (2001). *'Energy and Climate in the Urban Built Environment'*. James & James.
- Santamouris, M. & Kolokotsa, D. (2016). *'Urban Climate Mitigation Techniques'*. Routledge.
- Schiano-Phan, R. & Ford, B. (2003). *'Evaporative Cooling Using Porous Ceramic Evaporators – Product Development and Generic Building Integration'*. Proc. 20th Plea International Conference. Santiago, Chile. 9–12 November.
- StudioKahn (2010). *'Ecooler'* [Online]. http://studiokahn.com/portfolio/ecooler/. [Accessed 10/2018].
- Szokolay, S. (2008). *'Introduction to Architectural Science'*. Elsevier/Architectural Press.
- Vallejo, J., et al. (2017). *'Predicting Evaporative Cooling Performance of Wetted Decorative Porous Ceramic Systems in Early Design Stages'*. Proc. 33th PLEA International Conference. Edinburgh, Scotland. 2–5 July.
- Wood, A. & Salib, R. (2012). *'Natural Ventilation in High Rise Office Buildings'*. Routledge, pp: 64–73.
- Yannas, S., Erell, E. & Molina, J. L. (2006). *'Roof Cooling Techniques'*. Earthscan.
- Yoklic, M. & Thompson, T. (2004). *'Cooltowers, Passive Cooling and the Case for Integrated Design'*. Proc. 2004 American Solar Energy Society meeting, Portland Oregon, USA.

CHAPTER 3
CLIMATE APPLICABILITY MAPPING

At early stages in the design process, speedy and robust assessments of feasibility are facilitated by reference to reliable sources of weather data to determine the applicability of natural cooling strategies. Weather data, in the form of averaged hourly values for each environmental parameter, can be analysed and interpreted to establish the viability of different strategies. Linear plots of annual weather data together with psychrometric charts can provide an excellent basis for a quick and robust initial evaluation, for a particular location. Applicability maps also enable the evaluation of different passive cooling strategies, but at a larger geographical scale, without the need to access weather data for multiple locations.

This chapter reviews the interpretation of climate data for a particular location, and also the use of applicability maps, to evaluate the initial feasibility of different cooling strategies. These tools provide architects and engineers with quick and robust references to make early strategic decisions regarding the feasibility of different cooling options.

3.1. INTERPRETING WEATHER DATA

The local climatic and microclimatic characteristics (i.e. temperature, relative humidity, solar radiation, wind speed and direction, etc.) are clearly significant when considering the risk of overheating for a specific location and in determining the viability of a certain passive cooling strategy. Weather data describing the climate of a specific location can be obtained by reference to either local meteorological stations (airports have traditionally been a location for recording weather data), or to international databases distributed by professional institutions like CIBSE or ASHRAE and specialised websites and software applications such as, for example, 'Energy Plus', 'OneBuilding' or 'Meteonorm' *(EnergyPlus, 2018; OneBuilding, 2018; Meteonorm, 2018)*. The data, mostly provided in tabular format for periods of up to one representative year, can then be analysed using spreadsheet programs or available climate analysis tools like 'Climate Consultant', 'Climate Tool' or the recently developed software package 'Ladybug Tools' *(UCLA, 2017; ClimateTool, 2016; Ladybug, 2017)*.

From the point of view of cooling we are particularly interested in the summer period. However, the characteristics of the other seasons are equally important in determining an overall passive design strategy for the building. Monthly air temperature and humidity variation can help to reveal the opportunity (and limitations) of convective and evaporative cooling strategies, whereas solar radiation and rainfall data can indicate the importance of shading to protect vulnerable orientations and the opportunity for solar energy and rainwater harvesting.

As an example, using the climate data for Madrid, sourced from the Spanish Weather for Energy Calculations (SWEC), the applicability of natural cooling can be analysed (fig. 84). The weather data for Madrid is characterised by a large diurnal range, with night-time temperatures below 20°C during the summer months. This indicates that night-time convective cooling can potentially be exploited. Further, mean maximum temperatures remaining above 30°C over the summer period suggest the need for additional cooling strategies to overcome the risk of overheating.

> THE OPPORTUNITY FOR EVAPORATIVE COOLING IS TRADITIONALLY ASSOCIATED WITH DRY CLIMATES WHEN WET BULB TEMPERATURES ARE BELOW 22°C

The opportunity for evaporative cooling has been traditionally associated with dry climates when wet bulb temperatures are below 22°C *(Givoni, 1994)*. In Madrid, mean maximum wet bulb temperatures are below 16°C during the summer period. The principles and the opportunity for evaporative cooling can be better understood by using the psychrometric chart, first developed by Dr Willis H. Carrier in 1904 *(Gatley, 2004)*. The chart illustrates air moisture content in different units allowing proximity to saturation to be determined and comparison

CHAPTER 3 | CLIMATE APPLICABILITY MAPPING

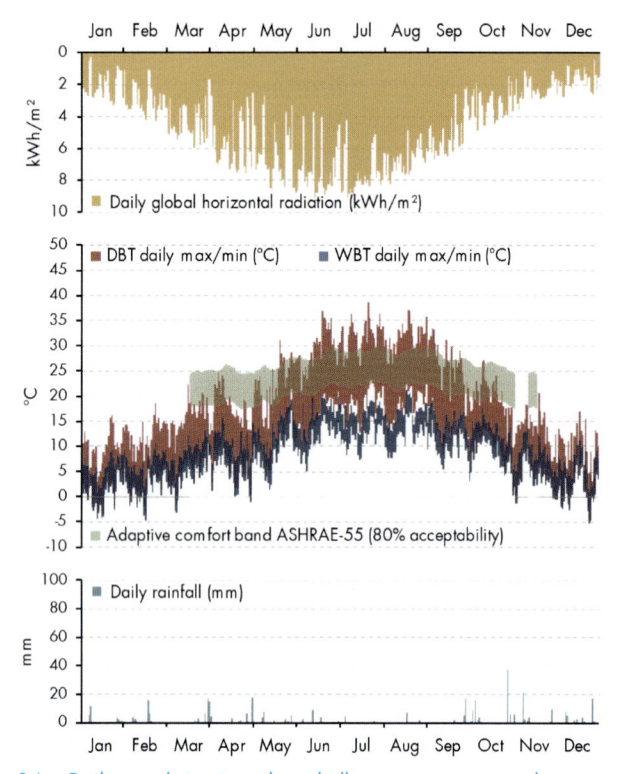

84. Daily max/min air and wet bulb temperatures, cumulative global horizontal radiation and rainfall for Madrid, Spain, ®Meteonorm.

of moisture levels against humidity thresholds for human thermal comfort. *Szokolay (2008)* suggests that evaporative cooling is not applicable above a threshold of 12g/kg of air moisture content, where human thermal comfort is of concern. If we compare the psychrometric chart for Madrid with that of Singapore (fig. 85), we see that in Singapore this threshold is exceeded for the whole summer season, whereas in Madrid it is rarely exceeded. Madrid is therefore (theoretically) an excellent location for evaporative cooling, whereas in Singapore a different strategy is necessary.

In winter, Madrid experiences mean maximum monthly temperatures below 15°C for three months. This suggests a significant heating load, but the climate is also characterised by substantial solar radiation per day in winter on both horizontal surfaces (2 kWh/m²) and south-facing vertical surfaces (3.5 kWh/m²). Therefore, the combination of passive solar gain and internal heat gains coupled with a high performance (well insulated) building envelope may reduce the residual heating requirement to below 15kWh/m² (the *'Passivhaus'* standard for heating energy).

Other strategic options for both summer and winter may include exploiting the heat capacity of the ground to provide both pre-cooling of ventilation air in summer and pre-heating in winter. The mean annual air temperature for Madrid is 14.5°C, equivalent to a constant soil temperature at a depth of about 10m as result of the thermal capacity of the soil.

CHAPTER 3 | CLIMATE APPLICABILITY MAPPING

During the midseasons of spring and autumn, mean monthly temperatures of 12–16°C suggest that comfort can be maintained within a house by natural ventilation alone. Occupant control of window openings should enable regulation of natural ventilation to remove any heat gains during the day, while set to 'trickle' ventilation at night to provide minimum fresh air.

3.2. APPLICABILITY OF PASSIVE COOLING METHODS

It is apparent that analysis of weather data, plotted on linear and psychrometric charts for a specific location, can promote rapid interpretation to support strategic decision making. Such plots can help to define the climatic basis of both the need for cooling and the opportunity for different passive cooling strategies. The combination of 'need' and 'opportunity' can provide the basis for determining the 'applicability' of a particular technique.

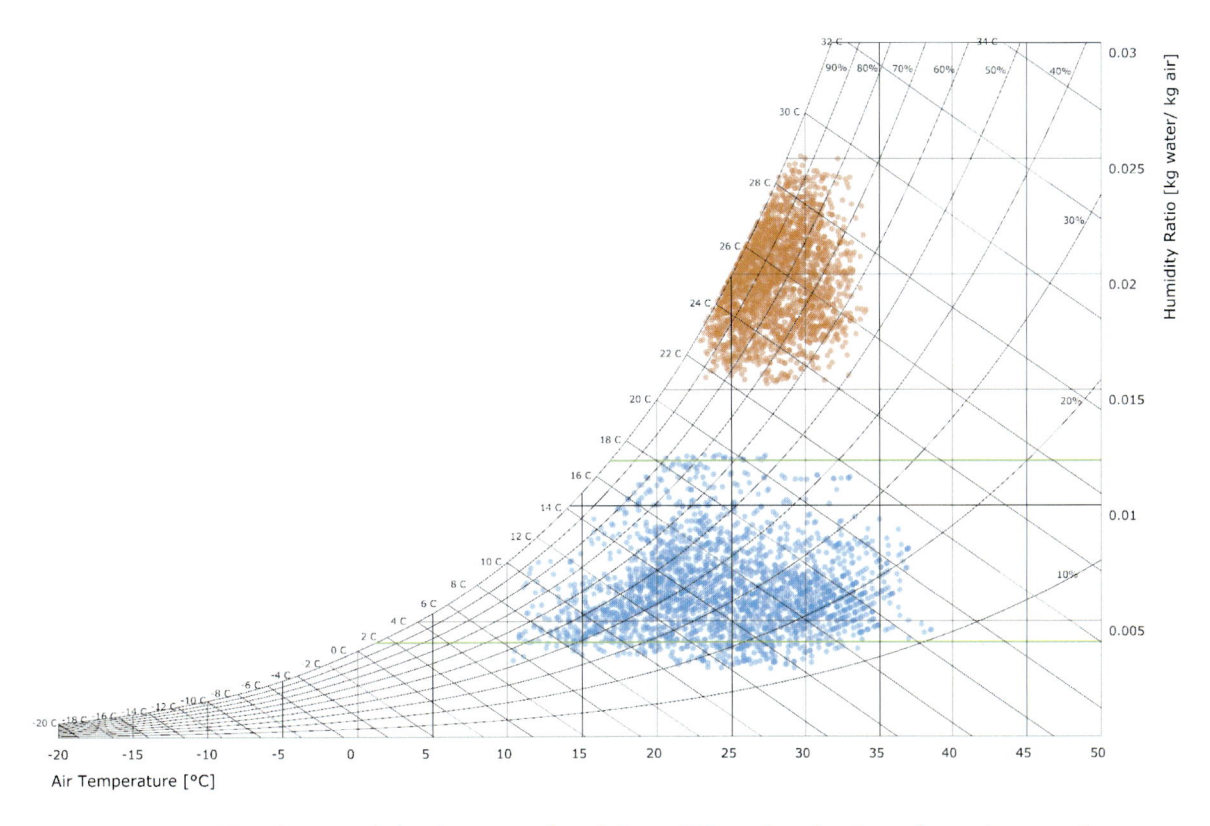

85. Air temperature and humidity in Madrid and Singapore from 1 June to 1 September plotted over the psychrometric chart.

CHAPTER 3 | CLIMATE APPLICABILITY MAPPING

The *'need'* (or demand) for cooling is based on a combination of climatic factors, and building design characteristics (uses, occupancy density, equipment & lighting). Preliminary assessments of cooling needs are often simply related to climatic factors and can be expressed as the number of cooling hours (CH) or cooling degree hours (CDH) for a location. The number of cooling hours represent the number of hours when cooling might be needed and can be determined directly from hourly weather data for the location, or from maps for the region.

Assessment of the *'opportunity'* of applying different passive cooling options strategies in a specific location will be determined by climatic factors alone (including dry and wet bulb temperatures and inside-outside temperature difference). The opportunity of a passive cooling strategy for a location can be expressed in terms of a temperature difference range (ΔT).

> THE 'APPLICABILITY' OF A PARTICULAR TECHNIQUE IS A MULTIPLE OF NEED & OPPORTUNITY

The 'need' for cooling in a location may be 'low' or 'high', just as the 'opportunity' for a particular passive cooling technique may be 'low' or 'high'. The 'applicability' of a particular technique can therefore be considered to be a multiple of 'need' and 'opportunity', and this is the basis for assessing the applicability of cooling by natural convection, evaporation and active downdraught. Essentially:

APPLICABILITY = NEED (CH) x OPPORTUNITY (ΔT)

NATURAL CONVECTIVE COOLING

Natural ventilation is an effective strategy to provide healthy and comfortable internal environments when outdoor conditions allow. In addition, the capacity to reduce indoor temperature through convection (convective cooling) is also widely appreciated and presents significant benefits against mechanical systems as discussed in Chapter 2.

The rate of (sensible) convective cooling is directly proportional to the volumetric heat capacity of air ($\approx 1.2 kJ/m^3K$), the air volume flow rate and the temperature difference between inside and outside (ΔDBT). For a natural ventilated space, this last parameter is the air temperature difference between the incoming (external) cool air and the average (indoor) warm air. This is further explained in Chapter 5.2. Assuming an indoor temperature of 26°C, equal to the upper limit of a thermal comfort zone for indoor environments with high humidity and high air velocity (*Givoni, 1994*), the climatic opportunity for convective cooling can be directly determined by the indoor-to-outdoor temperature depression, 26°C-DBT. The equivalent cooling is thus directly proportional to the indoor-to-outdoor air temperature difference and responds to the question: how much cooler is the climate with respect to indoor temperature?

Looking at the climate of Madrid (fig. 86), there is an evident opportunity to use natural convection to cool indoor spaces during the midseason (mean 26-DBT>6°C) and, if the internal gains are high, also during the winter season (mean 26-DBT>12°C). In summer, when mean ambient air temperature is high, the above index is low, indicating that alternative cooling strategies are needed to maintain indoor temperatures within a comfortable range.

A second index determining average daily temperature fluctuation can contribute to a deeper investigation on the feasibility of convective cooling strategies. Night ventilation is a useful strategy to release the heat received and often absorbed by the building mass during the previous day, and a higher temperature drop at night can increase convective heat exchange and internal heat losses. The average indoor-to-night temperature depression, 26°C-DBTmin, indicates the opportunity for a night ventilation strategy coupled to high thermal capacitance materials to reduce indoor peak temperatures. As in the previous index, a high temperature difference would lead to a higher opportunity for night ventilation cooling.

In Madrid during the warm season, mean 26°C-DBTmin for July is typically >8°C, indicating significant potential for night-time convective cooling (fig. 87). However, street noise and local heat island effects may influence viability. Also, while night-time convective cooling may potentially contribute to a passive cooling strategy in Madrid, it is unlikely to deal with the need for additional daytime cooling, and therefore alternative cooling strategies may be needed to address overheating risks during the day in summer.

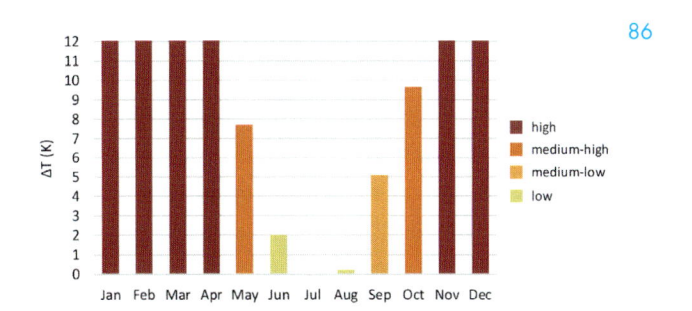

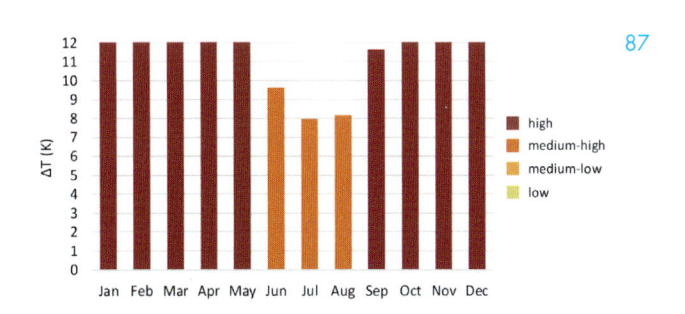

86. Opportunity for daytime convective cooling (26°C-DBT) in Madrid, Spain.
87. Opportunity for night-time convective cooling (26°C-DBTmin) in Madrid, Spain.

PASSIVE EVAPORATIVE COOLING

Chapters 1 and 2 have illustrated how the opportunity for evaporative cooling has been exploited in different ways, both historically and in contemporary applications, in many different locations around the world. Recent research has established the performance characteristics of different evaporative cooling techniques. Studies by *Givoni (1994)* confirmed that air temperature can be reduced by about 70% to 80% of the wet bulb temperature depression (DBT-WBT) depending on the type of system. Considering a typical indoor air temperature about 3 to 5°C higher than ambient due to internal heat gains, Givoni concluded that direct evaporative cooling can be applied only in regions where WBT remains below 22°C in order to satisfy indoor comfort conditions. Likewise, *Santamouris (2007)* discouraged the applicability to regions where wet bulb temperatures are too high for human thermal comfort. A 22°C WBT may be a reasonable numeric reference when the objective is to condition indoor spaces but, when the objective is to cool people in semi outdoor/transitional spaces as traditionally practised, this threshold could be increased up to 24–25°C. Still, this threshold is indicative and may be adapted to a local thermal comfort criteria and ambient conditions.

> THE OPPORTUNITY OR EFFICIENCY OF AN EVAPORATIVE COOLING METHOD DERIVES FROM THE WET BULB TEMPERATURE DEPRESSION (DBT-WBT)

The opportunity or efficiency of an evaporative cooling method thus derives from the wet bulb temperature depression (DBT-WBT) and responds to the question: how dry is the climate? This index can be adapted to different contexts:

- An average DBT-WBT between each hour of the analysis period that would broadly represent the humidity of the climate with no differentiation between day and night.

- An average DBT-WBT when DBT>26°C. This index represents the maximum opportunity by considering the wet bulb depression at the warmest hours of the day. It addresses evaporative cooling opportunity in the outdoor environment when cooling is most needed.

- Considering the previously used design indoor temperature of 26°C to determine the wet bulb depression. 26°C-WBT would indicate the opportunity to reduce cooling demand in indoor spaces with a passive evaporative cooling system that theoretically could supply air at wet bulb temperature.

Back to the climate of Madrid, the mean wet bulb temperature depression for each month indicates that there is a high opportunity for Passive Evaporative Cooling systems in July and August, both months presenting the maximum temperatures during the year (fig. 88). The high DBT-WBT confirms the dryness of the air and thus the air temperature reduction that could be achieved by evaporative cooling methods.

ACTIVE DOWNDRAUGHT COOLING

Active Downdraught Cooling becomes an environment-friendly solution in climates with warm and humid conditions presenting low Passive Evaporative Cooling applicability. It is achieved by using chilled water cooling coils or panels exposed to a warm internal environment, thus inducing a natural indoor air movement (downdraught). Cooling in ADC systems is achieved by convective heat exchange and no evaporation takes place. Although ADC is applicable for both humid and dry climates and air moisture content does not have a significant impact on the cooling delivered, the applicability assessment proposed in this book prioritises passive systems over active systems.

The opportunity or efficiency of an active downdraught cooling method is directly proportional to the temperature difference between the room and the coil temperature for a convective heat exchange. This characteristic makes ADC methods less coupled to climate and reaffirms its potential applicability for both humid and dry environments. To evaluate ADC opportunity the index coil-to-room temperature depression is determined together with a complementary index to prioritise ADC opportunity on humid climates. This second index responds to the question: how humid is the climate?

A first index would determine a potential maximum coil-to-room temperature depression. The room temperature is the design indoor temperature equal to 26°C. The coil temperature is set to the minimum

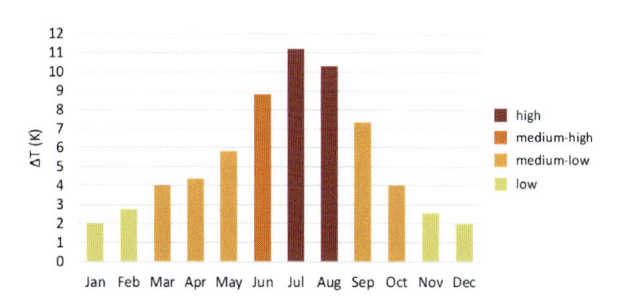

88. Opportunity for evaporative cooling systems (DBT-WBT) in Madrid, Spain.

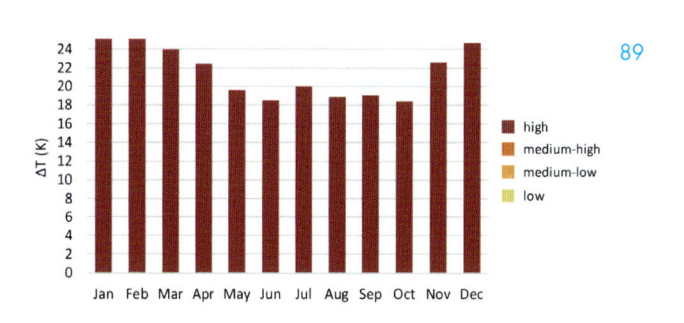

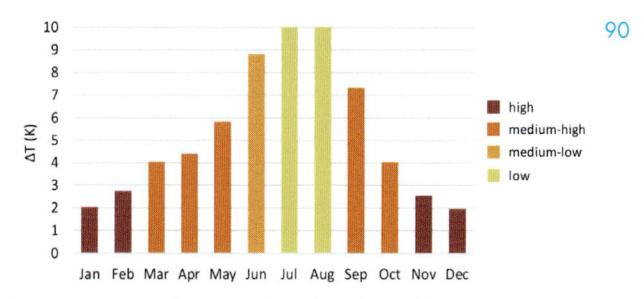

89. Opportunity for active downdraught cooling systems (DBT-DPT) in Madrid, Spain.
90. Opportunity for active downdraught cooling systems (DBT-WBT) in Madrid, Spain.

CHAPTER 3 | CLIMATE APPLICABILITY MAPPING

temperature at which condensation on the coil surface won't happen. In theory, the on-coil water temperature should be slightly above the dew-point temperature (DPT), but for simplicity it is considered equal to DPT. This first opportunity index is thus determined from 26°C-DPT.

The second opportunity index that complements the previous one is determined from DBT-WBT as determined in the previous section. In this case, and in order to prioritise ADC opportunity in humid climates, lower wet bulb temperature depressions are associated with high ADC opportunity.

ADC is mostly applicable in both dry and humid climates and, in the case of Madrid, the opportunity to reduce the room temperature by keeping the coil temperature close to dew point temperature is also high (fig. 89). However, the second index indicates a low applicability during the summer season (fig. 90), intentionally prioritising passive evaporative cooling systems in the hot-dry climate of Madrid.

> ACTIVE DOWNDRAUGHT COOLING BECOMES AN HYBRID SOLUTION IN CLIMATES WITH WARM AND HUMID CONDITIONS

3.3. PASSIVE COOLING APPLICABILITY MAPPING

At early stages in the design process, speedy and robust assessments of feasibility are enhanced by reference to reliable sources of weather data and an understanding of the building use. The direct relationship existing between climate and the applicability of convective and evaporative passive cooling systems allow the creation of applicability charts based on location-specific climate data for feasibility assessment. Moreover, weather data plotted on psychrometric charts can also promote rapid interpretation to support strategic decision making. Such plots can help to define both the need for cooling and the opportunity for different passive cooling strategies. The combination of 'need' and 'opportunity' can provide the basis for determining 'applicability' of a specific passive cooling technique.

Interactive psychrometric charts are accessible through web and desktop tools, mostly part of climate analysis software packages like Climate Consultant, Climate Tool or Ladybug Tools. By integrating the theory of psychometrics using Szokolay's methods *(Szokolay, 2008)*, these tools compare the climatic data against an 'extended' comfort zone for environments with evaporative cooling systems.

Applicability maps, instead, allow the evaluation of passive cooling techniques at a larger geographical scale without the need to access weather data for multiple locations. Previous work has published maps which have been constructed to communicate both the 'need' for cooling and the 'opportunity' for different climatic regions. A group at the University of Seville, Department of Energy Engineering, pioneered the definition of these maps, initially for Spain *(Ford et al., 2010)* and subsequently for the whole of Europe *(Salmerón et al., 2012)*. A similar approach has also been applied to map the applicability of different downdraught cooling options in China *(Xuan & Ford, 2012)* and more recently in the US *(Aparicio et al., 2018)*. In all these cases, climate data was obtained for most locations and post-processed to generate the applicability maps. When climate data is not available, applicability can be determined using the interpolation method described by *Sánchez et al. (2008)* by means of the geographical distance between the closest meteorological stations, latitude, altitude and proximity to the sea.

APPLICABILITY MAPS FOR EUROPE

Within Europe, each country may present multiple climatic zones. Typically, the national thermal regulations or building technical code of each country determines this classification. The developed Evaporative Cooling applicability maps by *Salmerón et al. (2012)* used zonal divisions based on primary administrative regions. For each region in Europe, the authors determined the mean DBT-25°C (proportional to the cooling degree hours) to evaluate the need for cooling, and the Dry Bulb-to-Wet Bulb Temperature depression (DBT-WBT) and Indoor-to-Wet Bulb Temperature depression (26°C-WBT) from June to August to determine the opportunity for using evaporative cooling techniques during the warm (summer) season.

The maps (figs 91–92) illustrate that most of the areas presenting higher demand are distributed in the central-south, and Mediterranean locations. In north Europe, the need of passive cooling methods can only be justified when indoor thermal comfort is compromised due to high internal gains, commonly associated with tertiary buildings with high density of occupation or equipment loads. The maps also revealed that most Mediterranean countries, and particularly central and southern Spain, present optimal conditions for the application of evaporative cooling systems.

> MOST MEDITERRANEAN COUNTRIES, AND PARTICULARLY CENTRAL AND SOUTHERN SPAIN, PRESENT OPTIMAL CONDITIONS FOR THE APPLICATION OF EVAPORATIVE COOLING SYSTEMS

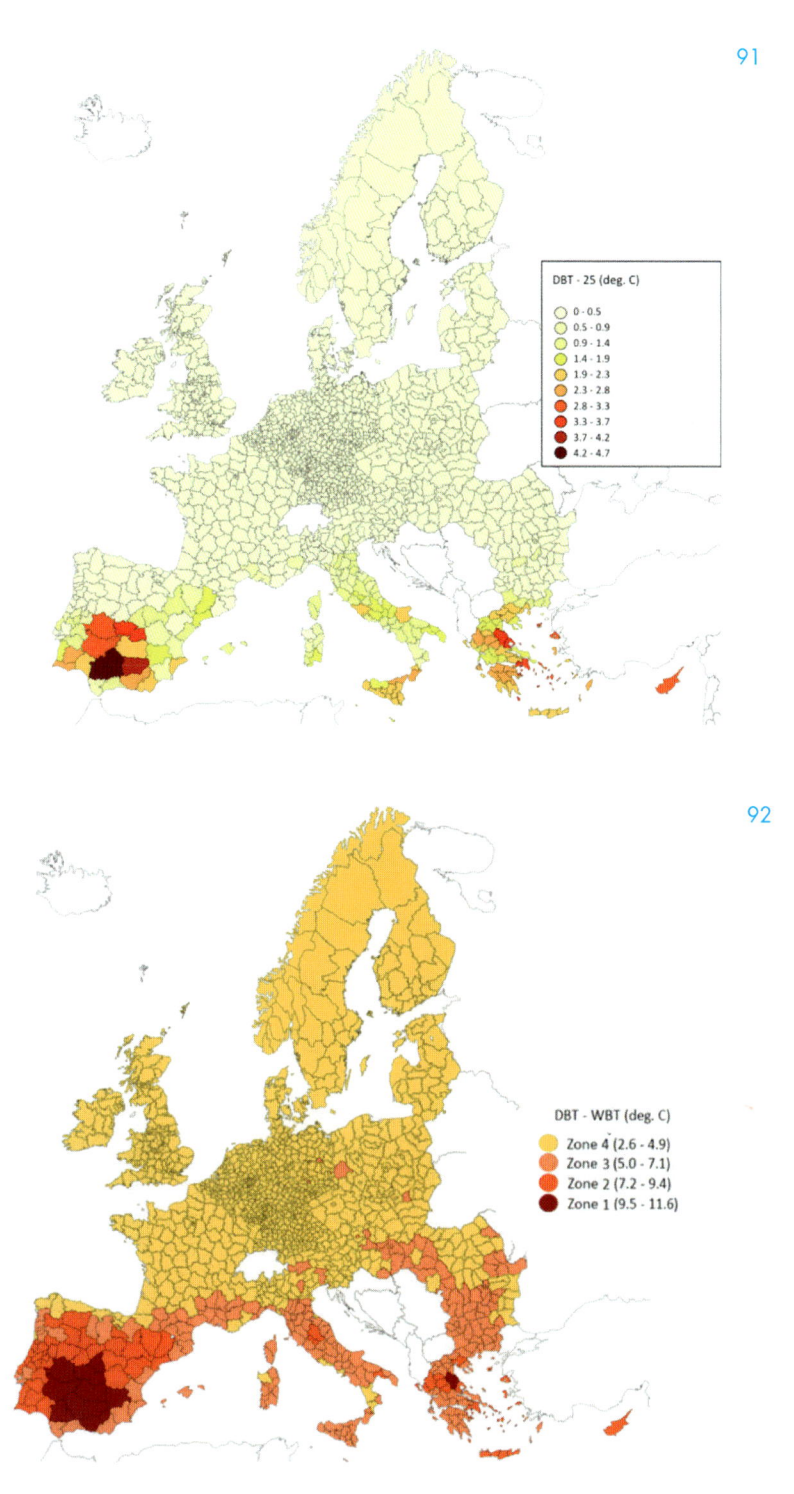

91. Demand for cooling (25°C-DBT) from June to August in Europe (Salmerón et al., 2012).
92. Opportunity for evaporative cooling (DBT-WBT) from June to August in Europe (Salmerón et al., 2012).

CHAPTER 3 | CLIMATE APPLICABILITY MAPPING   67

APPLICABILITY MAPS FOR CHINA

China is a large country with varied climate from subtropical zones in the south to temperate zones in the north *(Zhao, 1986)*. Climatic types or zones can be classified in various ways according to different criteria using different climatic variables and indices. The choice depends largely on the purpose of establishing such classification. In China, there are two major climatic zones classifications, the Code for Thermal Design of Civil Buildings (GB 50176-93, 1993) and the Standard of Climatic Regionalization for Architecture (GB 50178-93, 1996) which, apart from temperature, consider other climatic variables such as precipitation, relative humidity, diurnal temperature variation and wind speed.

The Evaporative Cooling applicability maps for China were developed by *Xuan & Ford (2012)* based on data for 267 locations around China. The authors developed a series of maps plotting cooling degree hours (demand) and DBT-WBT and 26°C-WBT (opportunity) for the warm (summer) season using linear interpolation techniques (figs 93–94). The maps revealed that about one third of China (the north and north-west provinces) presents optimal conditions for the application evaporative cooling systems; in another third (the south-eastern provinces) active downdraught cooling (ADC) is more applicable; and in the south-west and north-east cooling demand is very low, but could be met by natural ventilation.

> OPPORTUNITY FOR DOWN-DRAUGHT EVAPORATIVE COOLING IS LIMITED TO NORTH-WESTERN PROVINCES OF CHINA, AND OPPORTUNITY FOR ACTIVE DOWN-DRAUGHT COOLING IS MAINLY IN SOUTH-EASTERN CHINA

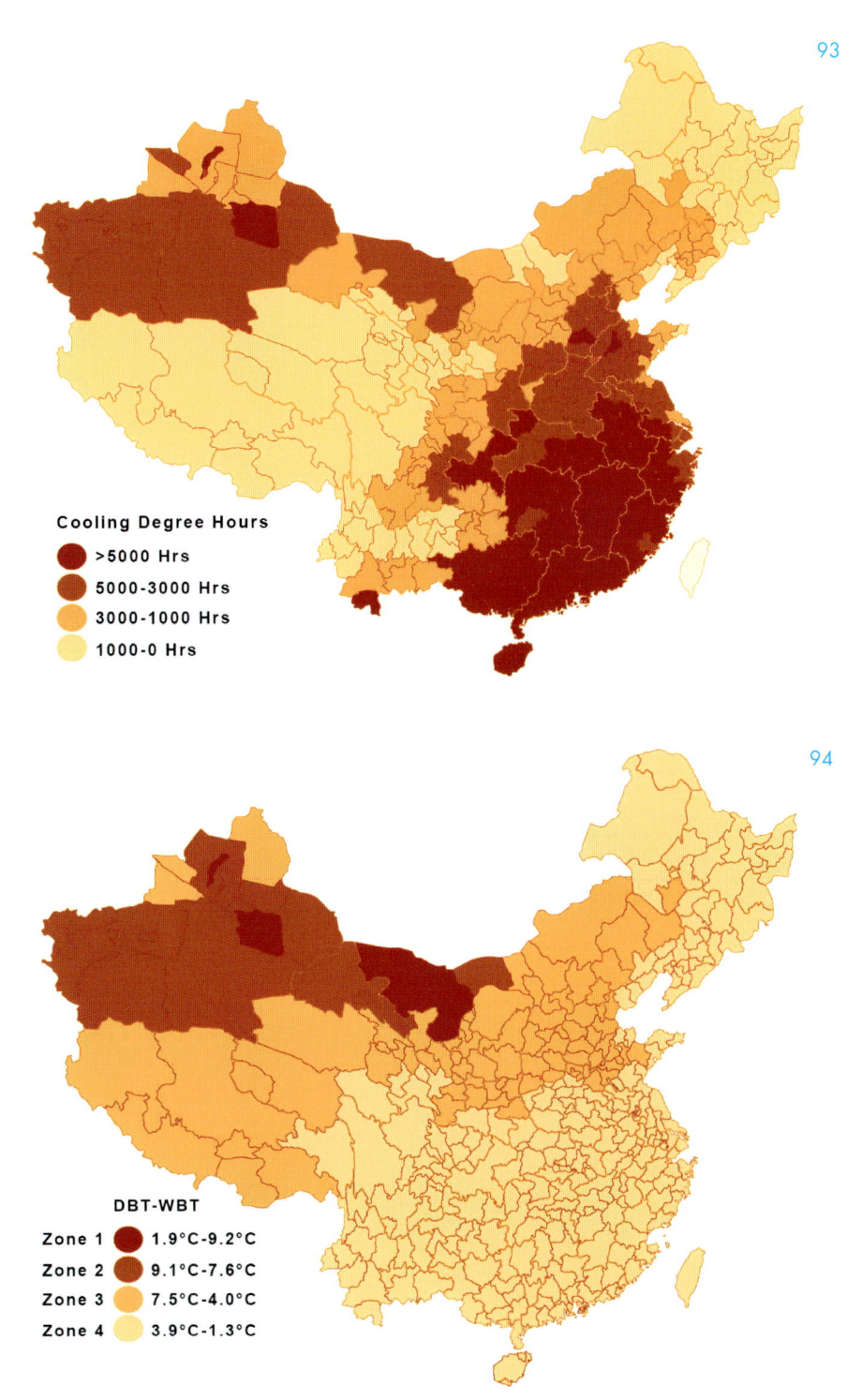

93. Demand for cooling (DBT>26°C) from June to September in China (Xuan and Ford, 2012).
94. Opportunity for evaporative cooling systems [DBT-WBT] from June to September in China (Xuan and Ford, 2012).

CHAPTER 3 | CLIMATE APPLICABILITY MAPPING   69

APPLICABILITY MAPS FOR THE USA

There are eight climate regions designated by the International Energy Conservation Code (IECC) and by the American Society of Heating, Refrigerating and Air-Conditioning Engineers (ASHRAE) of which seven are located in the continental areas of the USA. These eight zones can be further combined into six simplified climate categories: hot-humid, hot-dry/mixed-dry, mixed-humid, marine, cold/very cold, and subarctic.

The Passive Cooling applicability maps developed by *Vallejo et al. (2018)* used the third generation of Typical Meteorological Year climate data (TMY3), which derives from the 1961–1990 and 1991–2005 National Solar Radiation Data Base (NSRDB) archives, obtained for 1020 locations in the USA.

The need for cooling was determined from the number of cooling hours (CH) when DBT>26°C for a theoretical warm period from June to September. The resulting map (fig. 95) suggests a higher demand in areas with lower latitudes and altitudes, in other words, the south-east counties from Texas to Florida, Southern California and Arizona.

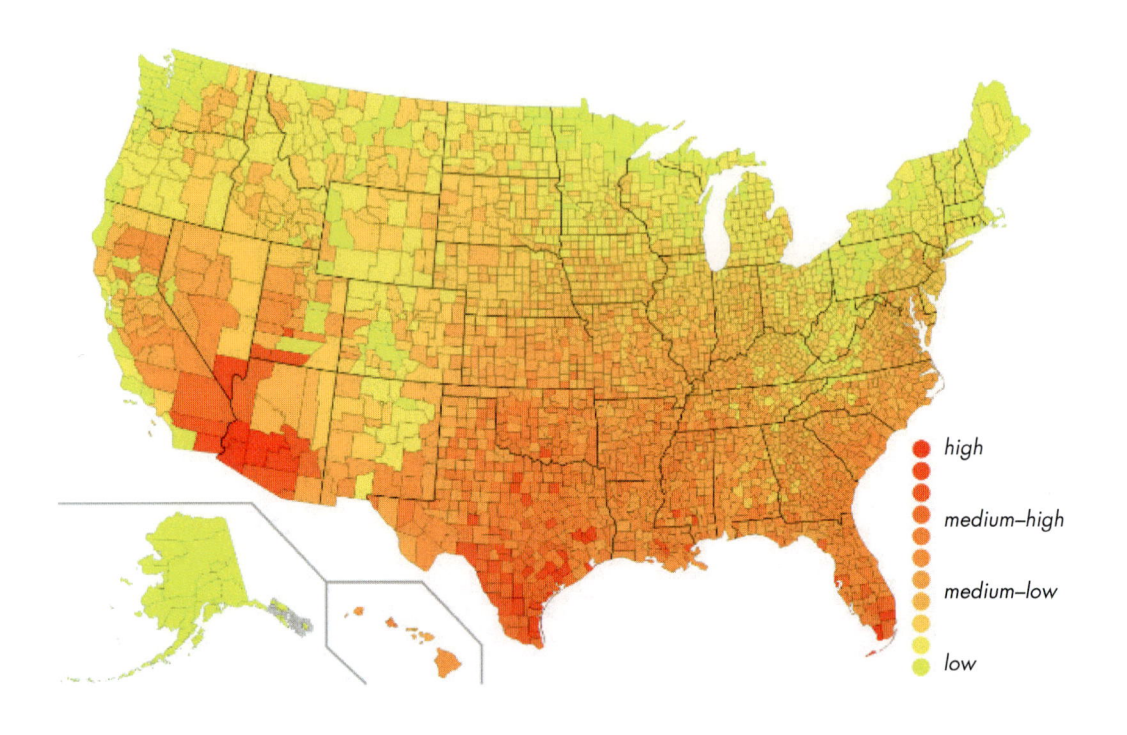

95. Demand for cooling [DBT>26°C] from June to September in the USA.

CHAPTER 3 | CLIMATE APPLICABILITY MAPPING

The opportunity for convective cooling was mapped using the index 26°C-DBT, determined for each hour of the analysis period (fig. 96). The map suggests a prevailing range of indoor-to-outdoor air temperature depression between 3°C and 9°C, with cooler areas referring to the northern counties and high altitudes. It is worth highlighting that 70% of the counties in the USA (presenting high applicability) could overcome overheating problems in buildings with a good natural ventilation strategy and without the need of mechanical systems.

The opportunity for evaporative cooling was mapped using three indices: DBT-WBT, DBT-WBT when DBT>26°C and 26°C-WBT. The developed map using the first index suggests a prevailing range of DBT-WBT between 3°C and 6°C, with drier areas referring to the western counties, and highlighting an evident relation with the altitude above the sea level. The combined results (need and opportunity) suggest a medium to high applicability in south and south-west regions in the USA for outdoor spaces and extended high applicability region towards the north for indoor spaces (fig. 97). The maps conclude that 30% of the USA counties present optimal climatic environmental conditions for the integration of passive evaporative cooling systems in architecture.

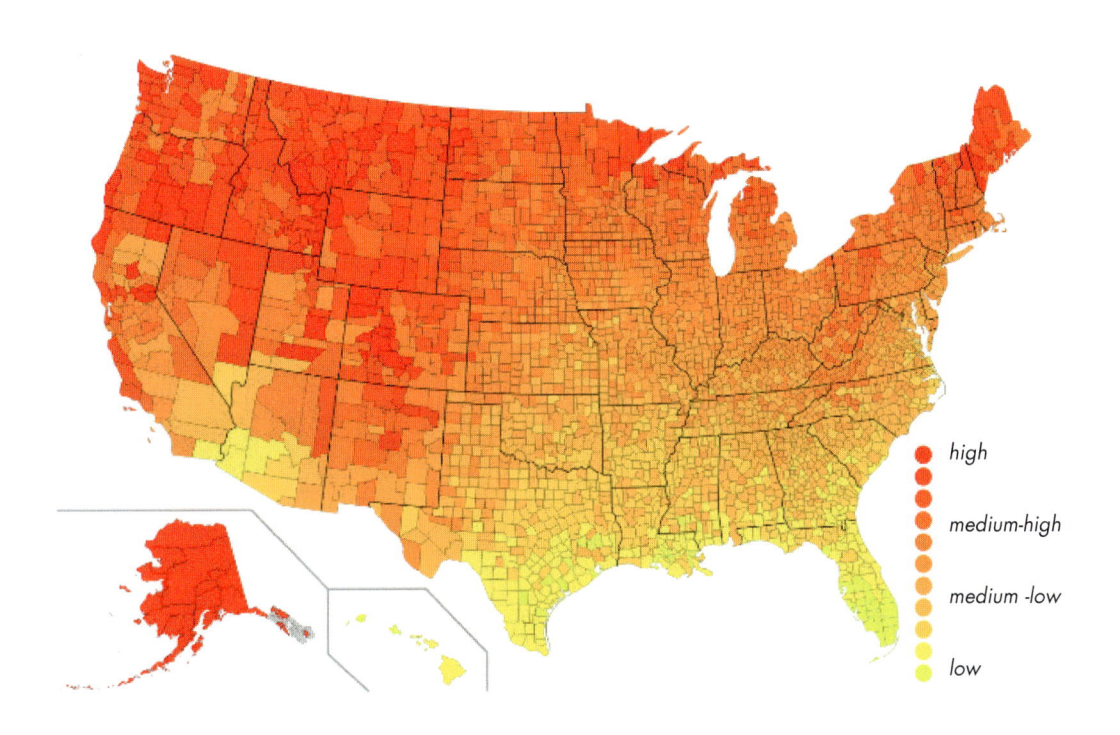

96. Opportunity for convective cooling [26°C-DBT] from June to September in the USA.

The opportunity for active downdraught cooling was mapped using the index 26°C-DPT, which determines a potential maximum coil-to-room temperature depression. The map suggested a mean range of coil-to-indoor air temperature depression between 10°C and 15°C. This is by about 4 degrees higher than a theoretical indoor PEC opportunity index (26-WBT) and its opportunity extends to most of the USA area. The combined results suggest that in principle, ADC is applicable in most of the USA, presenting the highest applicability in south and south-west regions. However, this strategy should be prioritised over PEC methods only in south-eastern regions where the opportunity for passive evaporative cooling is limited (fig. 98).

The results obtained are promising and suggest a large potential for the use of passive evaporative (PEC) and convective cooling solutions in the USA. In fact, from the climatic data available it can be concluded that more than 50% of the counties in the USA are eligible for the application of PEC methods and more than 70% of the counties could overcome overheating problems in buildings with a good natural ventilation strategy and without the need of mechanical systems.

> OPPORTUNITY FOR DOWN-DRAUGHT EVAPORATIVE COOLING IS HIGH IN SOUTH AND SOUTH-WEST REGIONS IN THE USA

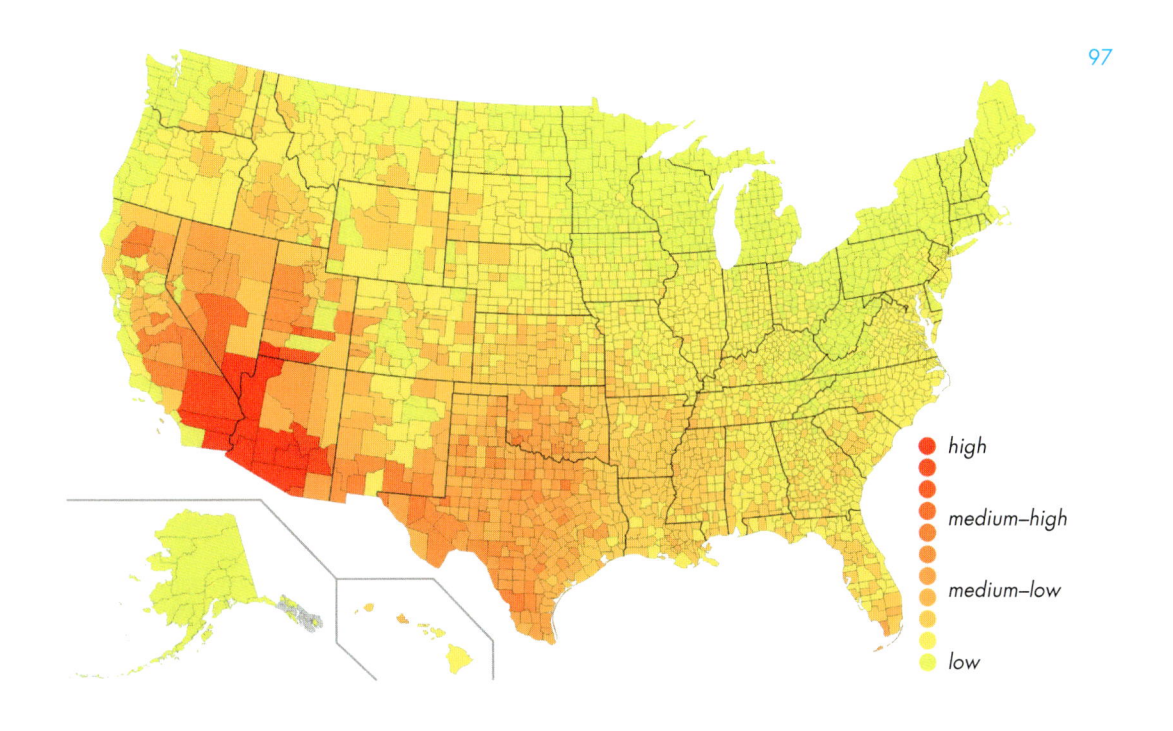

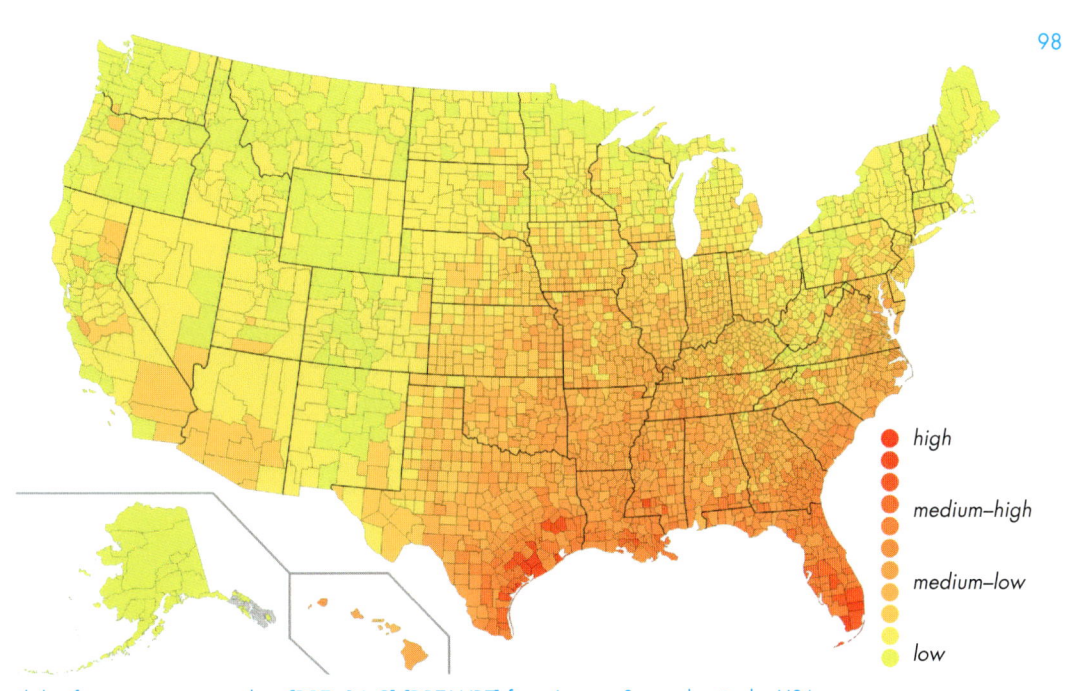

97. Applicability for evaporative cooling [DBT>26°C]·[DBT-WBT] from June to September in the USA.
98. Applicability for active downdraught cooling [DBT>26°C]÷[DBT-DPT] from June to September in the USA.

CHAPTER 3 | CLIMATE APPLICABILITY MAPPING

Climate analysis is essential for understanding the 'need' and 'opportunity' for the applicability of natural cooling strategies. Although this can be done in detail by analysing typical annual weather data for specific locations, large scale maps can graphically represent applicability indices for natural convective cooling, passive evaporative cooling and active downdraught cooling for entire continents and countries. Moreover, maps can assist architects, engineers and product designers with limited knowledge in this field to suggest the most suitable design approach or cooling strategy to overcome overheating problems, or to evaluate the market opportunity of a novel natural cooling product. The advantage of these applicability maps lies in the quick and accurate assessment of climatically appropriate design strategies at the early stages of the design.

Once this assessment is made, the detailed analysis of local weather data through linear graphs or psychrometric charts can help the designer to fine tune the seasonal and diurnal environmental design strategies. The next chapter explores how natural cooling strategies can form part of an integrated design approach to architecture and engineering.

REFERENCES

- Aparicio, P., Schiano-Phan, R. & Salmerón, J. M. (2018). *'Climatic Applicability of Downdraught Evaporative Cooling in the United States of America'*. Building and Environment 136: 162–176.
- ClimateTool. (2016). *'Climate Tool 2016'* [Online]. http://www.climate-tool.com/en/climatetool.html. [Accessed 12/2017]
- EnergyPlus. (2018). *'Weather Data'* [Online]. https://www.energyplus.net/weather/ [Accessed 08/2018].
- Gatley, D. P. (2004). *'Psychrometric Chart Celebrates 100th Anniversary'*. ASHRAE Journal, 41.
- Ford, B., et al. (2010). *'The Architecture & Engineering of Downdraught Cooling: a Design Sourcebook'*. Italy, PHDC Press.
- Givoni, B. (1994). *'Passive and Low Energy Cooling of Buildings'*. Van Nostrand Reinhold, pp.139–143.
- Ladybug. (2017). *'Ladybug Tools'* [Online]. http://www.ladybug.tools/. [Accessed 12/2017].
- Meteonorm. (2018). *'Meteonorm Software'* [Online]. https://www.meteonorm.com/ [Accessed 08/2018].
- OneBuilding. (2018). *'Climate Files'* [Online]. http://climate.onebuilding.org/ [Accessed 08/2018].
- Salmerón, J. M., Sánchez, F. J., Sánchez, J., Álvarez, S., Molina, J. L. & Salmerón, R. (2012). *'Climatic Applicability of Downdraught Cooling in Europe'*. Architectural Science Review 55(4): 259–272.
- Sánchez, F. J., Álvarez, S., Molina, J. L. & Gonzalez, R. (2008). *'Climatic Zoning and Its Application to Spanish Building Energy Performance Regulations'*. Energy and Buildings 40(10): 1984–1990.
- Santamouris, M. (2007). *'Advances in Passive Cooling'*. London, Earthscan.
- Szokolay, S. (2008). *'Introduction to Architectural Science'*. Elsevier/Architectural Press.
- UCLA. (2017). *'Climate Consultant 6.0'* [Online]. http://www.energy-design-tools.aud.ucla.edu/climate-consultant/request-climate-consultant.php/ [Accessed 11/2017].
- Vallejo, J., Ford, B., Schiano-Phan, R. & Aparicio, P. (2018). *'Passive Cooling Applicability Mapping. A Tool for Designers'*. 34th Plea International Conference: Smart and Healthy within the 2-degree Limit. Hong Kong, China, 10–12 December.
- Xuan, H. & Ford, B. (2012). *'Climatic Applicability of Downdraught Cooling in China'*. Architectural Science Review 55(4): 273–286.
- Zhao, S. (1986). *'Physical Geography of China'*. New York: Van Nostrand Reinhold.

CHAPTER 4
INTEGRATED DESIGN

Previous chapters have explored historical roots, contemporary opportunities, principles and strategies related to natural ventilation and convective cooling in buildings. Chapter 3 illustrated the feasibility of these techniques in relation to climatic characteristics. The potential benefits of reduced carbon emissions, capital, running and maintenance costs have been established, but delivery of these benefits in practice will require embedding this approach in the design (and procurement) process so that it becomes almost routine. This chapter discusses how this can become an integral part of everyday architectural and engineering design practice.

4.1. WHY INTEGRATED DESIGN?

Architectural design requires a very wide ranging 'holistic' approach, which today includes an understanding of environmental design principles. Thermal performance and passive cooling are part of this, and to be successful an understanding of the principles and strategies need to be embedded within the design and procurement process. It is not a technological *'add-on'*, but part of an approach to 'integrated design'. To achieve this, leading practice promotes close collaboration between the different members of the design team. The manipulation of plan and section, building form and fabric, the disposition of openings and the relationship between them (normally the province of the architect) have as big an impact on the feasibility of environmental design strategies as the specification of plant, actuators and controls (normally the province of the engineer). After nearly 30 years of developments in low-carbon design and the feedback on performance and occupant satisfaction provided in the ground breaking PROBE series of studies, Bill Bordass suggests: *"one key reason why buildings don't work well is their unmanageable complexity, so we advise designers to keep it simple and do it well"* (Bennetts & Bordass, 2007). This underlines the importance of a thorough understanding of basic principles.

Over this period building procurement and the role of the different construction professions have also changed. The architect's traditional leadership role has been challenged, and in many instances replaced by a project manager. The challenges have also resulted in greater specialisation: 'climate engineers', 'environmental designers' and 'sustainability consultants' have joined the team, as gaps in knowledge and experience have appeared, making collaboration and a shared understanding even more vital.

> IT IS WIDELY RECOGNISED THAT GREAT BENEFITS ARISE FROM ARCHITECTS AND ENGINEERS WORKING CLOSELY TOGETHER FROM THE EARLIEST STAGES OF THE DESIGN PROCESS

It is widely recognised that great benefits arise from architects and engineers working closely together from the earliest stages of the design process. Given an understanding of the basic principles, low carbon design can provide a springboard for invention and an opportunity for innovation, based on an holistic approach by both architect and engineer. This is very elegantly demonstrated in the Tjibaou Cultural Centre, Noumea by Renzo Piano Building Workshop and Arup, and more recently in their projects for the extension to the Kimbell Museum, Fort Worth, Texas and the California Academy of Sciences in San Francisco (figs 99–100). Alistair Guthrie (Arup director and long term collaborator with Renzo Piano Building Workshop) has said that Ove Arup *"…was convinced that an integrated architecture and engineering team could best provide the total building design or 'total architecture' to which he aspired"* (Guthrie, 2018).

CHAPTER 4 | INTEGRATED DESIGN

> *The mission of the California Academy of Sciences, since its inception in 1853, has been to explore, explain and protect the natural environment. It now promotes sustainability through its research, its exhibits and events. As well as offices, laboratories and scientific archives the new building includes an aquarium, a planetarium, a natural history museum and a 30m diameter glass rainforest enclosure. Renzo Piano with Arup have created a building which fundamentally embodies the values of the Academy.*
>
> *Although the aquarium and rain forest life support systems require a lot of energy, it has been designed to use a third less energy than comparable buildings that meet the US energy code. The main exhibition floor and research offices are naturally ventilated and daylit, and the building is constructed almost entirely from locally sourced, recycled materials (including fly ash substitute for cement in foundations and slabs).*

Part of the success of the Arup ethos is a recognition of the need for collaboration within the design team and with other consultants to achieve a successful 'whole'. "At Arup we have long realized that sustainability in the built environment is about total holistic design. Each part of the construction project, each element of the design interacts with each other and it is only as we consider the whole and its impact on the local environment and society that we can evaluate its future performance" (Guthrie, 2011).

At Arup an informal but strong commitment to seeking sustainable solutions for their clients has been formalised within the company by the implementation of a strategic approach throughout the company globally. "We needed to focus each project on a clear sustainability strategy at the outset of a project and get all the parties thinking

99. California Academy of Sciences, San Francisco [Renzo Piano Building Workshop], ©ARUP.

about how it might be achieved. It needed to be inspirational, strategic, visionary, imaginative as well as technically correct and economically feasible" (ibid).

This holistic approach is also shared by climate-engineers Transsolar, who state: *"Three pillars to the practice... are: To follow an integrated design process, knowing that this process leads to the most successful solutions; to constantly question assumptions and conventions, knowing that building performance can always be improved; to substantiate our design concepts with sound technical analysis"* (Transsolar, 2016).

While Transsolar have applied this approach in many projects around the world it is exemplified in their project with Selldorf Architects La Mechanique Generale et Les Forges in Arles in the south of France, in which the integration of architecture and engineering appears to be seamless.

Rab Bennetts (Founding Partner Bennetts Associates, London) pioneered a new generation of naturally ventilated office HQ buildings for PowerGen (now EON) in Coventry (1994), and for Wessex Water in Bath UK (2000). He has suggested that it is through collaboration that architecture of the highest quality can be achieved (more often than the rare genius of a talented individual).

"Design is at its best when the disparate authorship of great ideas is absorbed within a convincing whole...our collaborative model is the vehicle for complete integration between design & construction, cost & delivery, design quality & performance" (Bennetts, 2016).

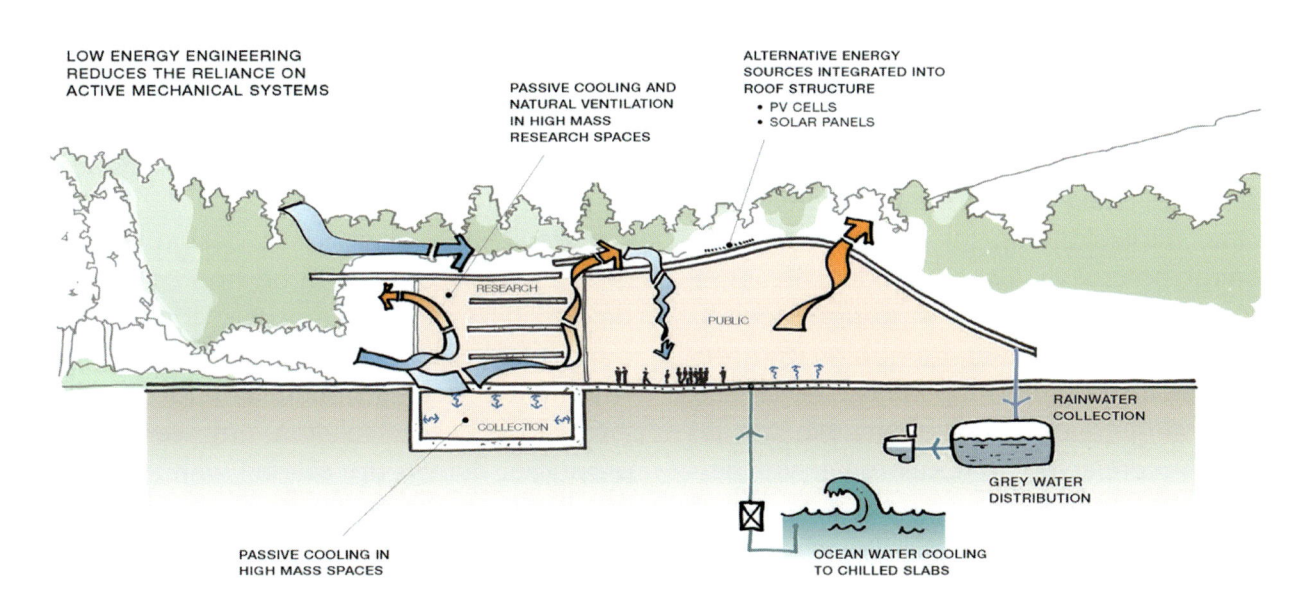

100. Sketch to show environmental strategy for California Academy of Sciences, San Francisco, ©ARUP.

CHAPTER 4 | INTEGRATED DESIGN

This is in a context of profound changes within the professions and the construction industry, including contractual and procurement changes which have challenged architects' traditional leadership position. It is therefore more important than ever to encourage understanding, mutual respect and collaboration within design and construction teams. (Bennetts' recent 'Potterrow' project in Edinburgh is discussed in some detail in section 4.4 below).

This book is concerned primarily with just one aspect of environmental and sustainable design, but because so many of these issues are interrelated, it is important to develop a shared understanding in order to prioritise and make sound judgements. Much is said about the potential benefit of passive design, as part of a sustainable design approach, but the benefits of an understanding of environmental design issues will only be realised if that understanding (of daylighting, solar control, natural ventilation, thermal response etc.) is embedded within the design process. This is because building form, plan and section, the disposition of openings, the choice and configuration of materials, etc. are intimately connected with environmental performance. But how can an understanding of environmental design issues be embedded in the design process if the strategic choices are not clear?

> **AN UNDERSTANDING OF PRINCIPLES ENABLES STRATEGIC OPTIONS AND ALTERNATIVES TO BE PROPOSED**

Sir Leslie Martin, in setting up the Centre for Built Form and Land Use Studies in Cambridge University in 1967, described the value of research in architecture in terms of *"clarifying the choices available to designers"* (Steadman, 2017).

The process of understanding the site, its physical, social, economic, cultural and environmental context, and the implications of the aspirations of the client, involves a process of analysis and evaluation which may be described as research, with the intention of 'clarifying the choices available'.

To some extent, familiarity with the context (repeatedly designing and building in the same location or region) may help to short-cut this process and reduce the range of choices. This is what occurred historically and has been referred to as the 'empirical tradition' in Chapter 1. However, in our contemporary world, change is so rapid and far reaching that proceeding on the basis of precedent, previous experience and intuition alone is rarely satisfactory.

It would be unlikely for a design solution developed historically to be appropriate today because the contemporary context is so different. However, the application of the environmental design principles underpinning the historic design solution may well be appropriate. But we are still left with the question: *how do we embed this understanding in the design process?*

An understanding of principles enables strategic options and alternatives to be proposed. These options can then be tested and evaluated to identify a preferred design proposal. At this point the design can be developed in more detail. It is therefore the case that the process of testing, analysis and evaluation of options needs to be built into the iterative process of design. At early design stages there is little time for testing and evaluation and so any analysis must be simple and quick. Rules of thumb, simple steady-state expressions and spreadsheet tools can be used, but importantly someone in the architect's office needs to understand and be able to apply these simple tools (there is rarely the time at this stage to always refer to an external specialist or consultant).

As the design develops, progressively more sophisticated analysis may be used to provide confidence in the decisions being made. This may require input from another member of the design team or an external specialist. Testing and checking to ensure that the assumptions built into the early design stages are achieved in the detailed design and construction of the building is vital to ensure that predicted performance is achieved in practice. Testing performance will also lead to robust conversations within the team as compromises may have to be made and as the demands and priorities of all the design team members need to be met. This stresses the significance of collaboration and understanding within the team throughout the life of the project. It is not just the design team, but the whole construction team who need to be aware of the importance of the detail to the success of the strategy in the completed building.

Increasingly the success of a new building is being judged in terms of occupant satisfaction. A standard procedure for post occupancy evaluation, developed by Building Use Studies as part of the PROBE studies carried out in the UK between 1995–2006, is providing a large data base of building performance. This provides not only useful insights for the designers and contractors involved, but also valuable feedback for clients and facilities managers, and contributes to benchmarking performance of different building types in many parts of the world.

The influence of the general physical context (characteristics of site and microclimate) has been discussed in Chapter 2. The applicability of different passive cooling strategies is of course influenced by the age and character of the urban form (morphology) and by the opportunity to reduce the cooling load through modifications to the existing building envelope. These issues

> THE APPLICABILITY OF DIFFERENT PASSIVE COOLING STRATEGIES IS OF COURSE INFLUENCED BY THE AGE AND CHARACTER OF THE URBAN FORM AND BY THE OPPORTUNITY TO REDUCE THE COOLING LOAD THROUGH MODIFICATIONS TO THE EXISTING BUILDING ENVELOPE

are discussed in more detail in this chapter, followed by a discussion of how passive and hybrid cooling can be integrated in the design of new buildings.

4.2. REFURBISHMENT OPTIONS FOR EXISTING BUILDINGS

In defining a passive cooling strategy for an existing building, it is important to establish how the building form and fabric can help to minimise the cooling load, and what the original strategy for providing light and ventilation was. Historically courtyards, lightwells and ventilation shafts frequently facilitated this. A study of urban morphology will often reveal opportunities to improve internal conditions within an existing building without reliance on mechanical conditioning. In buildings of high heritage value the avoidance of mechanical conditioning is also important in limiting the intrusion of plant, ductwork etc.

Foster & Partners' refurbishment of the UK Government Treasury in central London turns original courtyards into enclosed and glazed atria, simultaneously reducing the area of external walls (heat loss in winter) and promoting natural ventilation through the existing building in summer (figs 101–103).

The original building was completed in 1917 and is Grade II listed. It has a roughly symmetrical plan, with two parts linked by a cylindrical courtyard. The perimeter is punctuated by smaller courtyards and lightwells, some of which have been capped with translucent roofs to create five storey spaces that provide light and air to the surrounding offices, and at ground level a library, a café, training rooms and an entrance atrium.

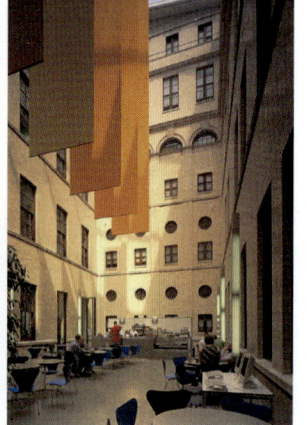

101. UK Government Treasury in London, UK [Foster & Partners].
102. Proposed enclosed courtyard.
103. Enclosed and glazed courtyard.

CHAPTER 4 | INTEGRATED DESIGN

Completed in 2002, the refurbished building set new environmental standards in Government buildings at the time. The reconfigured lightwells promote natural ventilation by the stack effect, drawing air through the office spaces. Fresh air is drawn into the building through the windows facing the internal courtyards and exhausted by vents at roof level *(Foster & Partners, 2002)*. This general strategy is applicable to a very large number of existing buildings where lightwells and ventilation shafts formed part of the original strategy. This is discussed in more detail below in the section on urban morphology.

URBAN MORPHOLOGY AND COOLING OPTIONS

The retention and refurbishment of the existing fabric of our cities will represent a significant part of all future construction activity, and it is vital that this process significantly improves the environmental performance of the existing building stock. This can be a challenge, particularly when the heritage value of the buildings is high (as in many parts of Europe).

Studies of urban morphology have been undertaken as part of the process of assessing the applicability of passive cooling strategies to non-domestic buildings in southern Europe *(Ford & Cairns, 2002)*. A preliminary generalised benchmark, which relates the applicability of passive cooling to the void/floor ratio, has been derived from the study of precedents (fig. 104). A method of determining the technical

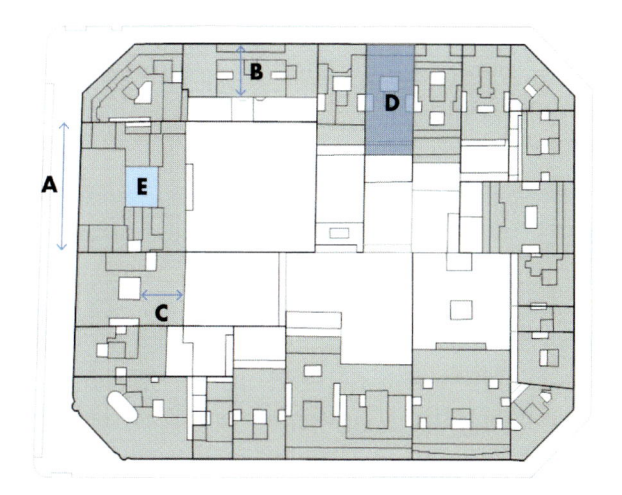

Minor intervention
Distance of void to external wall $C<12m$
PDEC applicability factor >0.75
Average of void areas E between $9m^2$ and $16m^2$

Intermediate intervention
Minor intervention is not suitable
Building depth $B<12m$
Can also be used when the void area E is $>16m^2$ and within 12m of an external wall (C)

Major intervention
Neither of the previous interventions has been chosen
Building depth $B>12m$

104. Passive cooling applicability study in European cities: urban block analysis methodology.

CHAPTER 4 | INTEGRATED DESIGN

applicability of convective cooling, based on relating a benchmark to sample areas of mainly commercial buildings in eight major cities in Spain, Portugal, Italy and Greece, was applied as part of an EU funded research project *(Ford & Moura, 2003)*.

The feasibility of integrating passive cooling as part of the refurbishment of an existing building will depend partly on whether an existing light/vent shaft or courtyard can be exploited to deliver convective cooling, or whether such a shaft will have to be added to the existing structure. Studies of European cities have revealed the high incidence of these features *(ibid)*. The cooling potential of a stack relates to the achievable volume flow rate, and hence the cross-sectional area. A rough proxy for the cooling potential within a given building can then be taken as the void to floor ratio.

From table 01 it is apparent that for recent buildings designed to incorporate downdraught cooling, the void: floor ratio varies between approximately 0.02 and 0.11 (i.e. the voids required for naturally driven convective cooling represent between 2% and 11% of the floor area of the building). The more recently designed buildings have a lower void to floor ratio, which suggests that experience, and improved performance modelling, has led to a lower 'safety margin' being applied. On this assumption, a benchmark ratio of 0.05 was chosen as the basis of a downdraught cooling 'Applicability Rating', which could then be applied to the analysis of the existing building stock. A method of determining the technical applicability of downdraught cooling, based on relating the benchmark to a sample area of commercial buildings in Athens, was applied as part of a pilot exercise, and subsequently applied to representative areas within a number of large cities in southern Europe. The method identifies three different levels of intervention: Minor; Intermediate and Major.

Table 01. PDEC Applicability rating for 5 buildings.

Project name	Footprint Area (m²)	Void Area (m²)	Served Area (m²)	Ratio Void/Floor	Prop V/F Ratio Against Benchmark (%)	PDEC Applicability Rating (0–1)
Seville Speculative Office	676.0	36.0	640.0	0.056	106.5	1.00
Catania Speculative Office	300.0	14.1	285.9	0.049	93.5	0.94
Torrent Research Centre, Ahmedabad	376.3	38.7	337.6	0.115	217.0	1.00
Malta Stock Exchange, Valletta	190.0	8.0	182.0	0.044	83.2	0.83
CII Centre of Excellence, Bangalore	317.7	6.8	311.0	0.022	41.1	0.41
PDEC Applicability Benchmark				*0.0528*	*100.0*	*1.00*

Minor intervention

A minor (least cost) intervention can be made when the existing building incorporates one or more lightwells, with a void plan area of approximately 5% of the floor area. For the purposes of the study, a three storey building, with an existing lightwell 3 to 4m across and 12m (or less) from the perimeter, was assumed (fig. 105). The minor intervention assumes that a supply/exhaust termination is added to the existing lightwell, incorporating glazed baffles or glazed louvres (and possibly a glazed roof) to minimise the reduction of light penetration, and to provide an effective free area for airflow equal to the void plan area of the lightwell. In downdraught cooling mode either misting nozzles or chilled water cooling coils may be provided within the top of the tower. In both updraught and downdraught cooling modes it was assumed that the existing window openings at each floor level provide sufficient free area for the required flow rate to be achieved. In practice of course, many of these assumptions may not apply, depending on location, building type and other factors. However, for the purposes of the study, these assumptions then provided a basis for an assessment of both the applicability and cost of applying a combination of updraught and downdraught cooling to the existing building stock.

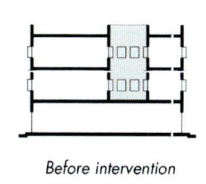

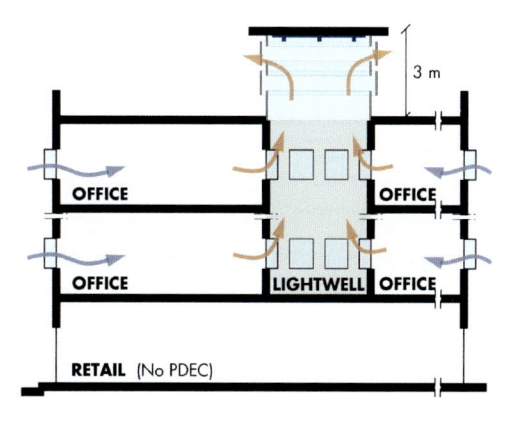

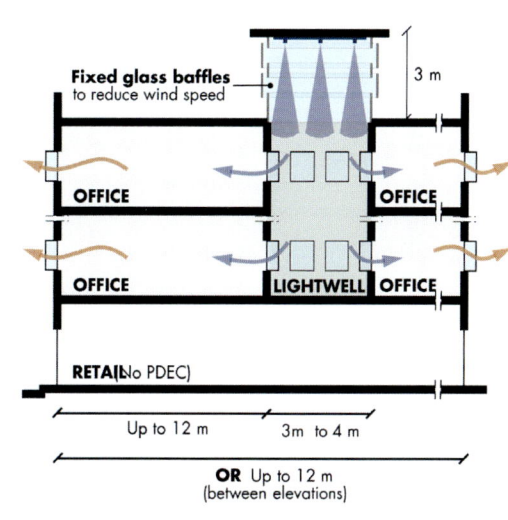

Before intervention

105. Generic Refurbishment: MINOR Intervention to promote 'updraught' and 'downdraught' convective cooling.

CHAPTER 4 | INTEGRATED DESIGN

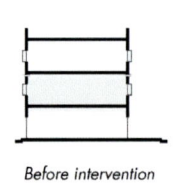

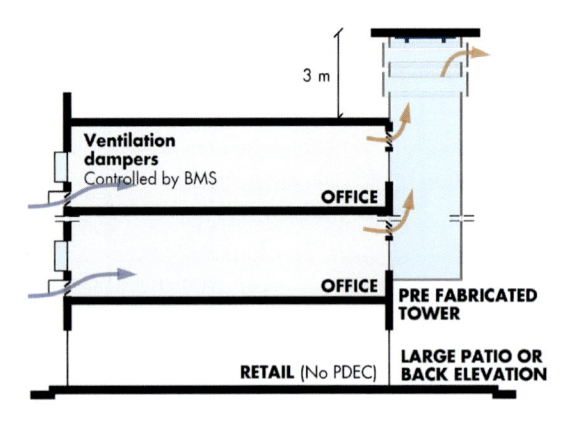

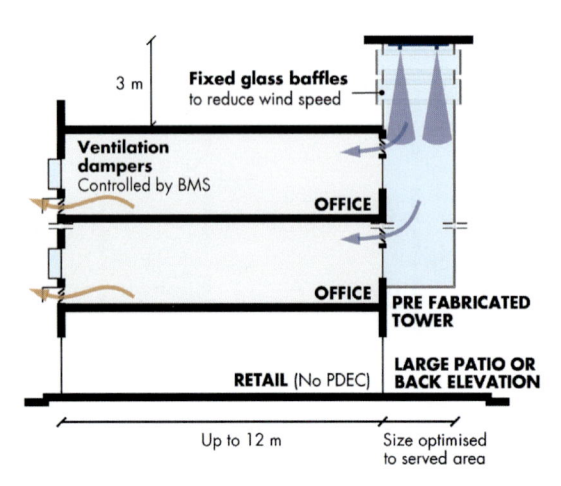

Intermediate intervention

The intermediate level of intervention (and cost) assumed the existing building does not incorporate an existing lightwell, and that a new pre-fabricated lightweight convective cooling tower or stack, equal in cross-sectional area to 5% of the floor area, is provided within the depth of the plan or attached to a rear elevation (fig. 106). This could alternatively be a continuous ventilated façade, providing thermal and acoustic buffering to the outside world, as well as a means to promote ventilation of the building. It also assumes an open floor plan and a maximum distance of 12m from the tower to the opposite external wall. The new tower incorporated glazed baffles or glass louvres and misting nozzles/cooling coils as for the minor intervention. For the intermediate intervention, however, it was assumed that the perimeter exhaust incorporated motorised dampers which are controlled by the building management system. This level of intervention is very flexible and could be applied to a wide range of existing buildings but is more expensive than the minor intervention.

Major intervention

During major refurbishments, involving the stripping out of existing services and extensive re-planning and re-modelling work, the creation of new light and ventilation shafts within a deep plan may be appropriate and would provide the opportunity to incorporate convective cooling. Such a scenario is assumed for the major intervention (fig. 107). In this case, in addition to the new

106. Generic Refurbishment: INTERMEDIATE Intervention to promote 'updraught' and 'downdraught' convective cooling.

enclosed voids created within the existing plan, automated control is provided on both supply and exhaust vent openings. This is clearly the most expensive of all the options but has the added benefit of increasing light penetration to the centre of a deep plan.

STOCK ANALYSIS IN SOUTHERN EUROPE

Results of the stock analysis referred to above has revealed the proportion of the building stock which is suitable for different levels of convective cooling intervention. Clearly, minor and intermediate interventions are likely to be most cost effective and will be taken up most quickly. Analysis has indicated that in Spain, 35 to 46% of all buildings are capable of having either minor or intermediate PDEC interventions, as described above. This compares with 22% in Italy, 27% in Greece and 33% in Portugal. If major interventions are included, then the overall applicability of downdraught convective cooling is very significant. The study indicates that, for all the cities examined, between 62% and 82% of all buildings are suitable for the application of naturally driven downdraught convective cooling. This indicates that the technical potential for such applications is very significant. However, local microclimate characteristics, market barriers, and the cost of refurbishment, will influence the final potential market penetration.

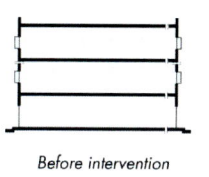

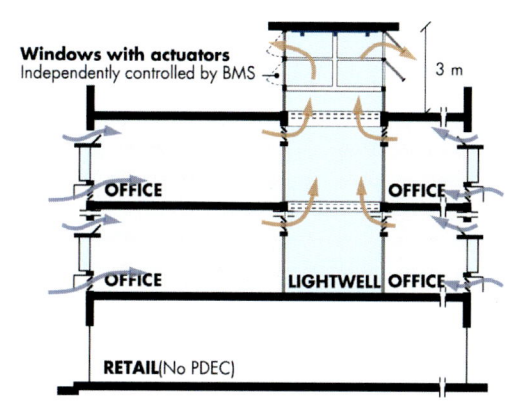

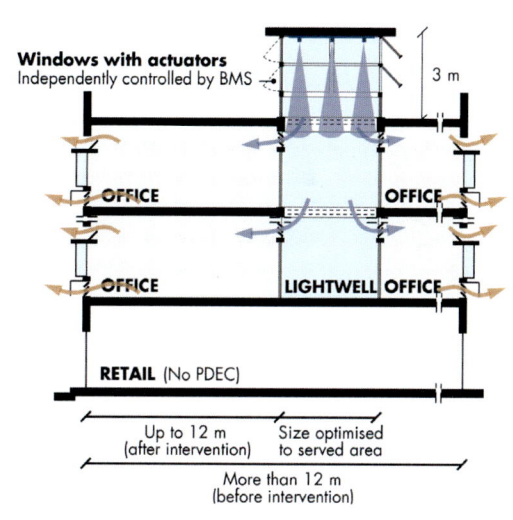

107. Generic Refurbishment: MAJOR intervention to promote 'updraught' and 'downdraught' convective cooling.

4.3. INTEGRATION IN EXISTING BUILDINGS

The context of existing buildings within urban environments is constantly changing, but opportunities may exist even where the constraints may initially appear overwhelming. Two major factors which influence opportunities for natural ventilation and convective cooling in many urban areas are the level of air borne pollution and traffic noise.

Located adjacent to a noisy and polluted dual carriageway within the city of Bristol, UK, the Temple Way House office complex built in the 1960s required refurbishment and complete renewal of the original air-conditioning system. It was the subject of a research project funded by UK Government to explore the feasibility of passive design solutions for non-domestic buildings (fig. 108).

The urban context of this project was particularly important. The dual carriageway ran past the east elevation, but the west side of the building was adjacent to the Bristol harbour area, which was significantly less polluted and quieter. It was therefore proposed to take supply air from the quiet, cleaner harbour-side, across the narrow section (12m) office floors, and exhaust air via sealed glazed shafts attached to the east elevation (figs 109–110).

CFD analysis demonstrated that it was technically viable strategy, and that provided the free area of the glazed stacks was

108. External view of Temple Way House, Bristol, UK [Short Ford Architects].
109. Proposed ventilation strategies for Temple Way House, Bristol, UK.

sized in relation to the internal heat gains, risk of overheating could be avoided even on the top floor. The glazed shafts provided a degree of thermal and acoustic buffering from the dual carriageway, as well as driving the natural ventilation strategy. This approach demonstrated the potential for low-cost, low-carbon refurbishment of 1960s city centre office buildings, but sadly the solution was not implemented.

However, a few years later Sauerbruch Hutton architects in close collaboration with engineers ARUP adopted a similar strategy in their extension of an existing office tower built in the 1950s in Berlin. Rather than providing individual glazed shafts, the strategy in the GSW Headquarters tower (1999) was to incorporate double-skin façades along the

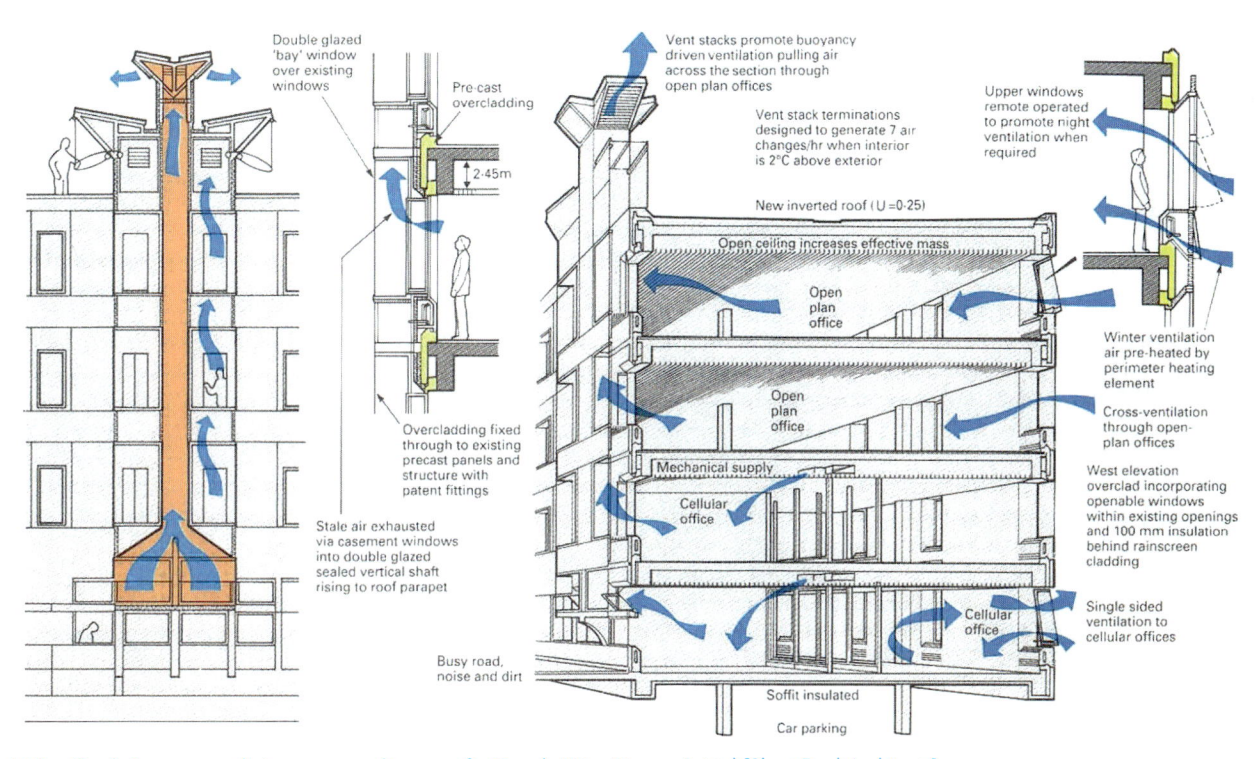

> **TWO MAJOR FACTORS WHICH INFLUENCE OPPORTUNITIES FOR NATURAL VENTILATION AND CONVECTIVE COOLING IN MANY URBAN AREAS ARE THE LEVEL OF AIR BORNE POLLUTION AND TRAFFIC NOISE**

110. Stack & cross ventilation strategy diagrams for Temple Way House, Bristol [Short Ford Architects].

CHAPTER 4 | INTEGRATED DESIGN

east and west elevations of the new tower. Air is drawn from the eastern façade, across the narrow section plan, and into the double skin western façade, which acts as a thermal flue (fig. 111). The prevailing westerly winds supplement the updraught by inducing suction above the exhaust opening at the top of the building by means of a *'wing'* canopy (Woods & Salib, 2013, pp. 64–73).

The natural ventilation strategy implemented in the GSW building varies according to the office configuration on each floor. Where cellular offices are on both sides of the building fresh air is drawn into the building via pivoting panels along the east façade which directly ventilate the eastern cellular offices and draw fresh air into the central corridor. The western cellular offices are naturally ventilated from the corridor through vents in the partitions and doors. Air is then exhausted through pivoting windows along the western double skin façade (fig. 113). This consists of an outer single glazed weather screen and an inner double glazed window enclosing a 1m deep cavity. The ventilated façade acts as a thermal and acoustic buffer between the offices and the outside world.

On the east façade the outer skin alternates between single-glazed window panels and full floor height fixed louvred screens that allow fresh air into the 200mm cavity (fig. 112). The inner layer of the east façade consists of double glazed window panels alternating with a pivoting panel, located directly behind the louvred screen, that allows the flow of fresh air into the interior.

111. Western double skin façade of GSW building Berlin [Sauerbruch Hutton], ©Annette Kisling.
112. Eastern fixed louvred screens to provide fresh air, ©Annette Kisling.

CHAPTER 4 | INTEGRATED DESIGN

While natural ventilation is the *'default'* mode for the building, the BMS starts a mechanical ventilation supply when the external temperature goes above 25°C or below 5°C. The BMS also controls the operation of brightly coloured moveable solar shading devices on the west façade and venetian blinds within the cavity of the east façade. Artificial lighting is also controlled in response to daylight levels (with occupant over-ride).

The exposed concrete ceiling of the intermediate floors stabilise temperatures within the offices in the summer, augmented by night ventilation which pre-cools the interior for the following day.

The use of double skin façades to provide both acoustic buffering and to drive natural ventilation has been widely adopted in urban office buildings. Double skin façades have also been integrated with mixed-mode ventilation and cooling systems.

The capacity for stack driven natural ventilation to successfully cool office and educational spaces with high internal heat gains has been demonstrated in numerous refurbishment projects over the last 20 to 30 years. The successful low carbon refurbishment of system built schools was spearheaded in the UK by Hampshire County Council in the 1980s. Hampshire involved Edward Cullinan Architects in a number of projects including Fleet Calthorpe School (1984) and Church Crookham School (1987). These buildings (characterised by ribbon windows and very poor insulation) tended to overheat

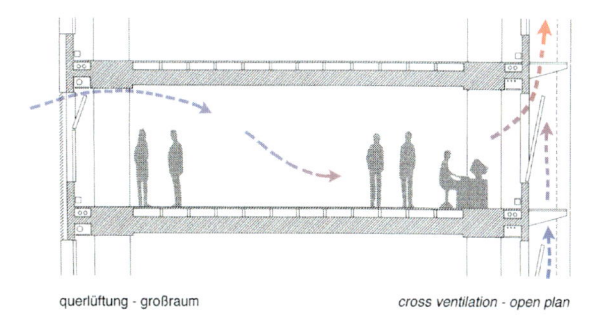

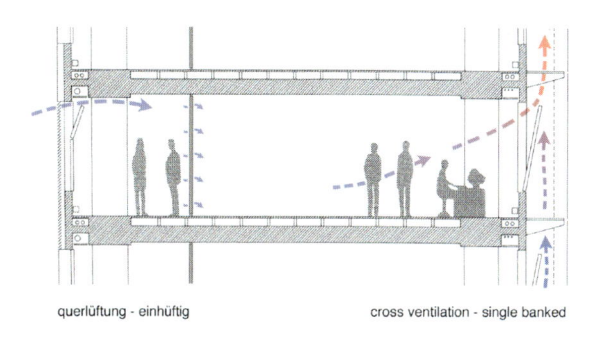

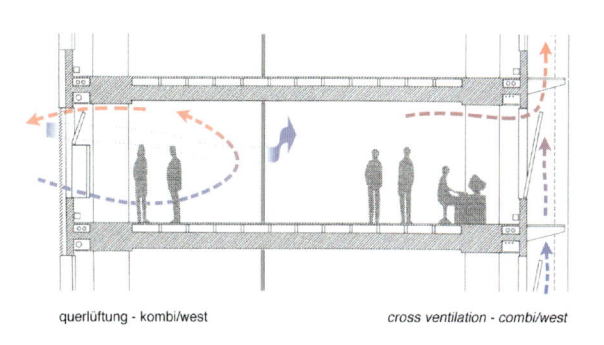

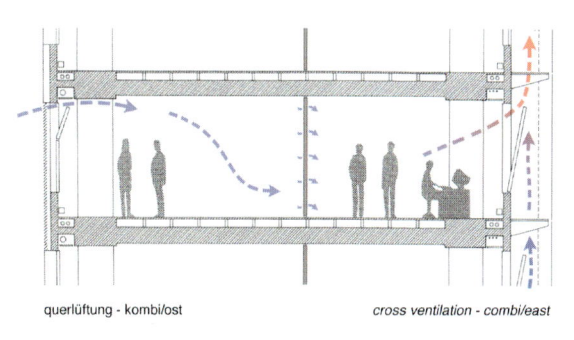

113. Natural ventilation options for GSW building Berlin [Sauerbruch-Hutton].

CHAPTER 4 | INTEGRATED DESIGN

114. South Façade, Fletcher Building, Leicester UK. Before & after refurbishment [Short Ford Architects].

> THE USE OF DOUBLE SKIN FAÇADES TO PROVIDE BOTH ACOUSTIC BUFFERING AND TO DRIVE NATURAL VENTILATION HAS BEEN WIDELY ADOPTED IN URBAN OFFICE BUILDINGS

excessively in summer and were cold in winter. The addition of insulation to walls and roofs and sun shading to vulnerable façades provided a dramatic improvement in internal conditions.

However, without improvements to the ventilation many classrooms remain stuffy. The refurbishment of the Fletcher Building at De Montfort University (a 1960s CLASP system building) by Short Ford Associates provided the opportunity to not only improve insulation levels and provide sun shading, but to also improve ventilation by increasing the free area of inlet and outlet openings and increasing the stack height between them (fig. 114). This strategy improved the internal environment while reducing the running and maintenance costs, acting as a model for refurbishment of similar lightweight buildings from the 1960s.

Regeneration within an historic urban setting presents particular challenges. In the renovation of an existing mixed-use building (the former Casa di Bianco) in Cremona, Italy, Mario Cucinella Architects (MCA) paid deference to the historic centre, restoring the 16th century façade and tower, while making significant alterations to the modern (1970s) block (fig. 115). The surrounding context is respected by using local building materials and elements, such as shutters and coloured façades interpreted in a contemporary way. Concerns for solar control, natural ventilation and daylight formed part of the process of design development and the different treatment of internal court and external façades *(Cucinella, 2004)*.

92 **CHAPTER 4** | INTEGRATED DESIGN

The renovation of the former Post Office complex in the Via Bergognone in Milan by MCA faced the additional environmental problem of noise and pollution from a busy traffic intersection. The complex consists of four office buildings dating from the 1960s and 70s that enclose a central courtyard. The west/south-west facing façade onto via Bergognone was particularly vulnerable to solar heat gain and to traffic noise, and a 'second skin' made from selective glass was designed to reduce solar gain and the need for air-conditioning. Chilled beams help to stabilise temperatures and provide comfort in summer while also reducing the size of the mechanical ventilation system. This project exemplifies the opportunities to improve internal environments while reducing energy and running costs.

The foregoing examples demonstrate the importance of an understanding of solar control, daylight, ventilation and thermal performance to the development of a satisfactory integrated solution.

Most of the buildings which we will have by the end of the century have already been built, and therefore the creative re-use and low carbon refurbishment of existing buildings is of great significance. However, the integration of environmental design principles in the design of new buildings provides possibly a greater opportunity for innovation and development.

> IN THE RENOVATION OF EXISTING BUILDINGS, THE FAÇADE BECOMES THE PRIMARY MODERATOR BETWEEN THE EXTERNAL CLIMATE AND THE INTERNAL ENVIRONMENT

115. Casa di Bianco, Cremona, Italy. Before & after refurbishment [Mario Cucinella Architects].

4.4. INTEGRATION IN NEW BUILDINGS: TEMPERATE CLIMATES

An holistic approach, collaboration within the design team and an understanding of basic environmental design principles are as applicable to the design of new buildings as they are for the refurbishment of the existing stock. However, there is even more temptation (particularly with large projects) to be led down the path of over-complexity. When asked how architects should approach the design of new buildings Bill Bordass said: *"Start with passive design to get rid of unnecessary requirements or technologies, install efficient equipment, control the remaining building efficiently and use low-carbon energy supplies. If you halve demand, double efficiency and halve the carbon content of the energy supplies, you're down to an eighth of emissions."* (Bordass, 2011).

In the design of new buildings there is much more flexibility and opportunity to respond to the demands of a particular location and the needs of the client. The characteristics of the site and its climate/microclimate are often the starting point for strategic thinking on environmental control options. This section on the integrated design of new buildings explores different typologies (higher education, offices, residential) located in different climate zones (temperate; hot humid; and hot dry). The intention is to identify common approaches to meeting differing demands.

NATURAL VENTILATION DESIGN

The UK climate has been characterised as being warm and wet in summer and cold and damp in winter. Average weather data reveals that temperatures in summer are rarely over 27°C, and that for almost the whole year external temperatures will be lower than those desired internally. This implies that external air can be used to remove internal heat gains throughout the year. Inspite of this, mechanical ventilation and air-conditioning was prevalent in non-domestic buildings in the 1970s and 1980s. By the 1990s natural ventilation was back on the agenda, but not widely applied due in part to a lack of 'know-how' within the professions.

Completed in 1993, the Queens Building at De Montfort University in Leicester (Short Ford & Associates) was described as the first of a new generation of naturally ventilated buildings in the UK (fig. 116). This building is described briefly in Chapter 2, and details of its performance in use were reported following surveys of occupant satisfaction and energy use in 1996 and 2006 as part of the PROBE series of studies *(Asbridge & Cohen, 1996; Bunn, 2006)*. The Queens Building demonstrated that daylit naturally ventilated buildings can reduce capital, energy and maintenance costs. It has also provid-

> THE CHARACTERISTICS OF THE SITE AND ITS MICROCLIMATE ARE OFTEN THE STARTING POINT FOR STRATEGIC THINKING ON ENVIRONMENTAL CONTROL OPTIONS

ed valuable feedback on some of the risks and detailed design issues which need to be addressed to improve performance. Other buildings from the 1990s and the first decade of the new millennium have engaged with the same issues and have demonstrated the benefits of passive design and natural ventilation, even where the final design has adopted a 'mixed-mode' approach.

The Queens Building at De Montfort University was perhaps a bit too closely tailored to the needs of its original occupants, and although subsequent changes have been accommodated they have sometimes been made at the expense of increased energy and maintenance costs (the 2006 PROBE revisit to the Queens Building indicated that both gas and electrical energy consumption had increased). Few would now question the need for new buildings to accommodate change, to be more resilient, more adaptable and less specific to their initial use. The common ground is to meet the needs of the occupants, and wherever you are in the world this includes providing and controlling light and air, thermal and visual comfort.

For Ted Cullinan, people and how they occupy space have been central concerns of his architecture. The cooperative practice he formed in 1965 continues to place people at the centre of their design process and believe that good buildings lift the spirit and create social value. It is not surprising that many of their projects are loved by their occupants.

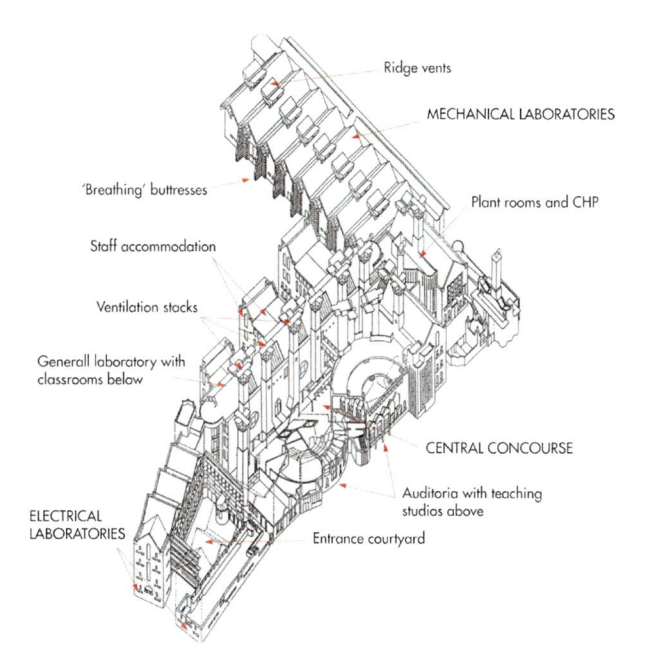

116. Axonometric of the Queens Building, Leicester, UK [Short Ford Architects].

THE QUEENS BUILDING DEMONSTRATED THAT DAYLIT NATURALLY VENTILATED BUILDINGS CAN REDUCE CAPITAL, ENERGY AND MAINTENANCE COSTS

CHAPTER 4 | INTEGRATED DESIGN   95

The Centre for Mathematical Sciences, University of Cambridge (Architect: Edward Cullinan Architects; Environmental Engineer: Roger Preston & Partners; Structural Engineer: Buro Happold) was completed in three phases between 2000 and 2003. It provides space for over 1500 students and staff within several departments bringing Pure and Applied Maths together into a single building to promote interdisciplinary research. This is achieved within a complex of seven pavilions which surround a central reception, cafeteria and administration building, and separate library and gatehouse (figs 117–119). Six of the pavilions are very similar in design, with offices and meeting rooms around a central lift and staircase topped by a glazed lantern and exhaust ventilator.

The client was averse to sealed, air-conditioned work spaces, preferring instead den-like spaces with openable windows. The design team considered various passive and mixed-mode options and decided to pursue natural ventilation for nearly all spaces within this large complex. Environmental control in the cellular offices was provided by single sided and buoyancy ventilation with provision for chilled beams should the need arise.

The project brief contained a challenging set of requirements from the client who wanted to avoid mechanical ventilation in offices, and the resulting design strategy incorporates solar shading, exposed thermal mass, buoyancy-assisted single sided natural ventilation, and the facility for automatic night cooling. However, the implementation of this strategy has been complicated by three key

117. Site plan for Centre for Maths & Sciences, Cambridge [Edward Cullinan Architects].
118. Pavilion F, Centre for Maths & Sciences, Cambridge.

96 **CHAPTER 4** | INTEGRATED DESIGN

factors: the potential 24h occupancy of the building, limitations on ceiling height, and the planning requirement for electric lighting at night not to disturb occupants of adjacent properties.

A survey of the occupants of Pavilion B in 2002 indicated that overall the occupants regarded the building as comfortable, and satisfaction with air quality and temperatures was excellent in winter, while occupants were forgiving of high temperatures in summer. Four years later a second survey confirmed that occupants of the pavilions have a high level of satisfaction with their environment except for summer air temperatures. However, lessons from the first survey had been taken on board and user controls for windows and blinds were significantly improved for the phase 2 pavilions (now described as "... clear in intent, well labelled and give instant feedback to the occupants" (ibid).

The high level of occupant satisfaction with the environment in the CMS highlights the quality of the architecture and the high level of collaboration within the design team to achieve an integrated approach to environmental design.

> **THE HIGH LEVEL OF OCCUPANT SATISFACTION WITH THE ENVIRONMENT IN THE CMS HIGHLIGHTS THE QUALITY OF THE ARCHITECTURE AND THE HIGH LEVEL OF COLLABORATION WITHIN THE DESIGN TEAM**

119. East Elevation of Pavilion B, Centre for Maths & Sciences, Cambridge [Edward Cullinan Architects].

CHAPTER 4 | INTEGRATED DESIGN

A similar high level of collaboration was achieved by architects Bennetts Associates and engineers Buro Happold in their proposals for the Potterrow Development in Edinburgh for the University of Edinburgh completed in 2018 (figs 120–121). Rather than tailoring the accommodation needs of the incoming faculty too closely, Bennetts proposed *"…a generic ribbon of office type space that could accommodate a wide range of University type functions over the long term"* (Bennetts, 2016). The University was persuaded that long-term adaptability should prevail over the strongly held opinions of the first occupier. The result has been hugely successful.

At Potterrow the original intention was that windows would be the principal means of ventilation, and mechanical ventilation would be back-up. However, full reliance on opening windows was not considered to be possible due to the level of traffic noise on Potterrow itself, and the University's poor experience of motorised windows at their Alexander Graham Bell building. Background ventilation is therefore provided by displacement ventilation from floor outlets with windows providing occupants with a level of local control (figs 122–123). The mechanically driven ventilation rate is not able to deliver significant convective cooling in summer, so pre-cooling is provided to air handling units from the main CHP unit. Inevitably perhaps the decision to provide mechanical ventilation with summer pre-cooling of ventilation air has resulted in higher electrical loads than originally predicted. On visiting the building it was found that occupants of

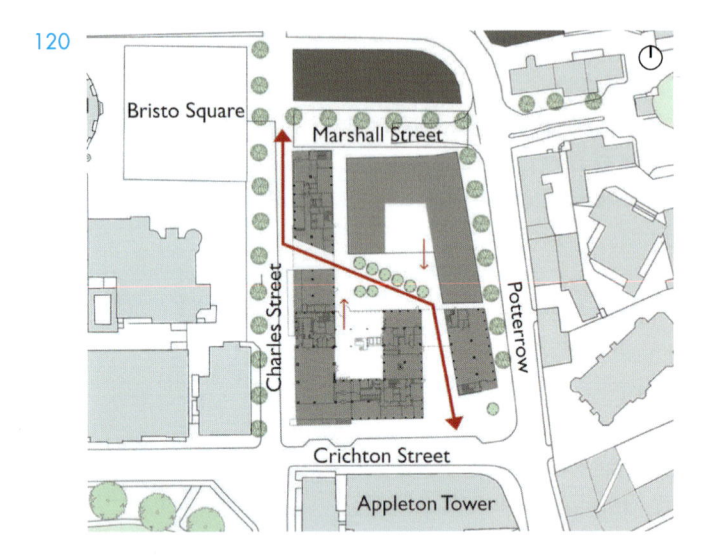

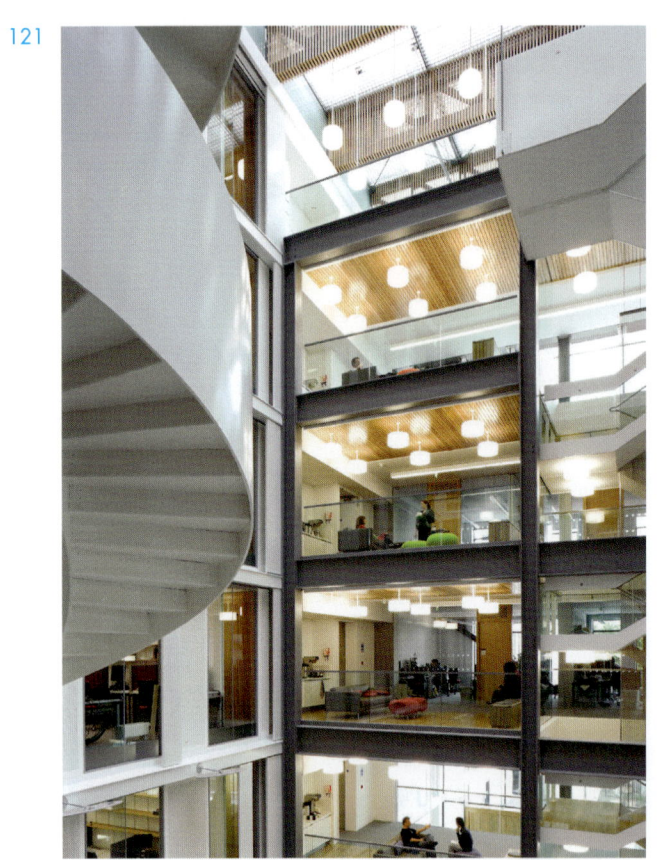

120. Site plan for Potterrow Development, University of Edinburgh [Bennetts Associates].
121. Interior of atrium, Potterrow Development, University of Edinburgh.

CHAPTER 4 | INTEGRATED DESIGN

rooms facing onto the open courtyard do open their windows in summer and are happy with conditions.

However, post occupancy surveys of occupant satisfaction revealed that overheating did occur in some rooms surrounding the atrium which have no direct openings to the outside. These investigations prompted action by the facilities managers to achieve improved control of louvres at the top of the main atrium. This alteration has enabled hot air within the atrium to be purged naturally without requiring manual operation, reducing the occurrence of overheating. Extract fans which operate in winter to recover

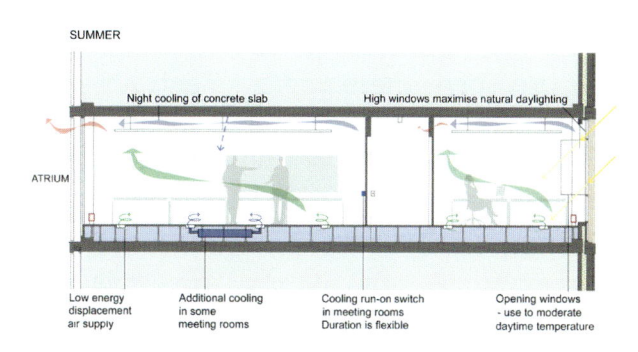

122. Mixed ventilation strategy for typical floor, Potterrow Development [Bennetts Associates].

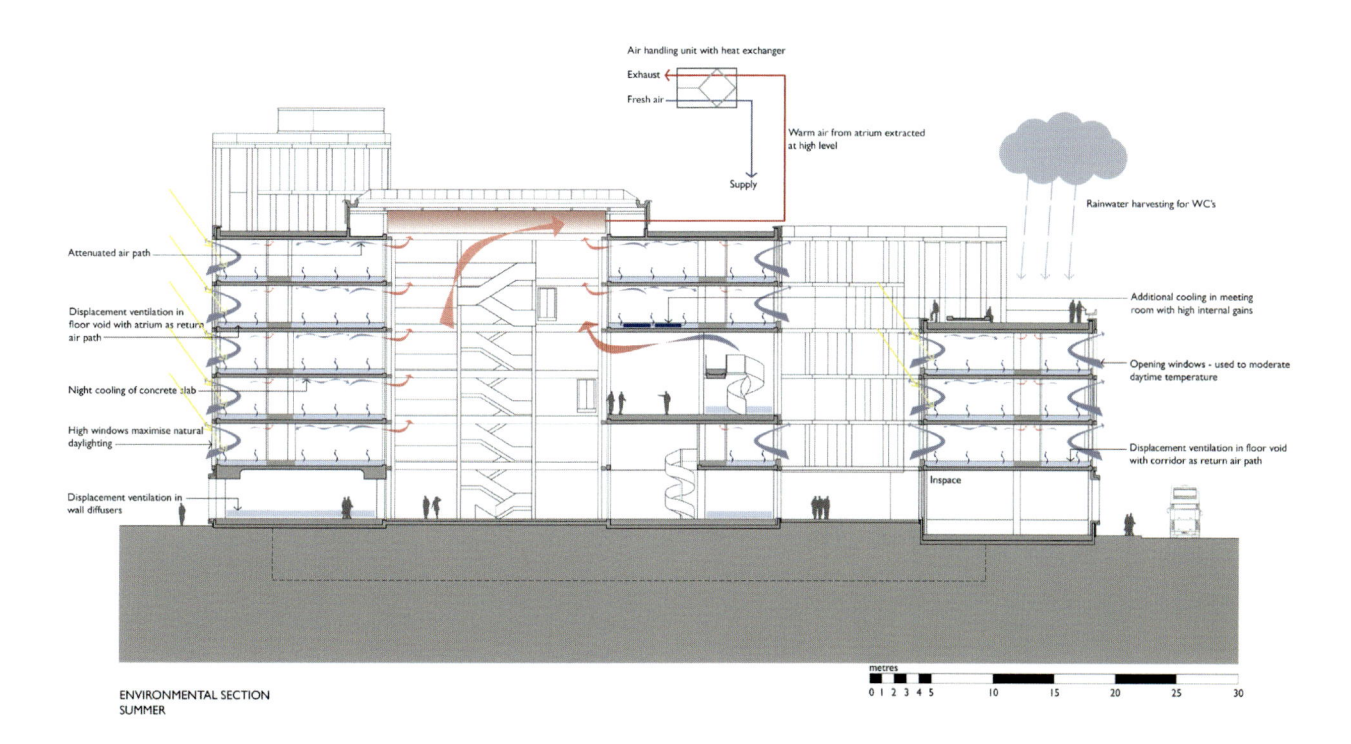

123. Site section through Potterrow buildings showing ventilation strategy [Bennetts Associates].

CHAPTER 4 | INTEGRATED DESIGN   99

waste heat are switched off when the louvres are open. These improvements have reduced fan energy consumption by 15kWh/m^2 and demonstrate that natural ventilation can have a role in this inner city location to purge hot air from the atrium.

Other recommendations coming from the surveys suggested that when designing airflow paths both air supply and exhaust should be easily adaptable and follow as direct a path as possible. Also, underfloor plena require a high standard of construction with rigorous quality control and testing. Air tightness can be improved if complex geometries are avoided.

The two reasons given for adopting a mechanical ventilation solution rather than purely natural ventilation in the Potterrow Development were the risk of street noise penetration, and the poor performance of window actuators. However, in similar urban areas it is sometimes possible to include acoustic attenuation on both supply and exhaust routes to achieve satisfactory noise reduction. It is also the case that robust, high performance vent actuators are available on the market, and provided the control system is equally robust and responsibilities within the design and construction team are clear, then natural ventilation can be viable in urban environments.

> IN SOME URBAN AREAS IT IS SOMETIMES POSSIBLE TO INCLUDE ACOUSTIC ATTENUATION ON BOTH SUPPLY AND EXHAUST ROUTES TO ACHIEVE SATISFACTORY NOISE REDUCTION

ACOUSTIC ATTENUATION ON AIRFLOW PATHS

The specification of acoustic attenuation on the supply and exhaust routes into and out of the building were part of the natural ventilation solution within the recently completed SPACE performing arts building for the Girls High School Trust in Nottingham (Architects Marsh Growchowski). During the planning review process, concerns were raised by nearby residents about the risk of noise break-out from the auditorium, and the conditions of planning approval included a requirement to meet certain noise reduction criteria. This had implications for the design of not only the natural ventilation system, but also the enclosing walls and roof structure.

The ventilation strategy therefore had to be capable of removing excess heat at the warmest time of the year while simultaneously providing sufficient acoustic attenuation to avoid disturbance to neighbouring residences, and avoiding noise from vehicles on the adjacent roads penetrating the auditorium. This flow of air had to be achieved without the assistance of the wind (i.e. by buoyancy alone).

The strategy introduces air primarily from the clean and quiet area on the south and west side adjacent to the arboretum. Air is supplied below the floor of the auditorium via a large plenum which also provides space for moveable wagons and floor equipment. Supply to this plenum was also initially anticipated on the east side, but an existing mature tree in this location mitigated against this. Even distribution of the air to all parts of the auditorium, and to the upper seating levels, would be required to avoid the risk of creating stagnant air pockets (figs 124–126). The need for the airflow to be promoted due to buoyancy alone implied exhaust at high level above the auditorium. Of course, the exhaust air terminations also had to function successfully when the wind was blowing, irrespective of the direction of the wind.

It was anticipated that both supply and exhaust air paths would have to include significant acoustic attenuation to prevent noise breaking in or breaking out of the auditorium. From an early stage it was assumed that sufficient acoustic attenuation would be provided if it resulted in a 50% reduction of the free area of the inlet and outlet areas. As the design developed, detailed acoustic calculations were made to ensure that the noise reduction criteria were met. An assessment

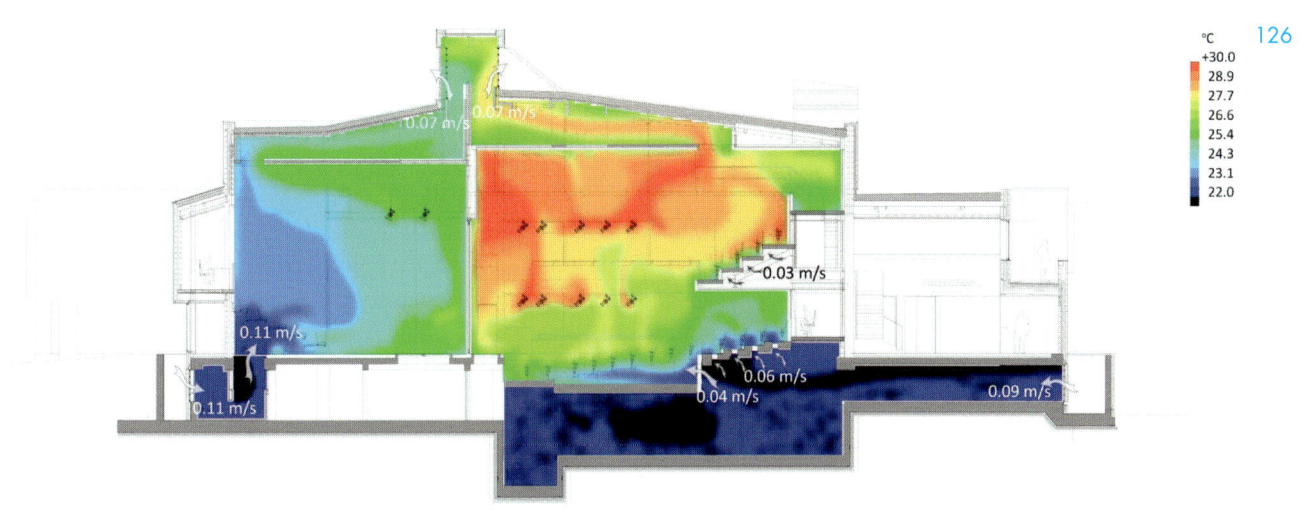

124. Site plan for the SPACE building, Nottingham [Marsh:Grochowski].
125. Interior of auditorium from stage area in the SPACE theatre.
126. CFD analysis for a summer evening performance in the SPACE theatre.

CHAPTER 4 | INTEGRATED DESIGN

of existing noise levels on site was undertaken and used as a benchmark for proposed limits for noise levels arising from the new building, with reference to adjacent residential buildings. These assessments were made in the context of a defined set of activities and pattern of occupancy for a 'typical' week set out by the client. Acoustic 'silencers' to all inlets and outlets were then specified to meet the proposed activity noise limits. It was found that to meet the prescribed noise levels the supply plenum to the north stage area had to be deeper than originally proposed. The noise reduction criteria also influenced the final geometry and detailed design of the exhaust plena from the main auditorium and from the stage area (the central outlet tower). The successful resolution of the numerous (sometimes conflicting) requirements for the satisfactory functioning of the central outlet tower involved numerous iterations and close collaboration of all the consultants in the design team.

This new building for the performing arts has been occupied since December 2016, and although a full post occupancy analysis has yet to be completed, the thermal and acoustic performance has broadly been in line with predictions *(Ford & Vallejo, 2015)* and results of a survey of opinions of visitors and staff (June 2018) have been very positive.

4.5. INTEGRATION IN NEW BUILDINGS: HUMID TROPICS

An understanding of the influence of humidity and air movement on thermal comfort is essential when designing for the humid tropics. Daytime highs of 28–32°C and night-time lows of 23–25°C are typical, with absolute humidity varying between 10–25 g/kg. In such conditions, small diurnal temperature variations are typical throughout the year in the tropics, with the only variation being the wind speed, humidity and the direction and intensity of rainfall. In Singapore the north-east monsoon winds bring very heavy rain and the 'empirical tradition' was to open façades and use slatted shutters to keep out the sun and rain while allowing significant air movement. During much of the 20th century façades were sealed shut and reliance placed on air-conditioning, and this is still the case particularly for work environments. An exception to this was provided by an office building designed by Geoffrey Bawa for the State Mortgage Bank in Colombo, Sri Lanka in 1976–78. *"The objective was to provide a working environment that could be lit and ventilated by natural means…. The main elevations face south and north …to reduce solar gain and catch the main breezes"* (Robson, 2002). Bawa led the way, and

> AN UNDERSTANDING OF THE INFLUENCE OF HUMIDITY AND AIR MOVEMENT ON THERMAL COMFORT IS ESSENTIAL WHEN DESIGNING FOR THE TROPICS

there is now a new generation of workplace and residential buildings in the humid tropics which combine sun shading and convective cooling to achieve thermal comfort.

Singapore based architects WOHA have gained international recognition for their integration of environmental and social principles at every stage of the design process. Their work addresses the need to reduce energy use and carbon emissions by responding to local context, culture and climate. Singapore climate is characterised by two main seasons, the north-east monsoon (December to March) and the south-west monsoon (June to September). Daily dry-bulb temperatures range from 23–34°C, and humidity is consistently high. While air-conditioning has become ubiquitous in Singapore, WOHA appreciate that by careful manipulation of plan and section natural ventilation can be viable for much of the time. Their 28 storey residential project at No1 Moulmein Rise (completed 2003) achieves cross ventilation to apartments by careful arrangement of openings in the plan. Projecting ledges and perforated metal cladding reduce solar heat gain, and bay windows incorporating a sliding aluminium shelf promote natural ventilation while preventing rain penetration. This important detail is a development of Bawa's 'monsoon window' which he included in the State Mortgage Bank in 1978, to solve the problem of how to provide for natural ventilation when heavy rain is accompanied by strong winds (figs 127–129).

A strong response to the environmental agenda is also reflected in the Singapore

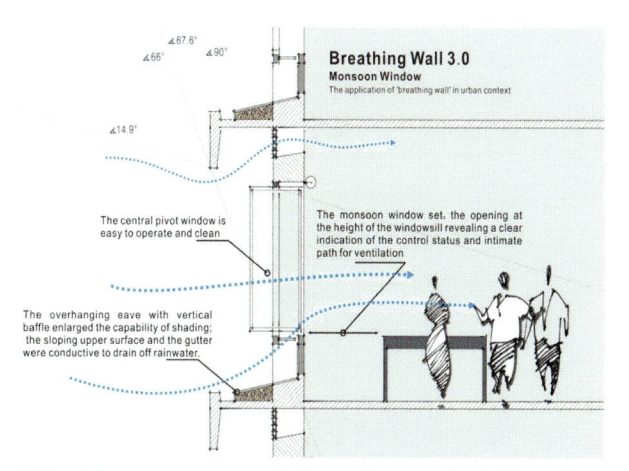

127. Monsoon window in the State Mortgage Bank, Sri Lanka [Geoffrey Bawa].

128. Bay window for 1 Moulmein Rise [WOHA], ©Tim Griffith.
129. Monsoon window for 1 Moulmein Rise, ©Tim Griffith.

CHAPTER 4 | INTEGRATED DESIGN 103

130

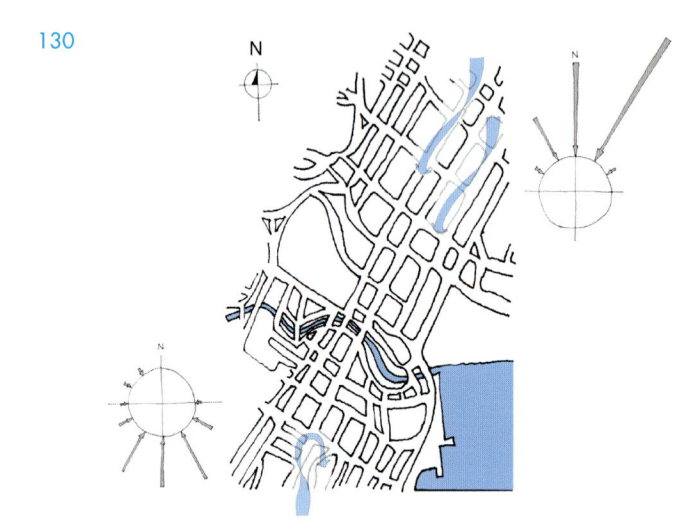

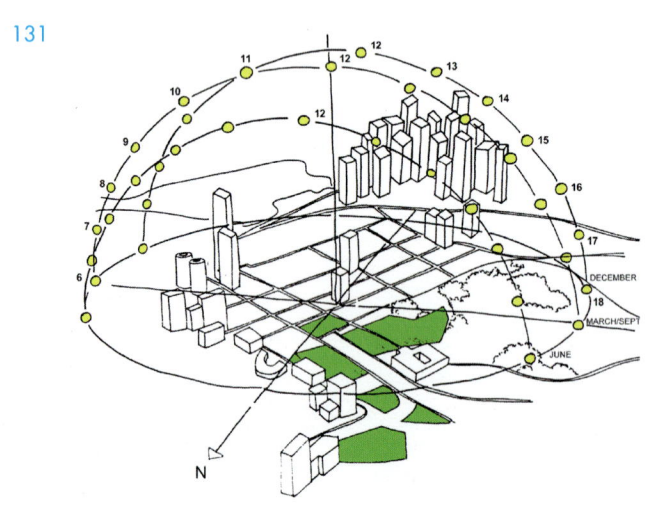

131

130. Singapore street pattern & winds.
131. Singapore: urban form & sunpath diagram.

Management University (SMU) by Edward Cullinan Architects (2000–2005). The competition brief called for the masterplanning of a new city centre campus, and the creation of seven new large departmental buildings. The relationship with the existing urban form and landscape was crucial, and part of this relationship is related to the prevailing microclimate. The competition stage design team (which included Structural Engineer Neil Thomas of Atelier 1; Landscape Architect Grant Associates; and Environmental Consultant Brian Ford Associates) developed a strategy to promote thermal comfort in external spaces as well as inside buildings, based on an understanding of local microclimate, the urban context and the contribution of building form, materials and landscape, to moderate the impact of the external climate.

Although seasonal and diurnal variation of temperature and relative humidity is small, the sunpath and direction of prevailing winds do vary seasonally, which informed the design team's response at both urban and individual building scales (figs 130–131). The orientation of the existing grid of streets originally helped to channel the north-east and south-west monsoon winds, alleviating the heat and humidity at street level. The new University campus is exposed to the north-east Monsoon winds, providing an opportunity to exploit this from November to March. Unfortunately, the tall buildings of the central business district obstruct the south-west monsoon winds (June to September) and promote a desire for more air movement at this time of year. The permeability of the ur-

CHAPTER 4 | INTEGRATED DESIGN

ban form has a major impact on street level air speed, and discomfort can be avoided provided there is adequate air movement. Therefore, as well as exploiting the existing urban 'grain', the planning of the different buildings incorporated gaps and 'perforations', particularly at ground level.

Of course, in external spaces thermal comfort is dominated by the sun heating surfaces which then continue to radiate heat, and a major objective has therefore been to provide both shade (to keep temperatures low) and air movement (to promote cooling). Shade from trees has been exploited, taking care in the choice of species, as this affects the width and height of the tree canopy, which in turn affects air movement at ground level. The campus plan includes both formal and informal shaded gardens along major pedestrian routes at ground level, and different landscape treatments from the sunken court to roof gardens (fig. 132).

In an equatorial climate, an east/west axis for buildings (i.e. north/south orientation) is most favourable, as sunlight penetration of building elevations can be controlled by simple horizontal shades. However, the existing street pattern in Singapore implies that most elevations will be vulnerable at different times (north-west and north–east elevations more vulnerable in June and south–west and south–east elevations more vulnerable in December). The major pedestrian routes through the project are shaded

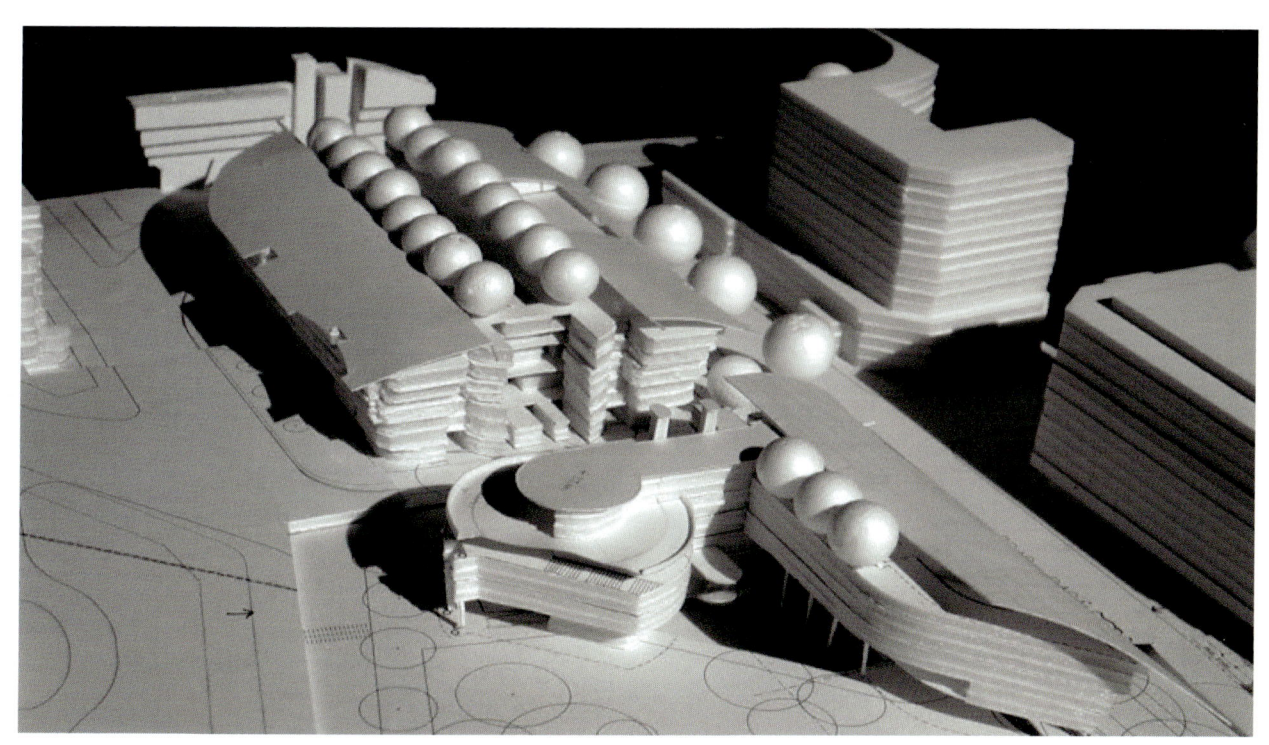

132. Singapore Management University Early Model for Heliodon Studies [Cullinan Studio].

CHAPTER 4 | INTEGRATED DESIGN

partly by surrounding buildings, and partly (from overhead sun) by the canopy of shade trees or fabric canopies.

With the sun *'overhead'* most of the time, the roofs of the buildings are particularly vulnerable. At competition stage solar heat gain through the roof was minimised by separating it into layers: an outer sun and rain 'parasol', and an inner air and thermal barrier. The 'parasols' are extended into large overhangs to protect the most vulnerable façades. A large air gap between the 'parasol' and inner roof would promote free air movement to remove the solar heat gain conducted through the outer roof (figs 133–135). Unfortunately this feature was not included in the final project.

A feature which was part of the original competition and which has been very successful has been the inclusion of shaded courts and 'breezeways'. The sunken pedestrian route, which runs from the MRT station north-west parallel to Bras Basah Road, and south-west parallel to Armenian Street, is protected by substantial shade trees at roof level. Elsewhere planted façades provide shade facing into the gardens, and 'raintrees' (Samanea Saman) shade the fabric of the street passageways and façades facing Bras Basah Road to reduce re-radiated heat build up and improve comfort at street level. For buildings to Bras Basah

> THE STEPPED SECTION OF THE SMU MAIN BUILDING ALSO PROMOTES THE USE OF DIFFUSE AND REFLECTED LIGHT TO PERIMETER SPACES

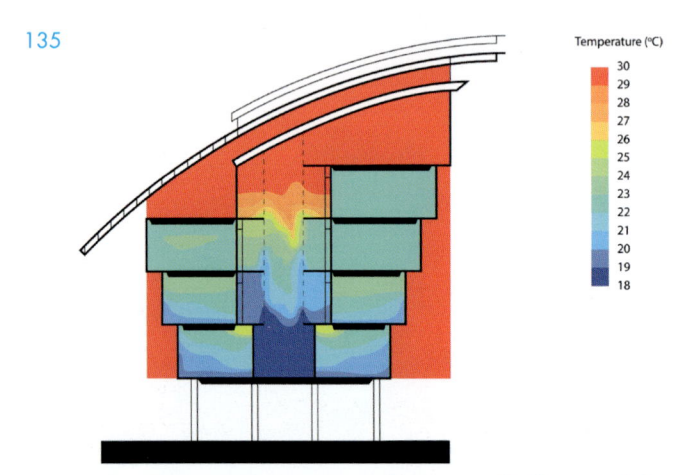

133. SMU central concourse and breezeway [Cullinan Studio].
134. SMU top floor central concourse.
135. CFD Temperature Plot to show departmental 'Cool Pool' using ADC (not implemented).

106   **CHAPTER 4** | INTEGRATED DESIGN

Road, the ground level presents a continuous barrier, preventing noise and pollution reaching the sunken pedestrian court. Air movement though the court at this point is promoted by ground floor openings to the south-west, and by the open 'slot' at roof level (fig. 136). The *'breezeways'* have proved to be popular informal social gathering places and are regarded as highly successful. The natural ventilation of these transitional spaces reduces energy use and also the thermal shock of moving from unshaded street to fully conditioned interior.

While solar control is vital, the buildings must allow diffuse and reflected daylight to penetrate most of the interior spaces. This is important to reduce electrical energy consumption from artificial lighting, and the associated cooling load, as well as maintaining contact with the rhythm of the outside world. Daylight is diffused by horizontal

> *Roddy Langmuir (Senior Partner, Cullinan Studios):*
>
> *The real success of SMU is the outcome of an integrated design approach, whereby the passive design measures also become responses to the context and the place-making objectives of a horizontally stratified campus. For example:*
> • *Breezeways are used as informal social gathering places.*
> • *Planted façades face into the gardens and present a softer, green backdrop seen from the park.*
> • *Raintrees shade the fabric of the street passageways to reduce re-radiated heat build-up and encourage their use.*
> • *The natural ventilation strategy for transitional spaces reduces energy use but also the thermal shock of moving from air con to baking hot street and back into air-con.*

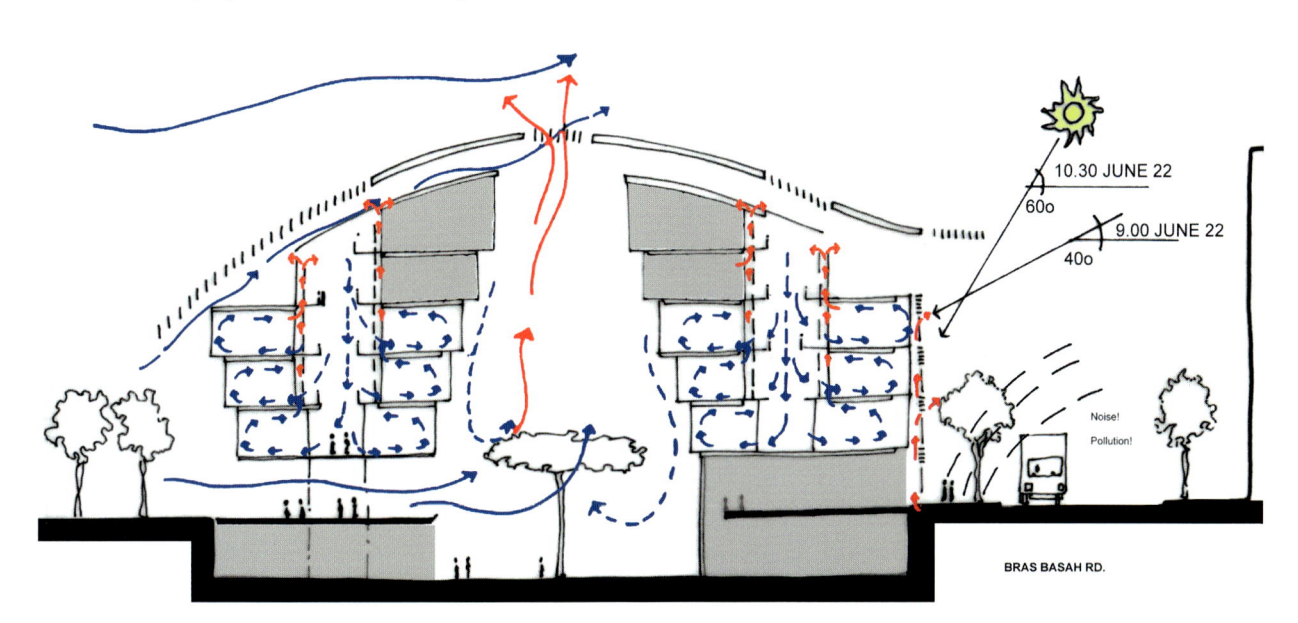

136. Sketch section of SMU Department adjacent Bras Basah Road, Singapore – environmental strategy.

CHAPTER 4 | INTEGRATED DESIGN

roof level louvres, and by a combination of horizontal and vertical louvres in parts of more vulnerable elevations. The stepped section of the main building also promotes the use of diffuse and reflected light to perimeter spaces. Elevations facing into the shaded *cool pools* and sunken courts are generally light in colour, to reflect and diffuse light to lower levels.

In the original proposals, the lightwells and interior circulation spaces to the separate departments were to be used as reservoirs of cool air. Chilled water cooling coils at high level were to induce 'Downdraught Cooling' (as applied in the Stock Exchange in Malta), creating a 'cool pool' from which fresh cool air was to be taken into perimeter classrooms on the lower three floors of accommodation. This technique would have substantially reduced fan power required to circulate air, achieving electrical energy savings of 30–40% of an equivalent fully air-conditioned building. Unfortunately, the client could not be persuaded to adopt this approach, preferring instead to adopt more conventional air-conditioning. Nevertheless, the other strategies which have been adopted will have reduced the cooling load by approximately 40% of a conventionally designed building.

Inspite of the fact that two of the five main passive environmental control strategies were not implemented as part of the final construction, the other elements have been successful, and illustrate the value of an integrated approach which is based on a careful analysis of urban form and microclimate.

4.6. INTEGRATION IN NEW BUILDINGS: HOT DRY CLIMATES

In hot dry climates where daytime ambient air temperatures regularly exceed the upper threshold for adaptive comfort, passive cooling can only be achieved by exploiting an ambient heat sink (as described in Chapter 1). Where the wet bulb air temperature is below 24°C, and a sustainable water supply is available, direct evaporative cooling can be exploited to reduce the supply air temperature.

In 2010 the Solar Decathlon Europe (SDE) competition took place in Madrid and provided the University of Nottingham an opportunity to explore the contribution direct evaporative cooling could make to summer time occupant comfort, as part of a broader passive environmental strategy. Designed and built by students this affordable modular terrace (fig. 137) included a high-performance envelope (U-value of walls & roof ≤0.17W/m²K), triple glazed windows, lobbied entrance and shading to glazed openings, all helping to minimise external heat gains. In addition, the specification of A-rated appliances minimises internal heat gains. The house was erected and tested in Madrid in June and the final evaporative cooling system installed in the house was evaluated for the week of the competition (19–26 June 2010).

The two bedroom, two storey house (floor area 75m²) was designed to promote passive downdraught evaporative cooling to maintain thermal comfort during the summer (fig. 138). Nozzles positioned at the top of the double height space generated a mist of water that evaporated in warm external air drawn through the roof light (figs 139–140). Evaporation of the water cools the air generating a plume that drops into the dining area and then divides, part flowing through the living room and exiting via a window on the south wall, and part through the kitchen, absorbing heat from any appliances that are operating and exiting via a window in the north wall. The nozzle system was controlled manually, with the occupants increasing or reducing the rate of cooling, according to their perception of how warm they felt.

Summer Strategy. During the hot, dry summers in Madrid, maximum daily external air temperatures can frequently be above 35°C. Protection of glazed openings from the high direct solar radiation is essential. Night-time

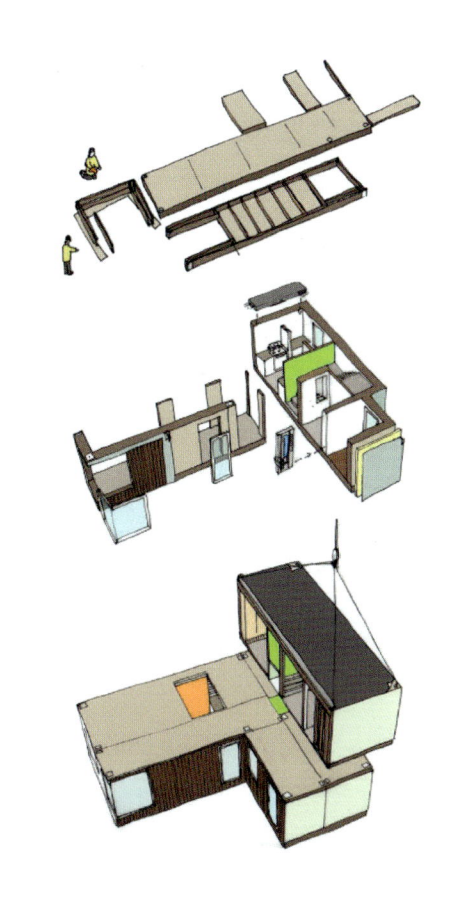

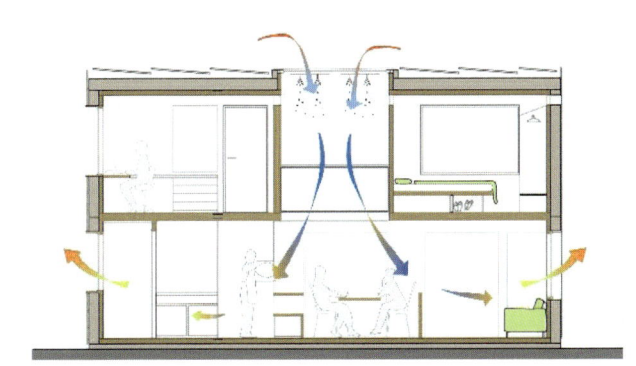

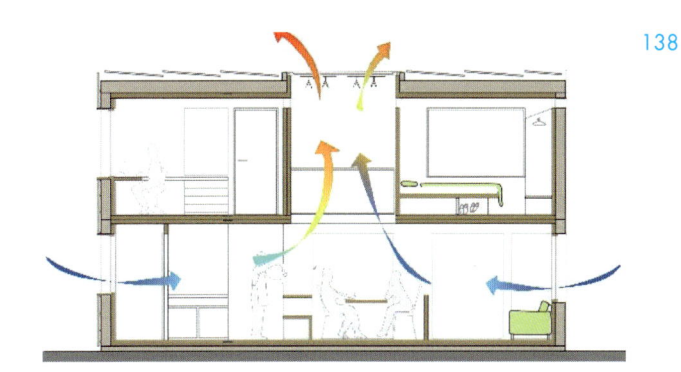

137. Volumetric prefabrication of Nottingham HOUSE, Solar Decathlon Europe, Madrid 2010.
138. Sections to show summer day & nigh-time airflow paths, SDE 2010.

CHAPTER 4 | INTEGRATED DESIGN   109

temperatures can drop below 20°C, providing an opportunity for convective cooling to be exploited.

Plotting data for Madrid on a psychrometric chart in Chapter 3 revealed that during the summer period, thermal comfort is achievable by evaporative cooling alone, even with dry bulb air temperatures above 40°C. This is because the external air is so dry – typically below 30% relative humidity in the afternoon. At night, convective cooling will be achieved by promoting buoyancy driven (updraught) natural ventilation. Shading of exposed perimeter openings is provided by a woven mesh, and solar gains are further minimised by the high performance envelope (U-value of roof = 0.1 W/m²·K, U-value of walls = 0.17 W/m²·K).

> NIGHT-TIME TEMPERATURES DROPPING BELOW 20°C PROVIDE AN OPPORTUNITY FOR CONVECTIVE COOLING TO BE EXPLOITED

This combination of strategies removed the need for conventional cooling. Monitoring dry-bulb air temperature and relative humidity inside and outside revealed that comfort conditions were achieved most of the time, in spite of the large numbers of visitors to the house in the afternoon period. Without the evaporative cooling system, internal air temperatures would have risen well above internal comfort conditions. In a typical house the evaporative system could run until 8pm keeping the indoor temperature below 26°C (Ford et al., 2012).

139. Misting nozzles in Nottingham HOUSE, Solar Decathlon Europe 2010, Madrid, Spain.
140. Exterior view of Nottingham SDE House.

The requirements of the competition limited the overall height of the house, which meant that a supply air 'tower' could not be included. However, results for the week of the competition suggest that the absence of an inlet tower is no impediment to the successful integration of the system within a two storey house.

While Madrid may provide an *'ideal'* climate for the application of evaporative cooling, this approach is also viable in other locations in southern Europe where the climate is intermittently hot and dry. In Catania, Sicily the climate is hot and dry in the summer months, although the wet bulb occasionally goes above 24°C, reducing the time during the summer when evaporative cooling is effective. The construction and testing of a full-scale prototype experimental building located at the Conphoebus Research Institute in Catania is described in Chapter 2. This revealed that evaporative cooling is viable for a significant proportion of the summer months in Sicily, and led to proposals for the design of a new office building by Mario Cucinella Architects to exploit evaporative cooling as part of a wider strategy to achieve passive environmental control (fig. 141).

The four storey building provides flexible open plan offices and a conference centre (fig. 142). To achieve uniform air speeds across the office spaces, the building has a number of glazed towers which connect the floor plates vertically. The evaporative cooling system in each tower can be switched on or off as needed, cooling only enough

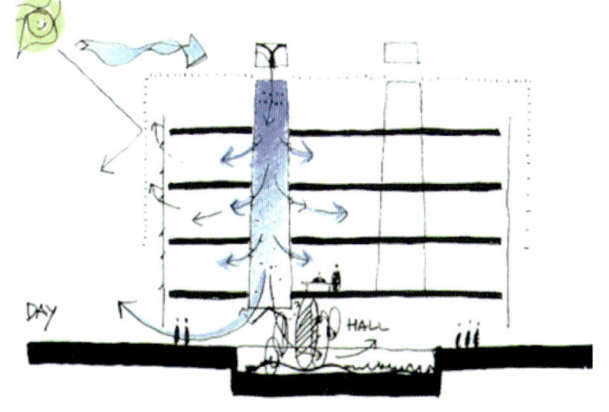

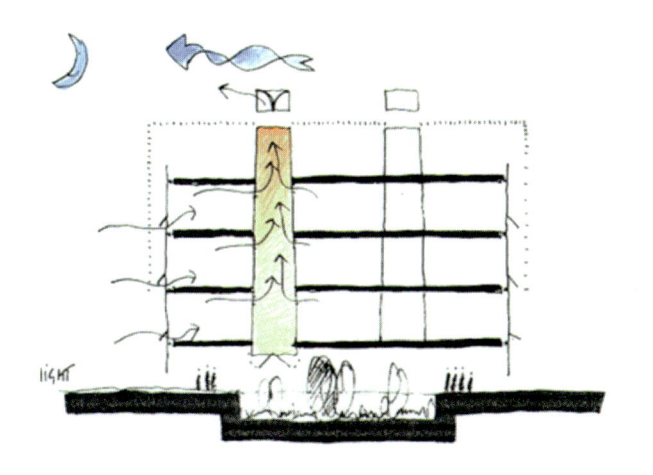

141. Day & night ventilation strategies for Office Building, Catania, Italy [Mario Cucinella Architects].
142. Model for proposed Office Building Catania, Italy.

CHAPTER 4 | INTEGRATED DESIGN 111

air as required by the zone served by that tower. This means that the amount of cooling can be controlled to respond to the varying demand in each part of the building. The towers also promote *'up-draught'* convective cooling during the midseasons and at night in the summer.

The towers were developed as an integrated architectural and environmental element by providing not only cooling but structure, ventilation and daylight, to each floor plate (fig. 143). Simulations of tower performance confirmed their potential in terms of cooling, comfort and energy efficiency. It was found that the openings in the tower with a low aspect ratio (width over height close to 0.1) reduced risk of uneven airflow at each level within the building. The simulations enabled refinement of the design to improve the uniformity of air velocity and temperature distribution throughout the office spaces (fig. 144). The overall energy consumption of this building was estimated to be 15% of that of a conventional building of the same size, significantly reducing energy running and maintenance costs.

> WHEN THE AMBIENT ABSOLUTE HUMIDITY GOES ABOVE 12G/KG OR WHEN THE INTERNAL RELATIVE HUMIDITY GOES ABOVE 65%, THEN DIRECT EVAPORATIVE COOLING IS NOT VIABLE

The structural scheme was generated by the geometry of the PDEC towers, providing design efficiency and construction economy, and further emphasising the value of an integrated design approach. The floor slabs

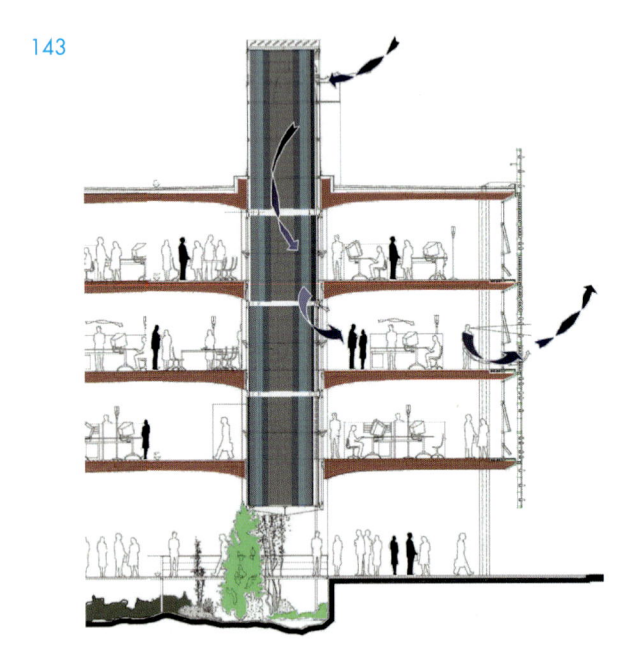

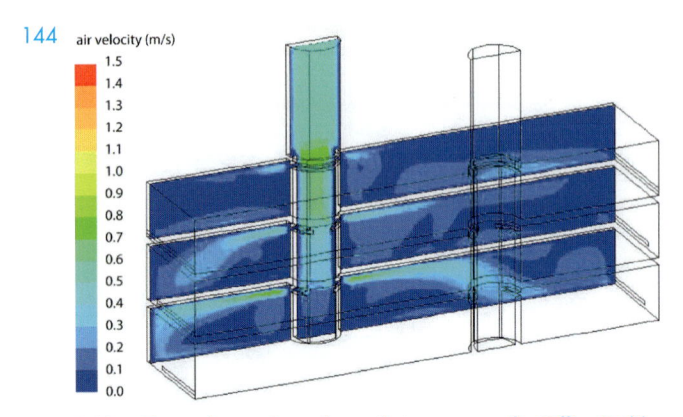

143. Tower design & stack ventilation strategy for Office Building Catania, Italy [Mario Cucinella Architects].
144. CFD plots showing air velocity contours for Office Building Catania, Italy.

sit on structural rings that are supported by four concrete pillars at each tower. The rings are horizontally braced between each other and are connected to the pillars by curved concrete ribs. Along the façade a series of single columns support the perimeter. This scheme concentrates all the structural elements around the cooling towers and the façade and leaves the plan free from columns. The glazed internal façade of each tower is supported from this structural system. Two service blocks at the ends of the building provide structural stiffness to the ensemble.

Each tower is composed of a *'body'*, that crosses all the floors, and a *'head'* that rises over the roof. The head acts as the air inlet and as a light catcher and diffuser. It also contains the ring of microniser and the evaporative zone (4.2m high). The air inlet is designed to control the wind effect and avoid turbulence inside the tower. The body distributes air and light to the offices. It is a glazed 3m diameter cylinder with a 200mm opening around its perimeter close to the ceiling at each floor.

The façades are composed of a double layer. The internal layer is single glazed with aluminium mullions. There are narrow openings for the air outlet at the bottom and openings for cross ventilation (free running situation) in the middle. The outer layer is composed of stone elements that provide shading to the internal layer, and to protect the façade from positive wind pressure, so that exhaust air is not forced back into the building. These shading elements are positioned horizontally with a variable density: more dense on the top floor and less dense on the lower floor, to balance the natural light provided by the towers. Daylight modelling and analysis found that the towers make a significant contribution to light levels in the centre of the building, reducing the need for artificial light and further reducing the electrical load *(Francis, 2000)*.

Catania (like much of Sicily and coastal southern Europe) has a composite climate in which summer time weather can switch from hot-dry to hot-humid. When the ambient absolute humidity goes above 12g/kg or when the internal relative humidity goes above 65%, then direct evaporative cooling is not viable and an alternative must be found to pre-cool supply air. This can be achieved by adopting *'active downdraught cooling'* as described in Chapter 2, and as applied in the Malta Stock Exchange. The application of active downdraught cooling in the Malta Stock Exchange is described in detail in Part 2.

This brief review of the integration of natural cooling strategies and techniques into the design of existing and new buildings illustrates the range of interrelated design issues which need to be resolved in collaboration with different specialist consultants. Environmental performance, structural efficiency and construction economy are all interrelated and provide significant benefits as part of an 'integrated' design approach. Reference has been made to the need to test strategies for natural cooling, solar control, daylighting etc., and a range of tools and analytic techniques are discussed in Chapters 5 & 6. It is also apparent that to be successful, design strategies must be carried through into detailed design, and Chapter 7 discusses the design of different components. The case studies in Part 2 of this book review the design and performance of buildings which have adopted direct evaporative and mixed-mode cooling as part of a wider environmental design strategy.

REFERENCES

- Asbridge, R. & Cohen, R. (1996). PROBE 4: Queens Building, de Montfort University, Building Services. Journal 35–41.
- Bennetts, R. (2016). 'The Way We Work'. Bennetts Associates: Five Insights. Artifice, London.
- Bennetts, R. & Bordass, W. (2007). 'Keep it Simple and Do it Well'. Sustainability Supplement to Building Magazine, 28 September. Digging Beneath the Greenwash, pp. 8–11.
- Bordass, W. (2011). 'Saving Money, Saving Energy, Saving Carbon'. Edge-CIBSE President's Debate, Arup Bristol, 8 September.
- Bunn, R. (2006). 'Project Revisit: Queens Building'. DeltaT, September, pp. 6–9.
- Cucinella, M. (2004). 'Works at MCA – Buildings and Projects'. The Plan – Art & Architecture.
- Ford, B. & Cairns, K. (2002). 'Market Assessment of Passive Downdraught Evaporative Cooling in Non-Domestic Buildings in Southern Europe'. EPIC Conference, Lyon, France. 2: 505–510.
- Ford, B. & Moura, R. (2003). 'Market Assessment of Passive Downdraught Evaporative Cooling in Non-Domestic Buildings in Southern Europe'. ALTENER II Project on 'Solar Passive Heating and Cooling'. European Commission – DG Reseach. http://www.phdc.eu/uploads/media/ALTENER_1_Final_report_extract.pdf.
- Ford, B. & Vallejo, J. (2015). 'Theatre Design and Natural Ventilation: A UK Case Study'. 32th Plea International Conference: Architecture in (Re)Evolution. Bologna, Italy.
- Ford, B., et al. (2012). 'Passive Downdraught Evaporative Cooling: Performance in a Prototype House'. Building Research & Information 40(3): 290–304.
- Foster & Partners (2002). 'HM Treasury' [Online]. http://www.fosterandpartners.com/projects/her-majestys-treasury-redevelopment/. [Accessed 10/2018].
- Francis, E. (2000). 'The Application of Passive Downdraught Evaporative Cooling (PDEC) to Non-Domestic Buildings: Office Building Prototype Design in Catania Italy'. 16th Plea International Conference. Cambridge, UK.
- Guthrie, A. (2011). 'The Challenge of Sustainability. Architecture and Urbanism'. A+U 487: 8–15.
- Guthrie, A. (2018). 'Engineering Innovation'. In Renzo Piano: 'The Art of Making Buildings'. Royal Academy of Arts, pp. 133–139.
- Robson, D. (2002). 'Geoffrey Bawa: the Complete Works'. Thames & Hudson, pp. 134–135.
- Steadman, P. (2017). 'The Legacy of Sir Leslie Martin'. Martin Centre 50th Anniversary Symposium.
- Transsolar (2016). 'La Mécanique Générale and Les Forges, Arles, France' [Online]. http://transsolar.com/projects/la-mecanique-generale-and-les-forges/. [Accessed 10/2018].
- Wood, A. & Salib, R. (2013). 'GSW Headquarters Tower. Natural Ventilation in High-Rise Office Buildings'. CTBUH Technical Guide. Routledge, pp. 64–73.

CHAPTER 5
TESTING THE STRATEGY

As soon as the location of a project is known then an analysis of the site and climate (Chapters 2 and 3) can proceed, informing strategic thinking even before an initial design proposition has been made. The feasibility of early strategic ideas can be evaluated using simple 'rules of thumb' or graphic tools informing the iterative process of developing outline design ideas. As the ideas become more clearly formed they can be tested using simple manual calculations and steady-state tools. These simple rules of thumb and analytic tools are as relevant to the architect as they are to the environmental and structural engineers. They are of particular value to the architect as very often they do not have the support of engineering advice at the initial stages of design or in smaller scale projects.

When applied to convective and evaporative cooling strategies, steady-state tools can provide architects with valuable information regarding the feasibility of different options. Once the strategic feasibility has been established then more detailed analysis can follow on with preliminary estimates of performance. Steady-state tools rely on the laws of physics or on empirical studies based on laboratory or field observations. They are quick and reliable and, in most cases, high level expertise in the field is not required.

5.1. HEAT GAINS

Preliminary sizing of the main elements required for a natural cooling system delivering a certain strategy enables its feasibility to be explored. As with any other cooling system, the *'size'* or capacity of the system will be determined by the estimated cooling load. The cooling load arises from internal and external heat gains which must be (totally or partially) removed in order to reduce the risk of overheating. Prior to sizing the passive cooling system, the first objective is to minimise this load. Complimentary strategies such as solar control or reduction of the occupancy density, equipment and lighting gains will contribute to this aim.

EXTERNAL HEAT GAINS

External heat gains arise from the prevailing ambient conditions (climate and microclimate) and are derived mainly from solar radiation (either directly or indirectly) and infiltration. The orientation and the solar exposure of glazed and opaque building elements have a significant influence on solar heat gains, and hence the importance of solar geometry and ensuring that most vulnerable surfaces are shaded.

To overcome overheating risk, it can be useful to make an estimate of the possible *'worst case'* external heat gains. This scenario, in a temperate climate, is usually a warm summer day under clear sky conditions. Heat gains through the external opaque fabric will be directly related to the heat transmittance (U-value) of the construction, and also to the surface temperature which is related to the surface *albedo* and to the amount of solar radiation falling on the surface. External heat gains therefore vary significantly depending on the time of the day, and are related to the thermal characteristics of the construction and level of exposure.

Conductive solar gains

If we take the example of solar radiation falling on a flat roof in Madrid in the summer, this may reach as much as 1,000W/m^2 at early afternoon and, depending on the nature of the surface material, the temperature of the surface could reach 50–60°C. In the case of the Nottingham HOUSE project located in Madrid (see Chapter 4), if the U-value of the roof was 0.2W/m^{2}K, then (assuming an internal temperature of 25°C and external surface temperature of 60°C) the heat gain through the roof would be 7W/m^2. With a total roof area of 75m^2, the heat gains through the roof could be up to 525W.

> THE 'SIZE' OR CAPACITY OF THE COOLING SYSTEM WILL BE DETERMINED BY THE ESTIMATED COOLING LOAD

The same rationale applies to walls with the difference that the incident solar radiation on vertical surfaces is strongly related to the surface orientation and time of day. Based on the foregoing assumptions, the total conductive heat gains at noon would be 1,035W for the Nottingham House in Madrid.

Direct solar gains

In Madrid, solar radiation falling on a south-oriented vertical surface can reach as much as 400W/m² at noon. Part of this heat can be directly transmitted to the indoor space through transparent (glazing) elements and the total amount of heat transmitted will be given by the glazing solar transmittance (G-value) and the area. With a south-oriented glazing area of 8m² and a glazing G-value of 0.65, the direct solar gains in the house would be 1,456W (considering that 70% of the transmitted heat is absorbed by the internal materials). This calculation can be replicated for the glazing areas exposed to the various orientations.

This simple calculation, based on a hot sunny day in summer at noon, is a steady-state *'snapshot'* of external heat gains. It is useful in providing an indication of the potential solar gains, which can represent a significant part of the cooling load. Whether it can be reduced further (by improving insulation, reducing glazed area, increasing shading, etc.), it can be explored as the design moves forward.

INTERNAL HEAT GAINS

Internal heat gains are derived from people, lighting and equipment used within the building (fig. 145). Internal gains from people will depend on their activity and their metabolic rate, but typically are about 80W (sensible heat gain) for a sedentary adult. Heat gains from artificial lighting can hopefully be minimised by ensuring that most spaces are adequately daylit (i.e. do not require artificial lighting during daylight hours), while preventing over-illumination

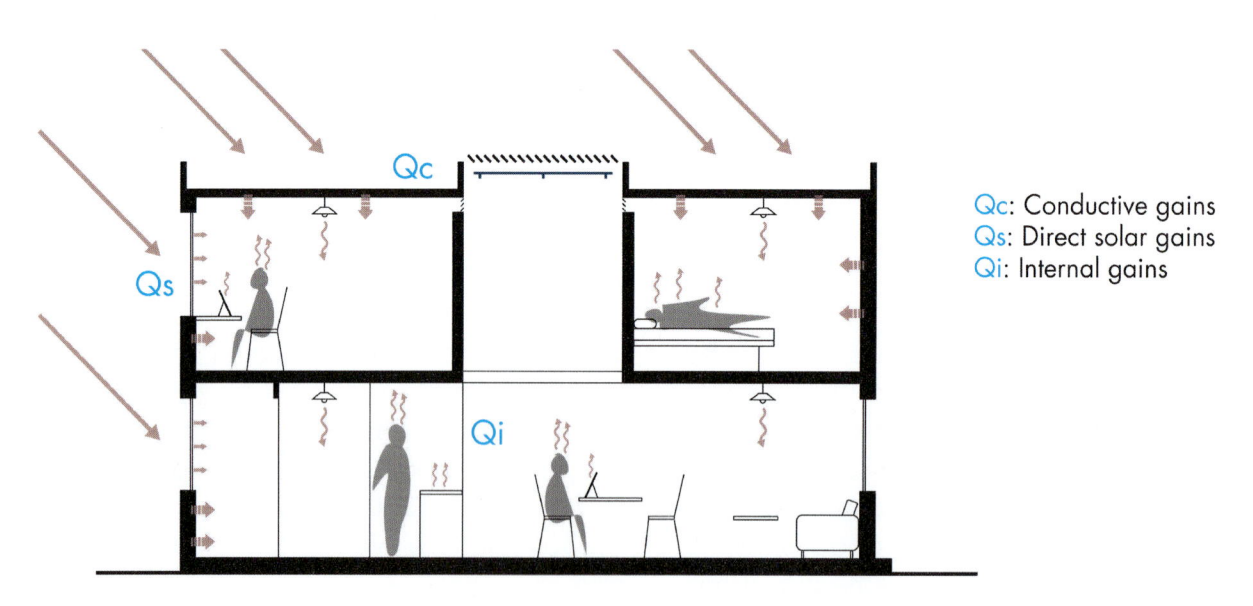

Q_c: Conductive gains
Q_s: Direct solar gains
Q_i: Internal gains

145. External & internal heat gains diagram for the Nottingham HOUSE project, Madrid, Spain.

118   **CHAPTER 5** | TESTING THE STRATEGY

and avoiding unwanted solar gains and glare. This is sometimes more difficult than it seems, and the phenomenon of *'blinds down – lights on'* is unfortunately very prevalent in many non-domestic buildings. Heat gains from equipment will depend on activity and the efficiency of the appliances *(CIBSE 2015, Table 6.2)*. For the 75m² four-person apartment in Madrid, typical internal heat gains (assuming 2W/m² of equipment and lighting gains) would be approximately 620W.

TOTAL HEAT GAINS

The sum of the estimated internal and external heat gains (calculated for noon on a hot day in summer) indicates the cooling load to be removed by the passive cooling strategy to avoid an internal air temperature rise. This simple calculation ignores dynamic effects within the building (thermal capacitance, occupancy pattern, etc.) but it provides a preliminary estimate to evaluate initial feasibility of a convective cooling strategy.

The heat gains to any space can be removed passively through natural convection using lower temperature external air, or (under hot dry conditions) by inducing evaporation within the airflow path. The cooling achieved by convection is directly proportional to the ventilation flow rate and the indoor-to supplied air temperature difference, thus an increase of either variable will lead to higher cooling achieved.

Conductive heat gains
Roof $(A_r \times U_r \times \Delta T) = 75 \times 0.2 \times 35 = 525W$
Walls $(A_w \times U_w \times \Delta T) = 170 \times 0.3 \times 10 = 510W$

Direct solar gains
Glazing $(ISR \times A_g \times G \times a) = 400 \times 8 \times 0.65 \times 0.70 = 1,456W$

Internal heat gains
People: $4 \times 80 = 320W$
Lighting: $2W/m^2 = 150W$
Equipment: $2Wm^2 = 150W$

Total External heat gains = 2,491W
Total Internal heat gains = 620W

TOTAL HEAT GAINS = 3,111W

THE SUM OF THE ESTIMATED INTERNAL AND EXTERNAL HEAT GAINS INDICATES THE COOLING LOAD TO BE REMOVED BY THE PASSIVE COOLING STRATEGY TO AVOID AN INTERNAL AIR TEMPERATURE RISE

5.2. DEFINING THE AIR VOLUME FLOW RATE REQUIRED

Convective cooling consists of the removal of heat by the movement of air, and the equivalent sensible cooling achieved can be estimated with the following expression from *ASHRAE (2013)*:

$$Qs = VHC \times q \times [Tin\text{-}Tout] \quad \text{Equation 01}$$

where Qs is the sensible cooling (W), VHC is the volumetric heat capacity of the air ($\approx 1,200 J/m^3 K$), q is the air volume flow rate (m^3/s), and $Tin\text{-}Tout$ is the temperature difference between the supply air temperature ($Tout$) and the maximum desired air temperature (Tin). From the above equation, the amount of air necessary to achieve the target cooling can be determined from equation 02 *(CIBSE, 2005)*:

$$N = \frac{Qs}{0.33 \times V \times [Tin\text{-}Tout]} \quad \text{Equation 02}$$

where N is the air volume flowrate now given in air change rate per hour (ac/h), Qs is the sensible cooling (W) and V is the room volume (m^3).

Returning to our example of the apartment in Madrid, we now wish to determine the air volume flow rate required to remove the estimated total heat gains. Since we know the cooling required (equivalent to the total heat gains of the space), the air volume flow rate (N) required to remove the heat gains can be obtained from equation 02.

For this example, the total heat gains (Qs) is 3,111W (determined in section 5.1), and assuming $Tin\text{-}Tout = 2K$ and $V = 225m^3$, N is found to be 21 air changes (or $1.3 m^3/s$).

In this way we obtained a quick estimate of the air volume flow rate required to remove internal and external heat gains from the apartment in Madrid, and we can now use this information to help size the opening areas required for stack driven convective cooling.

Air volume flow rate required

$$N = \frac{3,111}{0.33 \times 225 \times 2} = 21 \ ac/h$$

---

120   **CHAPTER 5** | TESTING THE STRATEGY

5.3. SIZING APERTURE AREAS FOR STACK-DRIVEN COOLING

Airflow in an indoor space is driven by thermal forces and (if available) wind forces. Thermal force alone (also known as buoyancy force) derives from the temperature difference between the room and the supply air temperature and is usually considered when sizing aperture areas for natural ventilation in early design stages. This represents a theoretical worst case scenario. The air is driven into the space as result of a pressure difference across the openings due to different stack pressure gradients of air temperatures (fig. 146).

The pressure gradient across the height of a building as result of thermal forces is given by the following expression:

$$\Delta p_i = \Delta p_0 + \Delta \rho_0 \times g \times z_i \qquad \text{Equation 03}$$

where Δp_i is the pressure gradient across the opening, Δp_0 is the hydrostatic pressure difference at ground level (Pa), $\Delta \rho_0$ is the density difference at ground level (kg/m³), g is the gravitational force per unit mass (m/s²) and z_i is the height of the opening above ground level (m).

The airflow rate achieved in a given aperture can then be expressed as function of this pressure gradient and the free area:

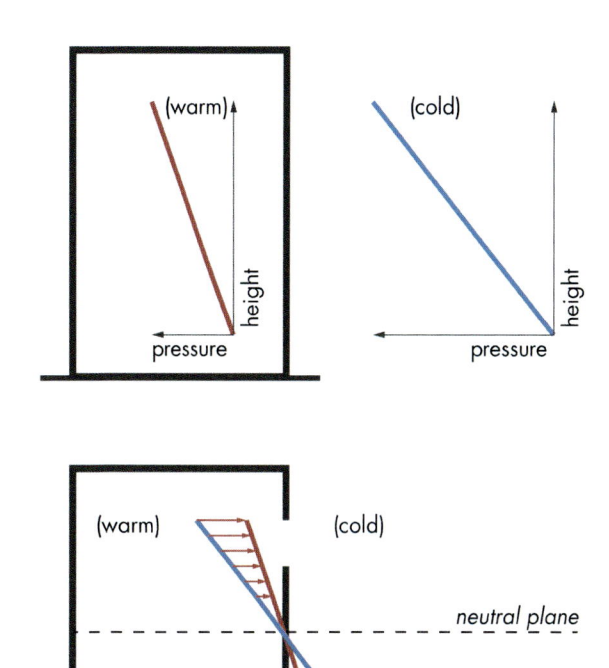

146. Representation of air pressure gradients & pressure differences across openings.

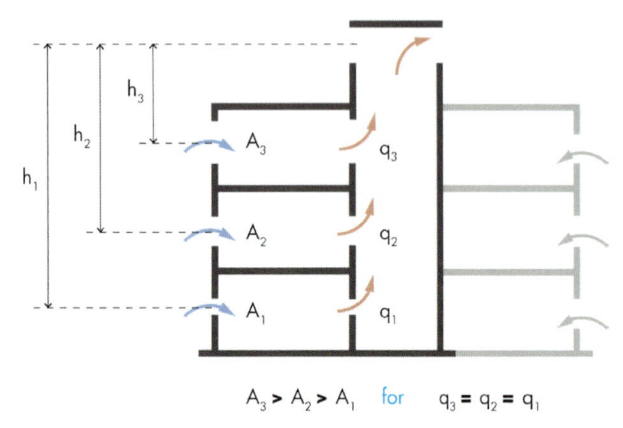

147. Direct implication of inlet-to-outlet height difference on opening areas for an even distribution of air in buildings with high stacks.

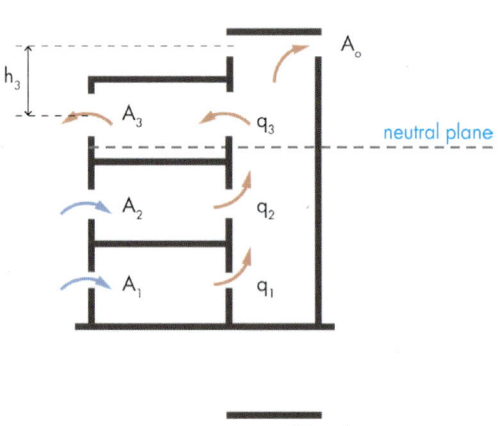

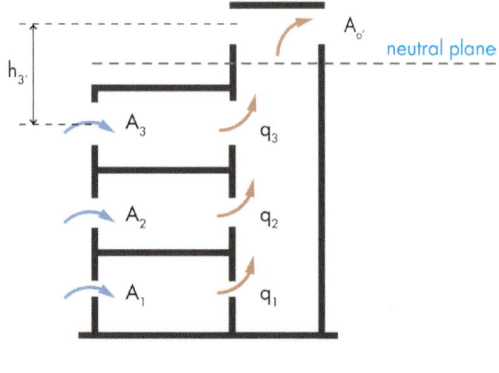

$A_3 > A_2 > A_1$ for $q_3 = q_2 = q_1$

148. Direct implication of inlet-to-outlet height difference and opening areas on the position of the neutral plane.

$A_{o'} > A_o$ and/or $h_{3'} > h_3$

$$q = Cd \times A \sqrt{\frac{2|\Delta p|}{\rho_0}} \quad \text{Equation 04}$$

where q is the flow rate through the opening (m³/s), Cd is a discharge coefficient (representing the opening's resistance to airflow), A is the area of the opening (m²), Δp is the pressure difference across the opening (Pa) and ρ_0 is the air density (kg/m³).

With stack driven airflow, as the supply air and the room air temperatures equalise, the driving pressures will approach zero, and no matter how large the opening, the flows will be very small.

Typically, it can be assumed that room air dry bulb temperature will be 1–3°C above the supply air temperature, and therefore large air change rates are generally required to achieve significant cooling of the building. *Cunningham & Thompson (1986)* reported air change rates of up to 30ac/h for their experimental building in Arizona under buoyancy forces alone, and at the TRC in Ahmedabad (see Part 2), air change rates of 6–12ac/h were recorded *(Ford & Hewitt, 1998)*. Experience suggests therefore that buoyancy driven cooling can achieve the large air volume flow rates.

From equation 03, the height difference between inlet and outlet openings has also a fundamental influence on the airflow rate achieved and on the area of openings required to deliver this air. In buildings with high stacks, openings higher up the tower will need to be larger than openings lower down the tower if an even flow rate is to be achieved on each floor of the building by natural convection. This is due to the difference in stack heights, and applies equally to updraught and downdraught airflow paths (fig. 147).

Inlet-to-outlet height differences and effective outlet areas will also determine the position of the neutral plane (height at which the pressure difference between the indoor and outdoor space is zero), so that the openings below and above this height will intake and exhaust air respectively. This implies the neutral plane should always be above the roof of the top floor in multi-storey buildings (fig. 148).

The above expressions (equations 03–04) can be simplified for scenarios presenting a single opening or identical inlet and outlet areas, so that the free area A (for each opening) required to provide a flow rate q for a specific stack heigh h is:

$$A = \frac{q}{Cd} \times \sqrt{\frac{Ti+273}{\Delta T \times g \times h}}$$ Equation 05

where A is the area of each opening (m^2), q is the airflow rate (m^3/s), Cd is the discharge coefficient (typically 0.6 for multiple openings and 0.25 for a single opening), Ti is the internal (room) temperature (°C), ΔT is the difference between the internal and supply air temperatures (K), g is the gravitational force per unit mass (9.8m/s^2) and h is the distance between openings or the sill-to-head height for a single opening (m).

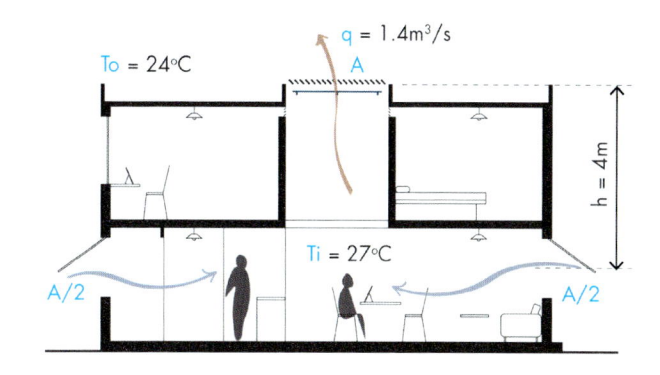

149. Natural ventilation strategy for the Nottingham HOUSE, Madrid, Spain.

Aperture area required

$$A = \frac{1.3}{0.6} \times \sqrt{\frac{27+273}{3 \times 9.8 \times 4}} = 3.5m^2$$

TYPICALLY, ROOM AIR DRY BULB TEMPERATURE WILL BE 1–3°C ABOVE THE SUPPLY AIR TEMPERATURE, AND THEREFORE LARGE AIR CHANGE RATES ARE GENERALLY REQUIRED TO ACHIEVE SIGNIFICANT COOLING

CHAPTER 5 | TESTING THE STRATEGY

Continuing with the example of Madrid house, the aperture area needed to deliver the required airflow rate of 1.3m³/s using a stack ventilation strategy as illustrated in fig. 149 can be determined from equation 05. For this example, the maximum acceptable air temperature in the room T_i is 27°C, the supply to indoor air temperature difference ΔT is 3K, and the height between openings h is 4m.

The required opening area for inlets and outlets is 3.5m², which of course can be distributed among multiple openings. For preliminary calculations, it is assumed that the total inlet area will equal the total outlet area, and that the cross-sectional area of the supply/exhaust air shaft will also be equal to the effective free area of the inlet area (of course it may be larger if it is also serving another function).

For each opening the pressure differences will of course be different, and so the building geometry has a profound impact on both the airflow pattern and the volume flow rate through separate floors. This further underlines the importance of defining a viable airflow strategy as described in Chapter 2.

> BUILDING GEOMETRY HAS A PROFOUND IMPACT ON BOTH AIRFLOW PATTERN AND VOLUME FLOW RATE THROUGH SEPARATE FLOORS

5.4. SIZING AN EVAPORATIVE COOLING SYSTEM

Whatever type of cooling delivery method is chosen, the viability of the envisaged airflow path, and hence the likely cooling performance which can be achieved, must be assessed as part of the design process. While a dynamic thermal and airflow model of the whole building is therefore likely to be required (certainly for large or complex buildings), for preliminary design sizing of the system elements a simple steady state model can be applied. The physics involved in convective cooling is to a certain extent manageable with simple calculations as demonstrated in previous sections. In the case of evaporative cooling, the physics involved are more complex and dependent on fluid dynamics. It is for this reason that preliminary sizing methods mostly rely on empirical data in tables or diagrams that are limited to certain systems and dimensions.

MISTING SYSTEMS

The preliminary design and sizing of misting systems can be based on estimates of the cooling load and airflow rates required as described above, and the rules of thumb derived from empirical data by *Givoni (1994)* for direct evaporative cooling via a tower. This suggests that the air temperature drop in the tower (determining the effective supply air temperature to occupied floors of the building) will be 80% of the dry bulb – wet bulb depression (Tdb-Twb). This same rule applies to cool towers using porous media to induce evaporation. However, the design of misting systems will vary from that of porous media in some important ways.

Hydraulic misting nozzles require a height of 2–4m for water droplets to be completely evaporated, depending on ambient conditions, water pressure and droplet sizes achieved. In the Torrent Research Centre in Ahmedabad (described in detail in Part 2), the central supply towers are above roof level to ensure that evaporatively cooled air can be supplied to the top floor. Initially the water pressure was below 25Pa and some droplets did not evaporate before reaching the walkway below. To avoid this problem and to increase the rate of evaporation the pressure was increased to 50Pa to achieve an average droplet size below 30 microns.

The alternative to hydraulic pressure alone is to use compressed air in combination with water at mains pressure to create very small droplets (<15microns). This can significantly reduce the height required to completely evaporate the droplets. The Nottingham Solar Decathlon House (see Chapter 4) avoided the need for a tower while still cooling the whole building by using pneumatic nozzles.

Area of openings

Preliminary estimates of the size of openings required for a misting system can be determined in the same way as for convective (updraught) cooling based on the air volume flow rate required to remove internal heat gains, as described above. In this case, the flow rate q is proportional to the difference

---

'RULES OF THUMB' FOR PDEC PERFORMANCE

Temperature depression
$T_t = T_{db} - 0.8 \times [T_{db} - T_{wb}]$
where:
T_t: tower delivery temperature
T_{db}: ambient dry bulb temperature
T_{wb}: ambient wet bulb temperature

(assume a room temperature: $T_i = T_t + 3°C$)

Cooling achieved
$Q_s = VHC \times q \times [T_i - T_t]$
where:
Q_s: sensible cooling
VHC: 1,200 J/m³K
q: airflow rate (m³/s)
T_i: room air temperature
T_t: tower delivery temperature

COMPRESSED AIR IN COMBINATION WITH WATER AT MAINS PRESSURE CREATE VERY SMALL DROPLETS, SIGNIFICANTLY REDUCING THE HEIGHT REQUIRED TO COMPLETELY EVAPORATE THE DROPLETS

in temperature between the room air T_i and the mean temperature of the tower T_t. For preliminary sizing this is normally taken to be 2–3°C to provide a conservative estimate of the useful cooling supplied. In order to achieve an even distribution of air to each floor, the free area of the vents at each level is inversely proportional to the square root of the stack heights of each opening as seen in section 5.3.

Number of nozzles

The determination of the number of nozzles required relates to an estimate of the amount of water required (H_2O) to reduce ambient air from the maximum dry bulb temperature anticipated to 1–2°C below the design internal air temperature. This is determined by establishing the difference in the absolute humidity of the external ambient air and the absolute humidity of the air in the tower, multiplied by the air volume flow rate (equation 06).

$$H_2O = \Delta MR \times q \times \frac{\rho_{air}}{\rho_{H2O}} \times [3{,}600] \quad \text{Equation 06}$$

where H_2O is the water required (l/h), ΔMR is the increment in air moisture content (g_{H2O}/kg_{air}), q is the airflow rate (m³/s), ρ_{air} is the air density (≈ 1.2 kg/m³) and ρ_{H2O} the water density (1,000 kg/m³).

The psychrometric chart shows that if we take ambient air at 35°C and 20% relative humidity it has a moisture content of 7.3g/kg. If the humidity is increased by evaporation (following the enthalpy lines on the psychrometric chart) to 12g/kg (regarded by *Szockolay, 2008* as the upper threshold) it will have a temperature of 23.3°C and result in an increment in air moisture content of 4.7g/kg (fig. 150).

Taking the example of the apartment in Madrid, in which the volume flow rate required to remove heat gains was estimated to be 1.3m³/s, then the volume of water required is estimated to be 26l/h.

Typical water consumption for a single nozzle is 5 litres/hour at 30 microns, so for the example given a total of six nozzles would provide sufficient evaporative cooling to achieve comfort conditions in the Madrid apartment. However, to provide a safety margin eight nozzles were provided, operating in pairs. Of course misting nozzles are not run continuously as they would tend to over-saturate the air, so they will normally be switched off when RH≥65%, subsequently reducing water consumption and also cooling. When nozzles are switched off the supply pipework drains back to the storage tank. These issues are discussed further in Chapter 7.

THE NUMBER OF NOZZLES REQUIRED IS DERIVED FROM AN ESTIMATE OF THE WATER REQUIRED TO ACHIEVE THE SUPPLY AIR TEMPERATURE AND THE RESULTING AIR VOLUME FLOW RATE

Water required

$$H_2O = 4.7 \times 1.3 \times \frac{1.2}{1,000} \times 3,600 = 26 \, l/h$$

Number of nozzles required

$$N = \frac{H_2O}{5} = \frac{26}{5} = 6$$

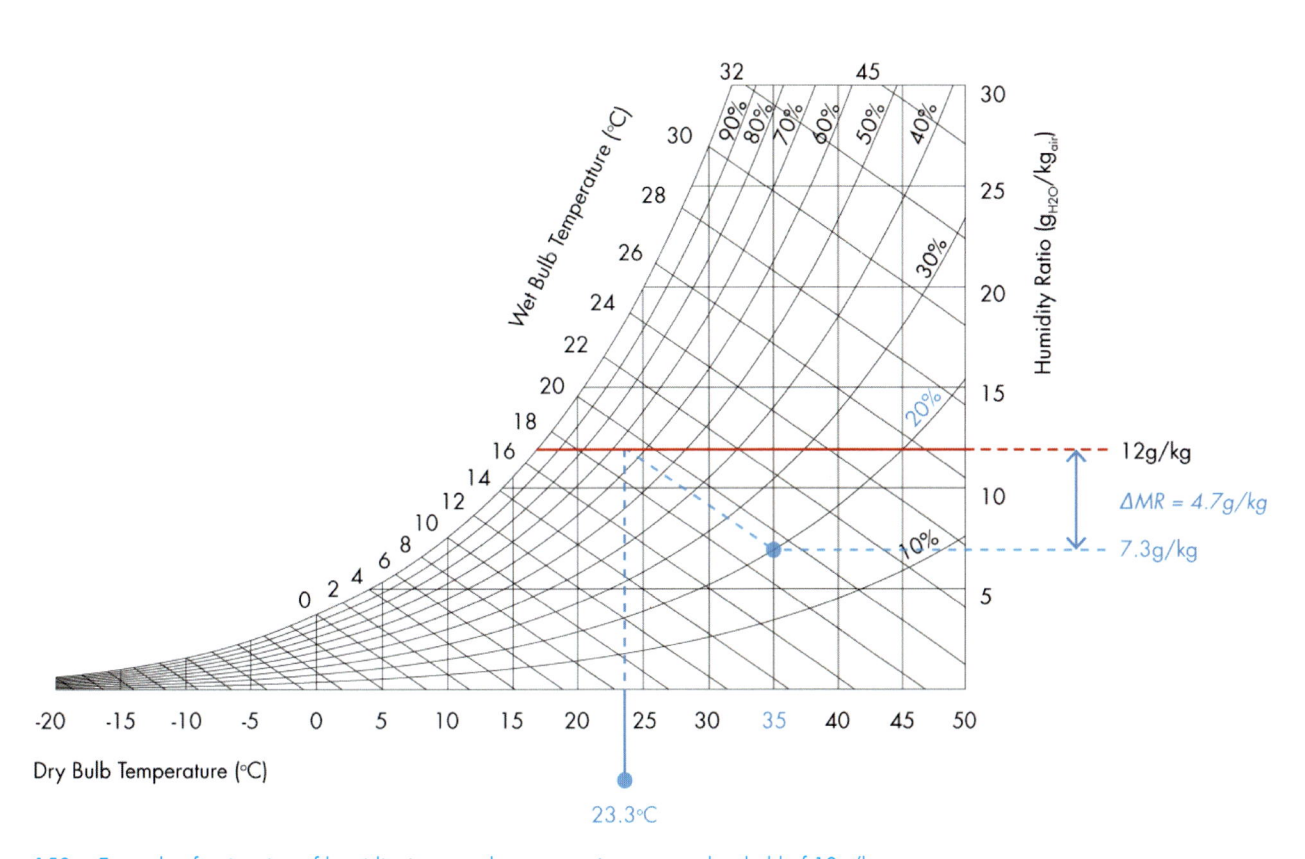

150. Example of estimation of humidity increase by evaporation up to a threshold of 12g/kg.

CHAPTER 5 | TESTING THE STRATEGY

151. Cellulose mats and spray nozzles mounted in test rig, Seville, Spain 1991.

152. Cool tower at Zion National Park Visitor Centre, Utah, USA [NPS Denver Service Centre Architects].

POROUS MATERIALS

Porous materials such as cellulose or fibrous mats, or porous ceramic elements, have been used historically to help cool building interiors (as described in Chapter 1) (figs 151–152). They allow water absorption and flow towards the external surface creating a thin water film that evaporates when exposed to the air stream. Given the appropriate climatic conditions, the technical applicability of a porous cooling system is dependent on the shape of the porous element, its arrangement and total surface area exposed to the air stream. Air movement over the porous surface is needed to renovate saturated air while contributing to the separation of water molecules from the surface to the air. It is thus important to consider the airflow path through the porous media and avoid major obstructions in the air stream in order to achieve maximum efficiency. The surface area of porous elements required to meet the space cooling demand will thus depend on these factors, and several studies exploring the cooling capacity of different porous media and arrangements have been developed in the past 30 years.

Cunningham and Thompson (1986) first suggested a method of relating the height of the cool tower and the area and 'efficiency' of the evaporative cooling matrix or pad, to the velocity (and thus volume flow rate) of the air within the tower. Later *Givoni (1992)* developed similar expressions from which exit air temperatures and airflow rates can be derived depending on tower height and pad area. More recently, *Kwok and Grondzik*

(2007) published a series of diagrams to size evaporative cooling towers and porous pads for building application by means of dry bulb air temperature, wet bulb temperature depression and tower height. The cool tower design procedure is as follows:

1. Find the design ambient dry bulb air temperature *Tdb* and the corresponding wet bulb air temperature *Twb* for the hottest time of the year for the building site, and derive the wet bulb temperature depression *Tdb-Twb*.
2. Estimate exit air temperature using ambient dry bulb air temperature *Tdb* and wet bulb temperature depression *Twb* by reference to fig. 153. The exit air temperature needs to be at least 1–2°C below the acceptable maximum internal air temperature.
3. Determine the airflow rate needed to remove the space cooling load from equation 01, where *Tin-Tout* would be the temperature difference between the room air and the supply (cool tower exiting) air.
4. Determine tower height and area of wetted pads required to provide this air volume flow rate from fig. 154.

The above methodology is now applied to the house in Madrid to evaluate the feasibility of a cooling tower using cellulose pads to remove the internal load of 3,261W. Maximum air temperatures in summer average 32°C and the subsequent wet bulb temperature depression is 16K. By using these values in fig. 153, it can be estimated that

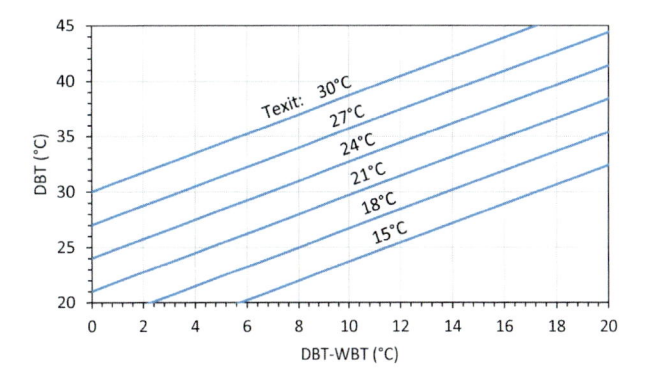

153. Cool tower exit air temperature as a function of wet bulb depression and outdoor dry-bulb air temperature, ©Kwok and Grondzik.

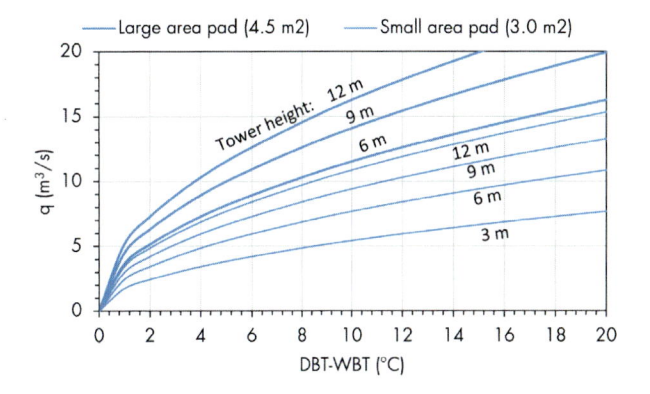

154. Recommended tower height and wetted pad area as a function of required airflow rate and wet bulb depression, ©Kwok and Grondzik.

> *Tower (exit air) temperature*
> Ambient dry bulb temperature: Tdb = 32°C
> Dry bulb to wet bulb temperature depression:
>   Tdb-Twb = 32-16 = 16K
> Tower temperature (fig. 153): Tt = 18°C
>
> *Airflow rate required*
>
> $$q = \frac{Qs}{VHC \times [Tin\text{-}Tt]} = \frac{3,111}{1,200 \times [26\text{-}18]} =$$
>
> $= 0.32 \; m^3/s$
>
> *Tower height & cellulose pad area*
> Recommended tower height (fig. 154): h<3m
> Cellulose pad area (fig. 154): a< 3m²

exit air temperature in the tower would be approximately 18°C, well below the upper threshold for thermal comfort and confirming the feasibility of this system.

Assuming a design indoor temperature of 26°C (an indoor temperature to be maintained and considered comfortable for the users), the resulting ΔT (Tin-Tt) would be 26-18 = 8K. By applying this temperature difference and the required cooling load on equation 01, the number of air changes needed to remove heat gains will be 5.2ac/h, equivalent to 0.32 m³/s.

Wet bulb temperature depression and required airflow rate can then be used to size the height and area of wetted pads using fig. 154. The optimal climatic conditions of Madrid to apply evaporative cooling systems suggest that the cooling load can be easily delivered with a small pad area of 3m² and tower height below 3m (which could deliver up to 5m³/s).

Other researchers (Schiano-Phan, 2004) contributed with additional diagrams to size porous ceramic cooling systems integrated in cavity walls by means of wet bulb temperature depression, ceramic evaporators' height and cooling capacity (figs155–156). This data, part of the EU funded 'Evapcool' project, include a performance chart (fig. 157) which correlates these parameters. By entering the wet bulb temperature depression at the peak hour of a typical summer day, it is possible to identify the system's specific cooling power in relation to system height. From the specific cooling power it is

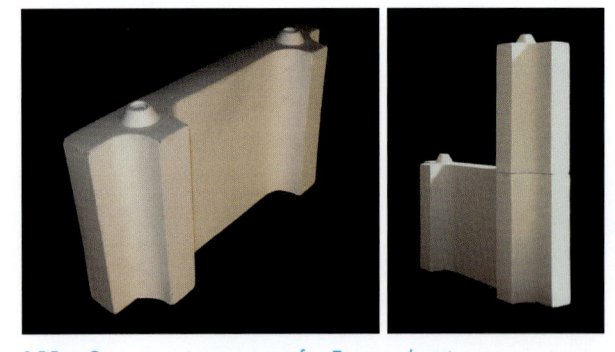

155. Cast ceramic prototype for *Evapcool* project.

possible to derive, by inverse formula, the height of the panels to be integrated into the building envelope.

For a dry-wet bulb temperature depression of 10K, for example, the specific cooling power of the porous ceramic panels ranges between 40 and 68W per square metre of surface area and unit of temperature differential between the supply air temperature and surface temperature of the ceramic. The variation in specific cooling output depends on the height of the system: shorter systems (h = 0.2m) offer greater specific cooling power due to increased efficiency; however, a greater width may be necessary to obtain the total cooling capacity required. On the other hand, taller systems (h = 1m), which offer smaller efficiency since the evaporative cooling effect is diminished by the saturation of the air along the system's height, can provide the required total cooling capacity with less width, which is easier to integrate into the envelope.

Recent research *(Vallejo et al., 2017)* explored the cooling capacity of porous ceramic screen walls applied in transitional spaces. The system consists of porous ceramic elements arranged over a vertical plane coupled with a top-drip irrigation system that distributes water over the ceramic surface, inducing evaporation when exposed to an airstream (fig. 158). Experimental and computational studies on different ceramic shapes output a steady-state tool to assist designers to size the system and predict cooling performance and water consumption for different ceramic designs. The cooling capacity

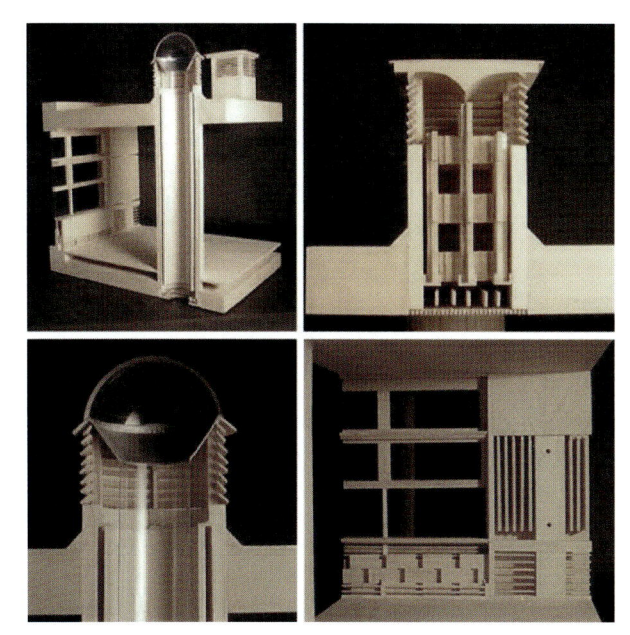

156. Generic Integration of the *Evapcool* systems in a typical office building.

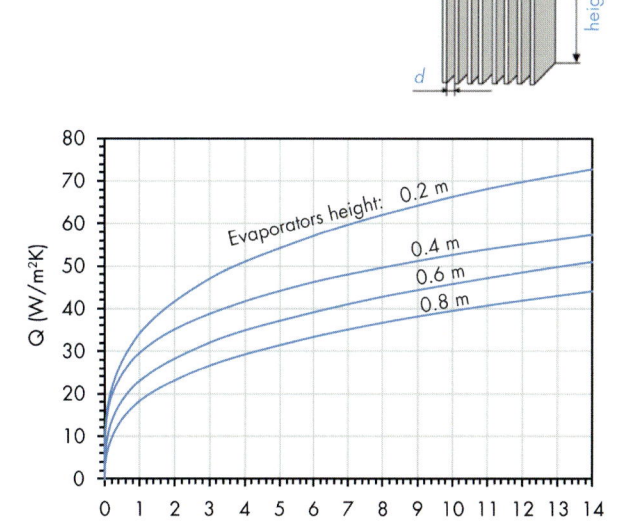

157. Specific cooling power per metre width and per metre height as a function of dry bulb-to-wet bulb temperature depression.

CHAPTER 5 | TESTING THE STRATEGY   131

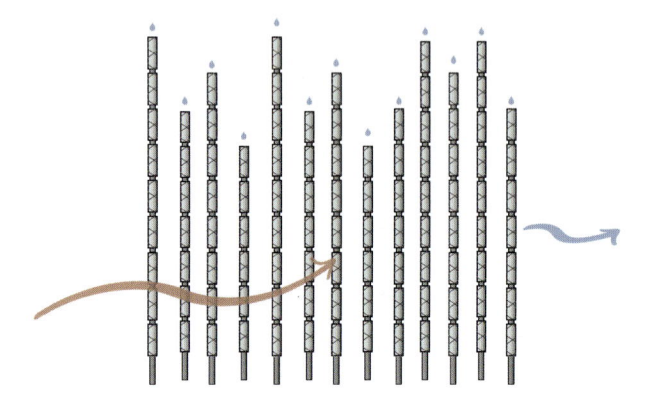

158. Evaporative cooling screen consisting in vertically arranged tubular-shaped ceramic elements.

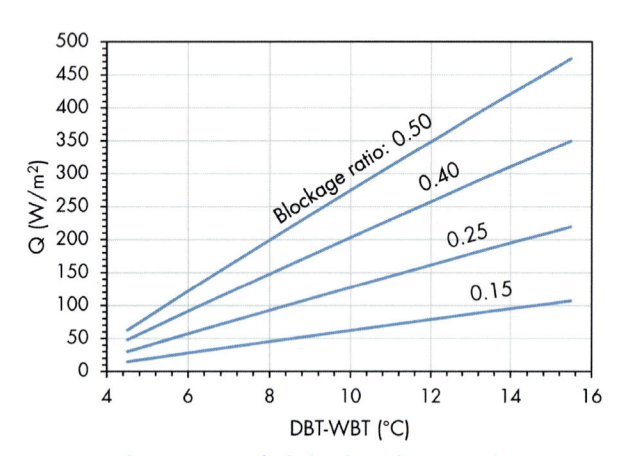

159. Cooling capacity of tubular-shaped ceramic elements (drag coefficient *Cd* of 0.5) exposed to an airstream of 0.5m/s for different blockage ratios as a function of dry bulb-to-wet bulb temperature depression.

of this system is determined by microclimatic and ceramic properties, including air velocity, wet bulb depression, ceramic shape (determined from the dimensions and resistance to flow of the sample) and the number of ceramic elements. Fig. 159 illustrates the relation between ambient wet bulb temperature depression and the cooling achieved by tubular-shaped ceramic elements (drag coefficient *Cd* of 0.5) exposed to an airstream of 0.5m/s for different blockage ratios. Blockage ratio determines the obstructed-to-free area ratio of the ceramic screen, and it's proportional to the number of vertical columns and the distance between them. Using the same context of the previous example, a dry-wet bulb temperature depression of 10K will give a specific cooling power ranging from 50 to 250 W/m² of ceramic screen. A denser screen (BR = 0.5) will give higher performance than screens with lower densities (BR = 0.15).

It is important, however, to design carefully the system as a whole, which should be understood as the combination of an airstream and the cooling device. A constant air movement is recommended to replace saturated air in the surroundings of the ceramic screen and this can be naturally achieved by strategically locating the system between spaces presenting different air pressure gradients.

> A POROUS CERAMIC COOLING SYSTEM SHOULD BE UNDERSTOOD AS THE COMBINATION OF AN AIRSTREAM AND THE COOLING DEVICE

CHAPTER 5 | TESTING THE STRATEGY

Water consumption of porous ceramic cooling systems can be determined from the increase in air moisture content obtained after the evaporation process. Unlike misting systems, porous ceramic systems present an evaporation media (evaporators) that may limit the final air temperature drop and humidity increase achieved. These two parameters, ΔDBT and ΔMR, are strongly related to the number of ceramic evaporators used, their geometry and arrangement among other factors and specific (empirical) performance data is needed for the estimation of water consumption. The calculation method from equation 06 (water consumption for misting systems) can be applied to porous ceramic elements when used with calculation graphs like figs 157 and 159 (ceramic performance data). Five steps are needed to determine the final water consumption:

1. Determine the cooling provided by the porous system (W) from figs 157–159 (ceramic performance data).
2. Using equation 01 (sensible cooling), obtain the equivalent temperature drop ΔDBT for the cooling and airflow rate provided.
3. Obtain the supplied air temperature by the porous ceramic evaporative system from $DBTin$ and ΔDBT, so that, $DBTsupply = DBTin\text{-}\Delta DBT$.
4. Using the psychrometric chart, locate the initial room temperature and relative humidity and, following the enthalpy lines (evaporation process), determine the increase of moisture content (mixing ratio) until reaching $DBTsupply$.
5. Use equation 06 (water consumption for misting systems) to determine water consumption by means of the estimated ΔAH and airflow rate.

ACTIVE DOWNDRAUGHT COOLING

Active downdraught cooling systems (or gravitational cooling systems) rely on buoyancy forces to recirculate air through chilled water cooling coils located at high level. Warm air at the top of a space passes through the chilled coils and is cooled down. The higher density of cooled air drives the air to the bottom of the space, which is then warmed up again by heat sources in the room, starting again the cycle. Chilled coils are typically located at high level within transitional spaces like pedestrian concourses or atria, or at ceiling level of small rooms such as office spaces. The cooling capacity of these systems is determined by the room-to-coil temperature difference, as well as the spacing of the fins and surface area.

> ACTIVE DOWNDRAUGHT COOLING SYSTEMS RELY ON BUOYANCY FORCES TO RECIRCULATE AIR THROUGH CHILLED WATER COOLING COILS LOCATED AT HIGH LEVEL

Cooling coils are normally manufactured for inclusion in air handling units where the airflow is driven by fans. Manufacturers provide information on sensible cooling delivered by these coils under high air velocities (3–7m/s) but this information is harder to find for buoyancy driven airflow. Analysis of fin spacing suggests that under a buoyancy driven airflow a spacing of 32mm would be most efficient, although most standard coils have a fin spacing of 10–12mm.

In the Malta Stock Exchange a total of 28 coils were installed and supported by a walkway below the central ridge ventilators. Each coil is 711x1,200mm and is fitted at an angle of 60 degrees from the horizontal, with a 200mm drip dray at the base (fig. 160). The on-coil water temperature was initially set at 17°C (much higher than the 6–10°C of most AC units). This reduces the level of condensation on the coil but increases the area of coils required.

> THE COOLING CAPACITY OF ADC SYSTEMS IS DETERMINED BY THE ROOM-TO-COIL TEMPERATURE DIFFERENCE, AS WELL AS THE SPACING OF THE FINS AND SURFACE AREA

160. Cooling coils attached to walkway below ridge vent in Malta Stock Exchange, Malta.

Other commercial products are available in the market and, in most cases, manufacturers provide technical data and diagrams to size the system. The example illustrated in fig. 161 applies to a commercial false wall system and relates the temperature difference between coil and room with the specific cooling capacity for different duct heights (fig. 162). It is evident that larger duct heights provide a higher pressure different across the intakes, increasing airflow rate and subsequent cooling achieved. For a room-to-coil temperature depression of 10K, for example, the cooling capacity will range between 300 and 750 W/m of wall for duct heights between 1.5 to 4m. These room-based re-circulation systems assume that a separate minimum fresh air supply is provided. Depending on location this fresh air supply may need to be de-humidified, or could be provided by trickle vents in the façade if serving perimeter spaces.

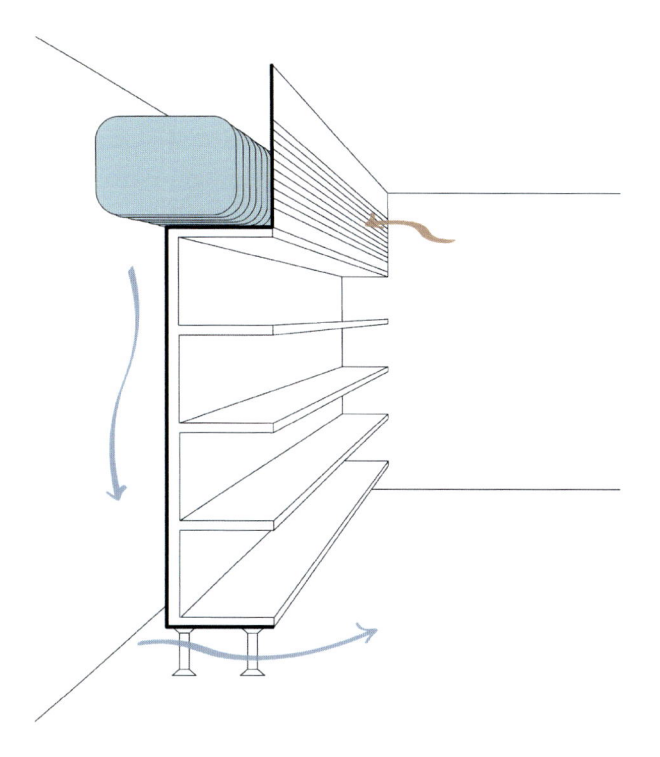

161. Example of integration of cooling coils in a commercial false wall system to induce a downdraught air movement, ®Gravivent.

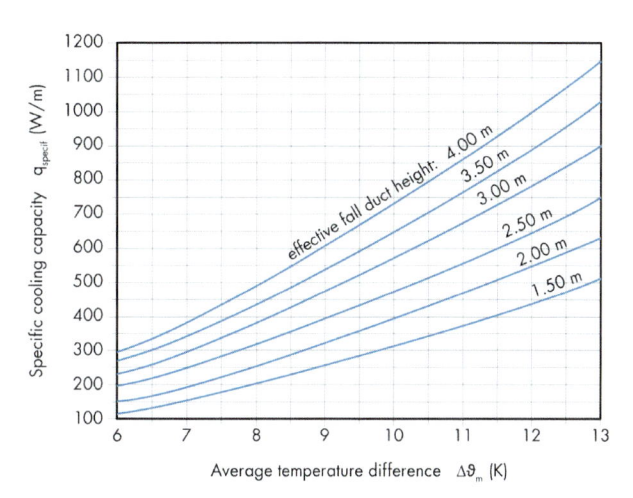

162. Specific cooling capacity for ADC systems for different duct heights as function of room-to-coil temperature depression ®Gravivent.

CHAPTER 5 | TESTING THE STRATEGY   135

This chapter has described simple methods to aid strategic decision making, to help preliminary sizing of components and to evaluate the feasibility of natural cooling options. A preliminary assessment of heat gains in the building is used to define the potential cooling demand, and this leads to the estimation of the air volume flow rate required to remove these heat gains. The different options for achieving natural cooling, natural ventilation, passive evaporative and active downdraught cooling have been reviewed and illustrated. Simple techniques to determine the supply air temperatures and the vent opening areas required have also been discussed.

At this stage it should be emphasised that, while actual performance will depend on a range of factors, these simplified tests are robust and reliable, and can provide the design team with the confidence that a particular preferred option is worth taking forward. As the design develops cost checks and other client requirements will influence the decision making process, and the different environmental control options may have to be revisited. Also, as the design is refined, questions will arise regarding many aspects of performance, and (depending on the scale of the project) more detailed analysis may be required to provide the design team and the client with a higher level of confidence that the strategic options chosen will indeed deliver the desired performance. Some of the tools and techniques available for more detailed analysis are reviewed in the next chapter.

REFERENCES

- ASHRAE (2013). *'ASHRAE Handbook: Fundamentals 2013'*. American Society of Heating, Refrigerating and Air-Conditioning Engineers.
- CIBSE (2005). *'Applications Manual 10: Natural Ventilation in Non-Domestic Buildings'*. London: Chartered Institution of Building Services Engineers.
- CIBSE (2015). *'Guide A: Environmental Design'*. London: Chartered Institution of Building Services Engineers.
- Cunningham, W. A. & Thompson, T. L. (1986). *'Passive Cooling with Natural Draught Cooling Towers in Combination with Solar Chimneys'*. Proc. 3rd PLEA International Conference. Pets, Hungary, The Hungarian Society of Sciences.
- Ford, B. & Hewitt, M. (1998). *'Cooling without Air Conditioning: The Torrent Research Centre, Ahmedabad, India'*. Renewable Energy 15, 1–4: 177–182.
- Givoni, B. (1992). *'Comfort, Climate Analysis and Building Design Guidelines'*. Energy and Buildings 18(1): 11–23.
- Givoni, B. (1994). *'Passive and Low Energy Cooling of Buildings'*. Van Nostrand Reinhold.
- Kwok, A. G. and Grondzik, W. T. (2007). *'The Green Studio Handbook: Environmental Strategies for Schematic Design'*. Oxford, Architectural.
- Schiano-Phan, R. (2004). *'The Development of Passive Downdraught Evaporative Cooling Systems using Porous Ceramic Evaporators and their Application in Residential Buildings'*. Proc. PLEA 21th International Conference. Eindhoven, The Netherlands.
- Szokolay, S. (2008). *'Introduction to Architectural Science'*. Elsevier/Architectural Press.
- Vallejo, J., et al. (2017). *'Predicting Evaporative Cooling Performance of Wetted Decorative Porous Ceramic Systems in Early Design Stages'*. Proc. 33th PLEA International Conference. Edinburgh, Scotland. 2–5 July.

CHAPTER 6
PERFORMANCE ANALYSIS

The last decade has witnessed a growing interest in building performance, leading to a wide range of simulation tools accessible to designers. Analytic tools can be classified according to the level of abstraction and complexity, including steady-state tools, dynamic tools and computational fluid dynamic (CFD) tools. This classification is of course proportional to the resolution of the digital model, the number of user inputs and the time required for a successful application. While steady-state tools can provide easy to interpret and quick design inputs to assist in decision making, dynamic and CFD tools provide detailed daily or seasonal performance and visualisations of energy flows over a two-or-three-dimensional domain that help to refine the design at later and more advanced design stages. This chapter introduces a range of different types of tool which can help designers to evaluate the performance of different options from outline to full scheme design.

6.1. PERFORMANCE CRITERIA: THERMAL COMFORT

One of the main objectives in environmental design is to achieve thermal comfort for the occupants of the building. Two different (and competing) models of thermal comfort have been developed over the last 50 years: a heat balance model, based on laboratory studies (largely undertaken by Ole Fanger *(Fanger, 1970)*, and an adaptive comfort model, based on field studies, initially put forward by Humphreys and Nicol *(Humphreys, 1976)*. The heat balance model treats the building occupant as a passive recipient of thermal stimuli, whereas the adaptive comfort model treats the occupant as an active participant, interacting and adjusting to their environment. A large number of field studies have shown that contextual factors have significant influence on perceptions of the thermal environment, and that the adaptive model provides a more accurate prediction of building occupant perceptions and responses in free-running (naturally ventilated) buildings than the heat balance model which is applied in mechanically conditioned buildings.

Adaptive comfort theory has gained widespread acceptance as a reasonable basis for defining the acceptable upper limit of temperature in buildings in different parts of the world. Indeed, the adaptive model has now been incorporated into major comfort standards around the world, including the European EN15251 and the American ASHRAE 55–2013 *(EN15251, 2007; ANSI/ASHRAE, 2013)*.

The range of acceptable temperatures defined by these standards derive from the prevailing mean outdoor temperature and responds to seasonal temperature variations. This comfort 'band' becomes in most cases the criteria for assessing the thermal performance of an indoor environment presenting an operative temperature relative to a particular space heat balance. Operative temperature is a simplified measure of human thermal perception derived from air temperature, mean radiant temperature and air speed. However, when mean radiant temperature may be similar, air temperature alone can be a reasonable indicator of thermal comfort.

> **ADAPTIVE COMFORT MODELS PROVIDE A MORE ACCURATE PREDICTION OF BUILDING OCCUPANT PERCEPTIONS AND RESPONSES IN NATURALLY VENTILATED BUILDINGS**

Recent developments by the Center for the Built Environment (CBE) – University of California Berkeley have led to the CBE Thermal Comfort Tool, a web-based graphical user interface for thermal comfort prediction in line with the above mentioned international Standards. The tool can be accessed through the following link: *http://comfort.cbe.berkeley.edu/*; and becomes a solid resource in early design stages.

6.2. STEADY-STATE METHODS

Natural cooling can be described as the dissipation of heat from buildings to a lower temperature environmental heat sink like air, water, sky or ground. This can be achieved through the design of the building (form and fabric) and the implementation of an airflow pattern in response to changing ambient conditions. To help designers have confidence to implement a natural cooling approach, reliable tools are required to evaluate feasibility and predict performance at different stages of the design process. Steady-state analysis allows quick approximations to be made to assist at the early design stage.

First, an understanding of site and microclimate (as described in Chapter 2) provides a sound basis to identify strategic cooling options, and whether natural convection is one of those options. In most cases, a preventive strategy like solar control is fundamental to the success of any natural cooling strategy, and the relationship between a site, its surroundings and the apparent movement of the sun should be a primary consideration. Natural ventilation may only be applicable at certain times of the year or of the day, so the first step is to identify the conditions under which it is applicable. The designer can then input those conditions into steady-state tools to establish preliminary feasibility. The simplified nature of these tools is often accompanied by intuitive graphic interfaces that invite designers to test multiple scenarios and design options in a matter of minutes.

> IN MOST CASES, A PREVENTIVE STRATEGY LIKE SOLAR CONTROL IS FUNDAMENTAL TO THE SUCCESS OF ANY NATURAL COOLING STRATEGY

SOLAR CONTROL

The need for cooling will depend on the characteristics of the site and microclimate (as discussed above), and the nature of the building (activities and density of occupation). In much of Europe, the historical tradition of masonry buildings with small windows meant that the need for cooling was rare, and when it arose could be dealt with by natural convection whenever the external air temperature was below the desired internal air temperature. Shutters prevented solar gain during the day in summer, while continuing to allow ventilation (fig. 163), and the high thermal capacitance interior helped to stabilise internal temperatures.

The contemporary context and expectations are completely different. The thermal performance of the external 'skin' of buildings has improved significantly as solar and internal heat gains from people and equipment increase the risk of overheating and therefore the need for cooling. The need for cooling has often been exacerbated by the design of the building itself, through excessive (and avoidable) solar heat gains. In the recent past this has led to the widespread use of mechanical air-conditioning, as discussed in the preface.

An effective shading strategy for vulnerable surfaces of a building should be the first stage when considering passive cooling strategies. Wherever we are in the world, the design of appropriate sun-shading devices for buildings requires a basic understanding of solar geometry, climatic variation, occupant needs and available design tools.

In the example below, the main source of solar gain to a complex of single storey kitchens and restaurant areas was through the roof, so in the refurbishment a canopy of shading louvres was proposed, radically reducing the need for cooling (figs 164–167).

The geometrical relationship between the earth (the building) and the sun is determined by the tilt of the earth (23.47°) as it spins on its axis, relative to the plane of the earth's orbit around the sun (fig. 168).

163. Shutters in the Stock Exchange, Valetta, Malta [Architecture Project].

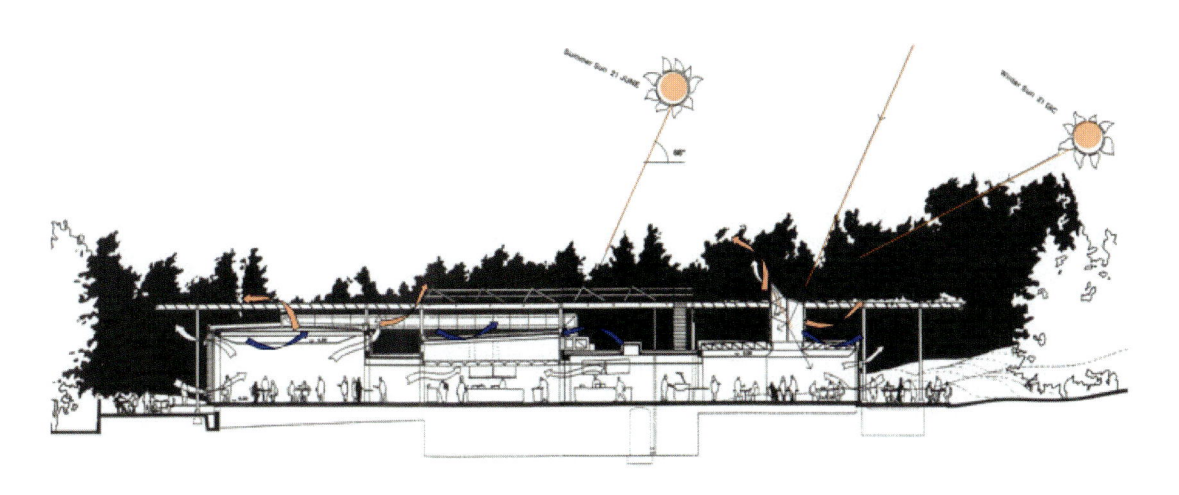

164. Section through canteen complex at Joint Research Centre ISPRA, Italy [Mario Cucinella Architects].

CHAPTER 6 | PERFORMANCE ANALYSIS

The latitude of the building's location (north and south of the equator) is therefore a fundamental determinant of the diurnal and seasonal variation in the sun's apparent movement around the building. The closer we are to the equator, the higher the sun will be in the sky at the middle of the day. This means that in many parts of the world the roof of a building receives the most solar radiation, leading to the use of protective 'parasol' roofs on many buildings within the tropics. Outside the tropics the amount of solar radiation received on different surfaces will vary significantly depending on the season. Generally, throughout the world, east and west facing building façades are more difficult to protect from unwanted solar gain than north or south facing surfaces. However, buildings are rarely oriented to the cardinal points of the compass, and these generalisations are not sufficient to determine an appropriate shading strategy. It is therefore necessary to understand the relation between solar geometry and climatic variation in more detail.

Solar geometry and climatic variation

The earth's orbit around the sun determines the broad climatic variations experienced around the world, but seasonal variation is rarely symmetrical around the solstices. For example, in Europe the summer solstice (21 June) is often referred to as mid-summer's day, but the hottest summer months are normally July and August. Similarly, the winter solstice (21 December) is not the coldest time of year (normally in February). This dislocation between solar geometry and heating and cooling seasons creates a

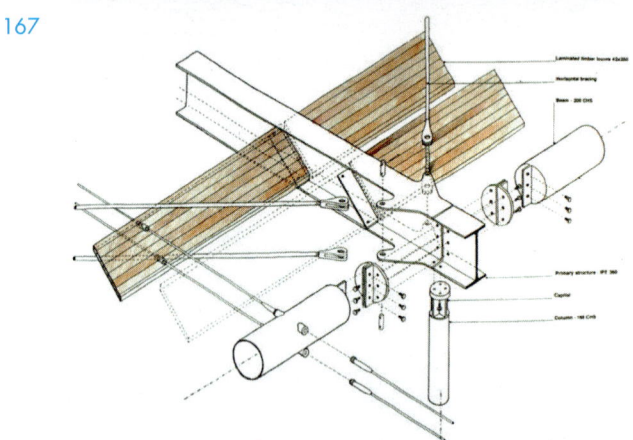

165. Exterior view of Joint Research Centre ISPRA, Italy [Mario Cucinella Architects].
166. Exterior view of shading canopy at Joint Research Centre ISPRA, Italy.
167. Detail shading to roofs over canteen complex at Joint Research Centre ISPRA, Italy.

problem for designers when determining an appropriate *'cut-off'* angle (horizontal or vertical shadow angles) for fixed shading devices. Indeed, this is also why some designers opt for moveable shading devices, in spite of high capital and maintenance costs, and frequent subsequent failure. Generally, however, fixed shading devices, if carefully designed, can be very effective in reducing unwanted solar gain and avoiding the associated cooling load.

In the northern hemisphere, south facing vertical glazing is generally assumed to be vulnerable, but for most locations, south-facing façades may be protected by means of simple horizontal shading devices. A horizontal shade designed to cut out the sun at noon on the equinox will prevent any direct radiation entering the façade throughout the six summer months. However, a horizontal shade will not be effective in preventing solar heat gains to east and west elevations at any time of year. Hence the use of vertical shades, and sometimes *'egg-crate'* shading devices, on east/west elevations. Within the tropics, east/west façades are vulnerable to solar gain even if they are opaque, and any opening may become vulnerable to intense reflected radiation. Within North Africa and the Middle East the traditional *'mashrabiya'*

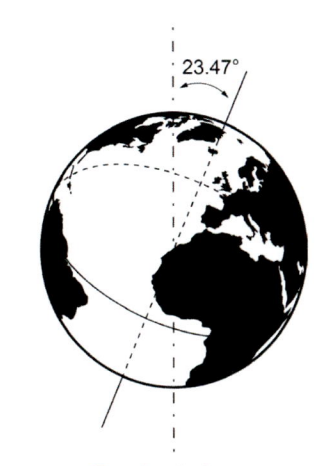

> IN THE NORTHERN HEMISPHERE, A HORIZONTAL SHADE DESIGNED TO CUT OUT THE SUN AT NOON ON THE EQUINOX WILL PREVENT ANY DIRECT RADIATION ENTERING THE SOUTH FAÇADE THROUGHOUT THE SIX SUMMER MONTHS

168. Representation of Earth's declination.

169. Mashrabiya in Sidi Saiyyed Mosque, Ahmedabad, India.

CHAPTER 6 | PERFORMANCE ANALYSIS

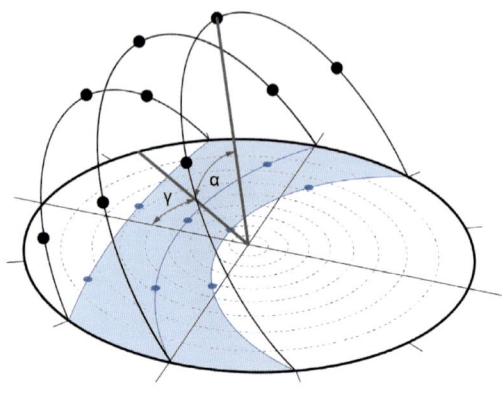

170. Schematic view of sunpaths.

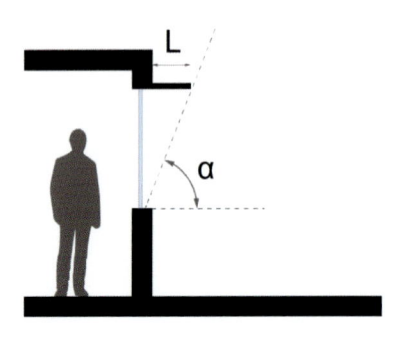

Overhang depth

$$L = \frac{h}{\tan(\alpha)}$$

In the context of Madrid, the sun altitude on the equinox at noon is 42°, and the required overhang depth to shade a south-facing window height of 1.2m throughout the six summer moths is:

$$L = \frac{1.2}{\tan(42)} = 1.30m$$

171. Relation between the solar altitude at noon on the equinox and the required overhang depth to prevent solar gains on a south façade in Madrid, Spain.

screen evolved to protect the building occupants from both solar heat gain and glare. Similar *'jali'* screens evolved in India (fig. 169).

What design tools are now available to analyse and respond to these considerations? In the past architects have analysed sunlight penetration using physical models in conjunction with sun-dials or heliodons. Physical modelling can still be a quick and effective way to evaluate the effectiveness of a shading strategy. Photographs (and animations) can record the apparent movement of the sun around and into a building. At early design stages, sketches and drawings can be evaluated by reference to 2D sun-path diagrams, which provide solar altitude and azimuth angles for any latitude. Solar altitude is the angle between the horizontal plane and the sun, whereas solar azimuth is the angle between the sun and the North, measured clockwise around the observer's horizon. 2D and 3D sunpath diagrams for any location can be obtained from Andrew Marsh's website: http://andrewmarsh.com/software/sunpath2d-web/. Solar altitude is often used to determine the required overhang depth to prevent excessive solar gains in south or north façades, whereas the sun azimuth can inform the design of vertical shading devices on east and west façades (figs 170–171).

Solar geometry has of course now been digitised and incorporated in a number of different software packages. New tools have emerged to meet the needs of environmental designers and engineers. Most of these tools now integrate the benefits of parametric design and building information modelling (BIM), making them more appealing to designers. The recently developed software package Ladybug Tools (*Ladybug, 2018*), highly supported by industry and academics, performs, among other capabilities, solar radiation, sun-hours and shading studies based on annual hourly climate data. The dynamics of the sun position measured by the solar angles, number of sun hours or amount of solar radiation can provide information for shaping shading devices, placing buildings and addressing outdoor comfort in the urban context (fig. 172). Incident solar radiation falling on horizontal, vertical or tilted surfaces is given by the angle on incidence (angle between the line perpendicular to the surface and the earth-sun line). The actual amount of direct solar radiation falling on the surface is proportional to the beam irradiance and the angle of incidence. It is evident that for vertical surfaces small angles of incidence result in higher irradiation, making East and West orientations especially vulnerable to solar gains and, at low latitudes, also horizontal surfaces (fig. 173).

NATURAL VENTILATION: OPTIVENT TOOL

Optivent is a simple steady state tool for the evaluation of natural ventilation options in buildings at an early design stage, which has been developed and refined over a number of years, through a process of application to built projects and peer review. It was developed originally as an in-house tool for use by a practice of consultants in the UK

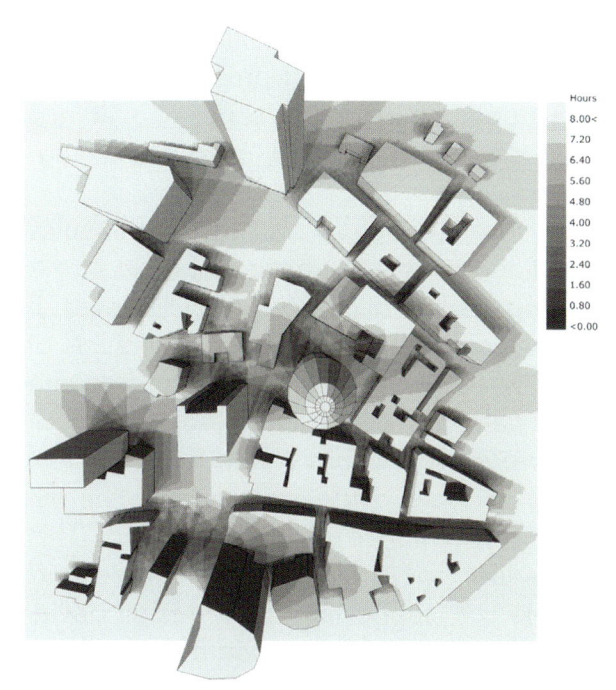

172. Example of sun hours and shadow analysis at summer solstice (10am-6pm) using Ladybug for Grasshopper for Rhino.

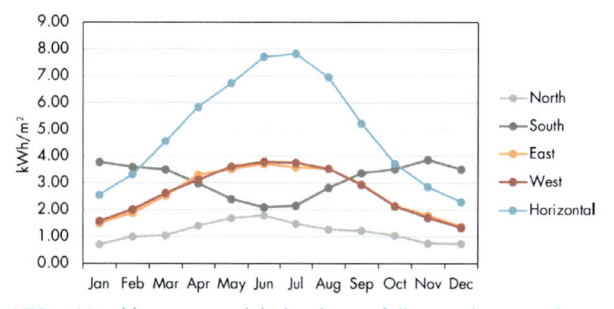

173. Monthly average global radiation falling on horizontal and vertical surfaces in Madrid, Spain.

CHAPTER 6 | PERFORMANCE ANALYSIS   145

and was applied in evaluating options, airflow path strategies and preliminary sizing of vent opening areas in a series of buildings in Europe, North America, India and China.

The tool is a *'snap shot'* of a particular condition and the dynamic response of the buildings is not accounted for. However, it was successfully applied in these projects at an early stage, and as the design became more defined, other (more sophisticated) tools were used to assist with refinement of details and to provide confidence in the dynamic performance of each building. This simple tool has proved to be reliable and quick to use, supporting rapid design development and providing a good basis for later stage, more detailed analysis.

Limitations and potential improvements emerged over time, and a revised version (Optivent 2) was developed and peer reviewed prior to being issued in 2015. The new version expands the range of generic airflow strategies that can be explicitly evaluated, extends the geographic applicability of the tool, and incorporates a more user-friendly graphic interface. It has been released on a web-based platform which is available from *http://naturalcooling.co.uk/optivent.html*

> OPTIVENT 2 SUPPORTS RAPID DESIGN DEVELOPMENT AND PROVIDES A GOOD BASIS FOR LATER STAGE DETAILED ANALYSIS

The tool assists with the definition of a reliable ventilation strategy by determining the ventilation rates required to provide a healthy environment and avoid the risk of overheating. The provision of adaptive opportunities for the occupants of buildings are also placing more demands on designers in relation to the health and comfort of building users. All this must be achieved while minimising carbon emissions. In addition to informing designers of the feasibility of a natural ventilation strategy, Optivent 2 provides additional information on thermal comfort and design considerations for the successful development of the strategy.

Design tool workflow

The most widely used engineer-oriented natural ventilation design guide is the UK CIBSE Applications Manual 'Natural Ventilation in Non-Domestic Buildings *(CIBSE AM10, 2005)*. Optivent 2 tool is based on the steady state expressions found in AM10, but goes further in providing a format within which feasibility can be quickly evaluated. It targets engineers, architects and designers through an approach that requires the user to describe the building characteristics and ventilation strategy (user inputs) in order to obtain the achieved airflow rates (software output) for evaluation. In the process, the areas of ventilation openings and stack heights can be varied to explore different design options. This results in a quick input process as in most cases the user is more familiar with the building layout than with airflow rates required.

The methodology is a five-step process where the user is asked to select a ventilation strategy, define the building layout and identify internal gains and solar exposure in order to estimate the airflow rates required to provide fresh air and for removing excess heat. Finally, a set of outcomes including airflow rates, air velocities and thermal comfort prediction are plotted in comparative charts to give the user sufficient information to evaluate the feasibility of the selected strategy.

The ease of use and quick input process also invites the user to explore other strategies under different environmental conditions. This also helps the user to understand the impact of different factors (stack height, aperture areas, internal-external temperatures) affecting the buoyancy driven and wind driven airflow rates achieved.

Natural ventilation strategies

Most ventilation strategies available for testing are in line with *CIBSE AM10 (2005)* document and are described in Chapter 2. The tool allows the evaluation of:
1. Single sided and cross ventilation options through a single aperture or two open vents in an isolated space (single cell),
2. A single-cell building with multiple inlets and a stack or atrium connected to wide open occupied spaces, and
3. A multi-cell building with multiple inlets and occupied zones connected to a well-defined chimney or ventilated façade, defining an additional zone.

A downdraught scenario is also available to allow feasibility studies on the application of passive and active downdraught evaporative cooling systems. These different scenarios are illustrated in fig. 174.

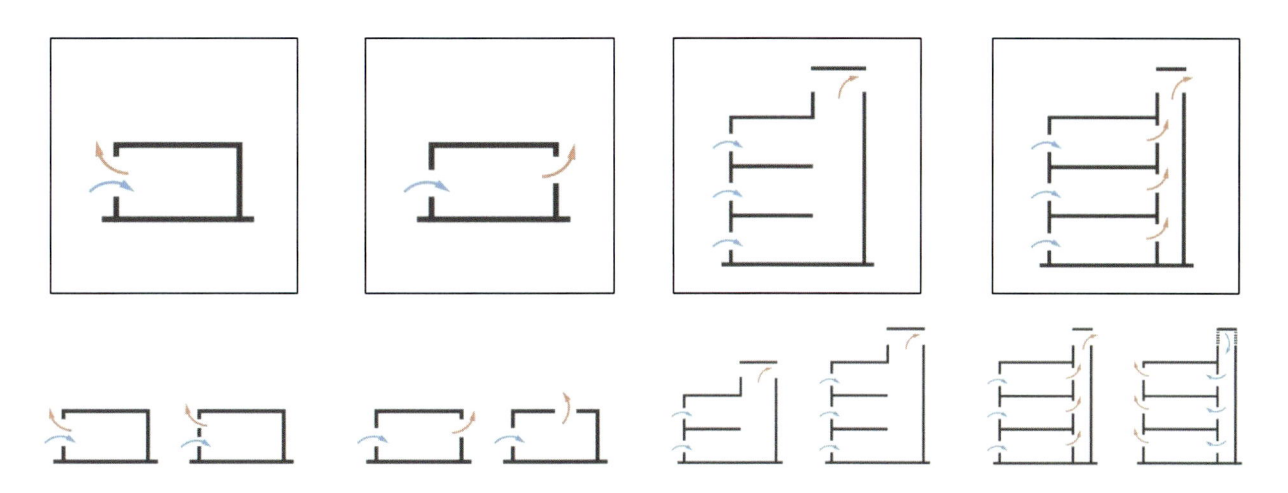

174. Optivent 2: Ventilation strategies that can be evaluated.

CHAPTER 6 | PERFORMANCE ANALYSIS 147

Graphic interface

The graphic interface of the tool enables most of the user inputs to be entered within diagrams and images that help understanding of each value. The diagrams also provide a simple record of input assumptions. Aperture heights, areas and temperatures are displayed as shown in fig. 175. The effective area of each aperture is also considered and a range of values are suggested to the user according to the way the window opens and the surrounding head, sill and jamb details.

Calculation methods and results

The model assumes the flow of air into the building equals the flow of air out of the building (the principle of mass conservation). This is applied in each envelope flow model *(CIBSE AM10, 2005, Equation 4.9)* and the airflow rate through each opening is expressed as a relationship between the pressure difference across the opening by means of the discharge coefficient and the specified effective aperture area *(CIBSE AM10, 2005, Equations 4.10–4.11)*.

Discharge coefficients and wind pressure coefficients have been set to default values optimised for each airflow model. Wind pressure coefficients have also been taken from *CIBSE AM10 (2005), p.54*.

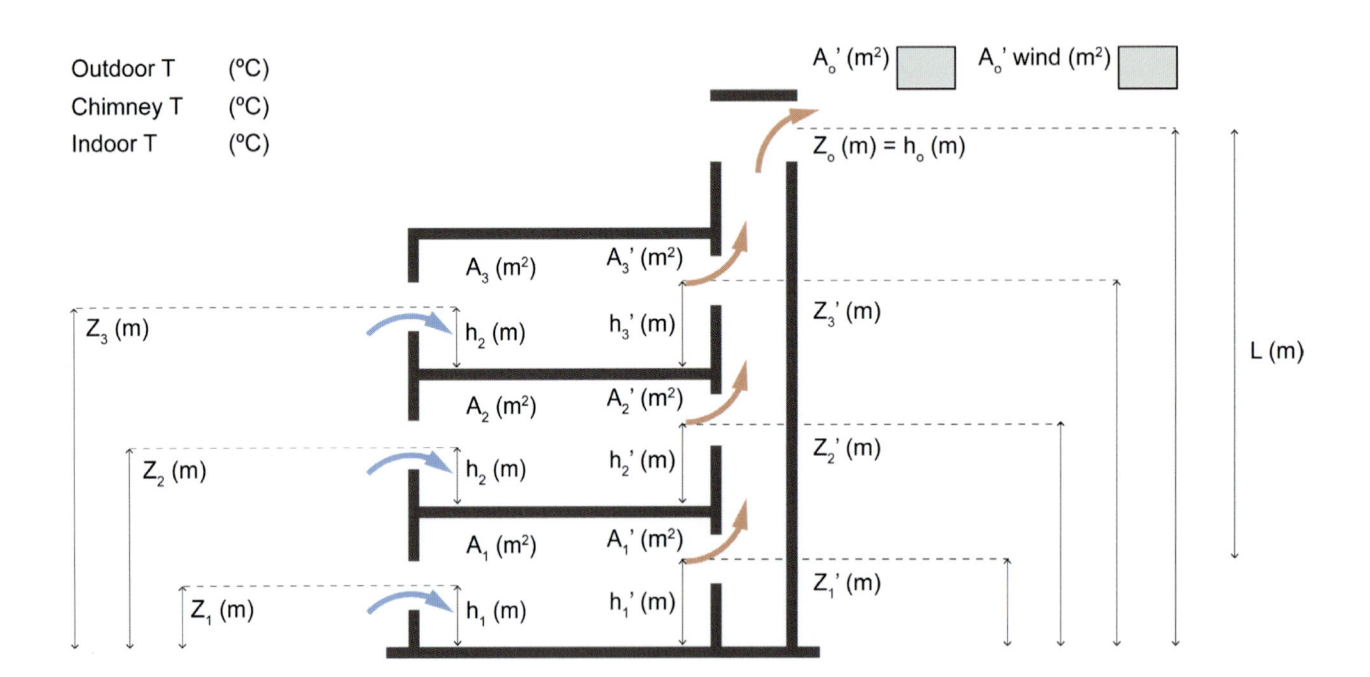

175. Optivent 2: Airflow data input interface.

CHAPTER 6 | PERFORMANCE ANALYSIS

The calculation process outputs airflow rates driven by buoyancy alone, and driven by buoyancy + wind. Buoyancy alone would represent the worst case scenario and must be considered during the feasibility assessment. Buoyancy + wind driven airflow rates will inform the user about the impact (and benefit) of wind forces in the chosen natural ventilation strategy.

Three charts are plotted to provide a general overview of the performance of the natural ventilation strategy. The first graph indicates the airflow rates achieved in the given scenario, and compares them with the airflow rates required for the supply of fresh air and for removing the total heat gains generated within the space (required for cooling). The results are given in m³/s, air changes per hour (ac/h), and Wh/m² by means of the space volume and indoor-outdoor temperature difference (fig. 176).

> BUOYANCY ALONE WOULD REPRESENT THE WORST CASE SCENARIO AND MUST BE CONSIDERED DURING THE FEASIBILITY ASSESSMENT

The second chart plots air velocities in m/s at inlets and outlets, together with the average room air velocity. The graphs also relate the air velocities achieved with the equivalent physiological temperature reduction provided according to the operative temperature expression given in *CIBSE (2015)*, pp. 1–3. Under hot conditions, air movement is a valid strategy to improve comfort conditions of occupants and may also be

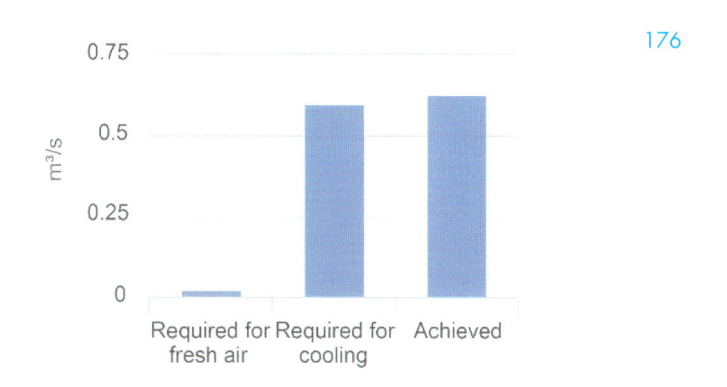

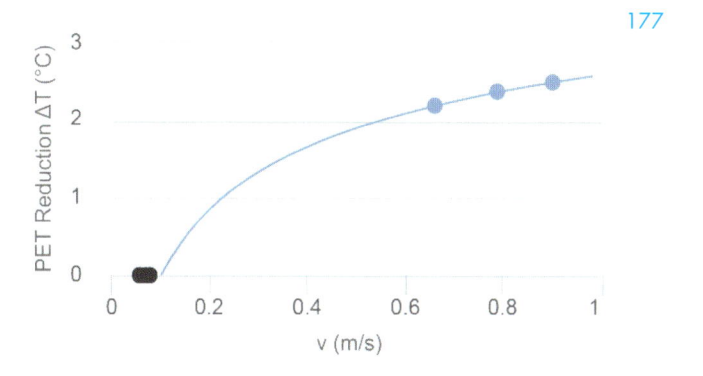

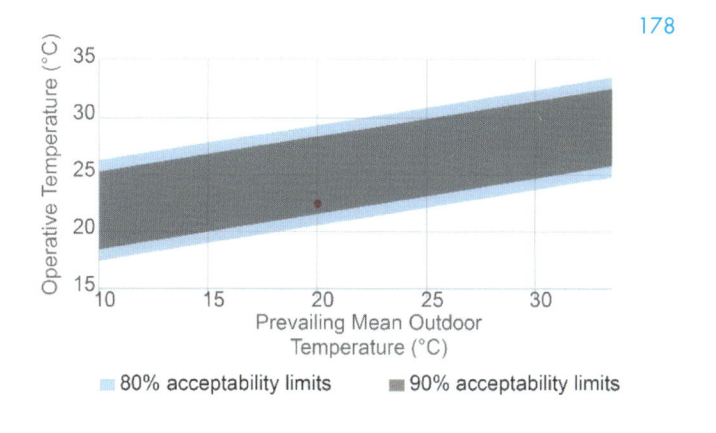

176. Optivent 2: Results on airflow rates achieved in the given scenario.
177. Optivent 2: Results on air velocities at inlets and outlets, together with the average room velocity and their equivalent psychological temperature reduction provided.
178. Optivent 2: Results on predicted comfort conditions based on the adaptive comfort method ANSI/ASHRAE Standard 55-2013.

CHAPTER 6 | PERFORMANCE ANALYSIS   149

considered when assessing the benefits of a natural ventilation strategy (fig. 177).

The third chart predicts comfort conditions within the occupied space based on the adaptive comfort method from ASHRAE Standard 55–2013. The chart plots the operative temperature over the predicted comfort bands referring to 80% and 90% of user acceptability limits in a similar manner as CBE Thermal Comfort Tool. The operative temperature is calculated by the software after considering the external and internal temperature, the total heat gains generated within the space and the heat losses achieved with natural ventilation (fig. 178).

Interpretation of the results will determine what design changes (like increasing vent area) may be required to improve performance (if necessary), and provide a basis for further testing of the viability of the design proposal, prior to the use of more sophisticated (and time consuming) analytic techniques.

> THE EVALUATION OF CONVECTIVE COOLING SYSTEMS USING DYNAMIC THERMAL SIMULATIONS IS ESPECIALLY EFFECTIVE AS THE BUOYANCY FORCES DRIVING AIR INTO THE ZONES ARE ESTIMATED FOR EACH HOUR OF THE ANALYSIS PERIOD WHILE RESPONDING TO THE DYNAMIC BEHAVIOUR OF THE BUILDING

6.3. DYNAMIC METHODS

As building design is developed in more detail, the passive cooling system may require a comprehensive level of definition. The system can be refined using dynamic thermal simulation tools, able to evaluate building performance on an hourly basis and indicate the achieved airflow rates and cooling delivered at different time periods and seasons throughout a typical (or hot) year. Dynamic thermal simulations take into account internal conditions (in addition to the climatic context), and consider most heat exchange processes involved in building physics: convection, conduction, radiation and, partially, evaporation. Further, commercial software like IES and ESDL Tas *(IES, 2018; EDSL, 2018)* have recently integrated advanced & dynamic daylight-responding user controls to achieve a higher level of definition in thermal modelling. Others, like the extensively used and validated EnergyPlus simulation engine *(EnergyPlus, 2018)* allow the customisation of user inputs through the also open-sourced interfaces OpenStudio *(OpenStudio, 2018)* and Honeybee *(Honeybee, 2018)*. The evaluation of convective cooling systems (natural ventilation) using dynamic simulations is especially effective as the buoyancy forces driving air into the zones are estimated for each hour of the analysis period while responding to the dynamic behaviour of the building. Wind force is also considered in most dynamic simulation software, but its integration in the calculation process is often simplified by means of typical wind pressure coefficients determined from the relation between a wind direction perpendicular to the surface of the building envelope, the meteorological wind speed, and the terrain type *(CIBSE AM10, 2005, p. 54)*. Although solution outcomes like air stratification and distribution within a three-dimensional domain are restricted, processed results can provide valuable and detailed information regarding the dynamic performance of the natural ventilation strategy. For example, peak, average and time related airflow rates for different building geometries.

The application of dynamic simulations in the design process is exemplified in the Kuwait School project by Mario Cucinella Architects (figs 179–180). The school, a three storey building with central courtyard originally designed as the first low carbon school building in Palestine, was located in

179. Mashrabiyas & ventilation shafts in SW elevation, Kuwait School, Gaza Strip [Mario Cucinella Architects & UNRWA]

180. Proposed natural ventilation strategy in Summer for the Kuwait School, Gaza Strip [Mario Cucinella Architects & UNRWA].

CHAPTER 6 | PERFORMANCE ANALYSIS

the Khan Younis refugee camp and hosted 32 classrooms distributed on three floors for a total capacity of 2050 children. In this project, a thermal model was used to evaluate among other strategies important aspects of the design:

- The effect of an external 'parasol' shading a concrete deck below.
- The capacity of high conductive materials applied on vertical ventilation shafts and exposed to the sun to improve natural ventilation flow rates.
- The need of thermal mass to minimise overheating risk in classrooms.

When using dynamic simulation tools, comparative studies between multiple design iterations often provide reliable information regarding the impact of each design solution. This method was applied for each of the studies performed in this project.

A graphical representation of data fluctuation for a typical warm week of the cooling season is often a good way to interpret results. A quantification of solar gains in the top floor classrooms for a shaded and unshaded roof scenario (for a prescribed roof surface reflectance and U-value) is illustrated in fig. 181. The graph suggest a considerable increase in heat gains in the exposed classroom, subsequently increasing the internal temperatures during the observed week. The comparative study confirmed that these classrooms could benefit from a 2°C drop during occupancy hours by having an external parasol shading the roof.

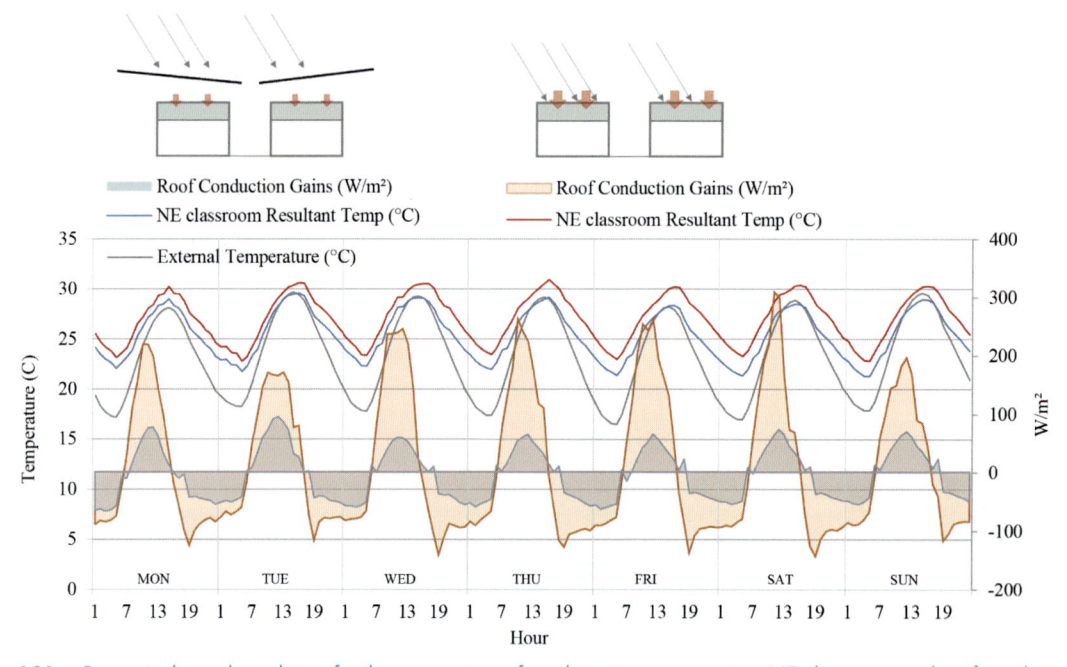

181. Dynamic thermal simulation for the comparison of resultant temperatures in a NE classroom and roof conduction gains between two scenarios (with and without a shade canopy) for a typical hot week in June.

CHAPTER 6 | PERFORMANCE ANALYSIS

A closer look at the natural ventilation flow rates achieved throughout the year allowed the typical airflow path in classrooms to be identified: flowing in from the courtyard, across the classrooms and out through the façade, mostly driven by buoyancy forces. The thermal model also allowed multiple ventilation strategies to be applied responding to the occupancy pattern and internal air temperatures. These strategies included single sided ventilation, cross ventilation and enhanced ventilation using vertical ventilation shafts. Average airflow rates could be determined for each scenario allowing a better understanding of the natural ventilation strategy. In this way, the capacity to dissipate heat from the occupants and from the building mass at night was confirmed (fig. 182).

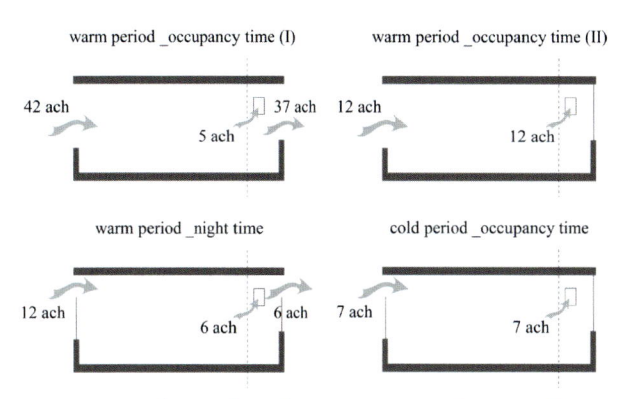

182. Proposed natural ventilation strategies and mean airflow rates predicted in classrooms using dynamic thermal simulation tools, Kuwait School.

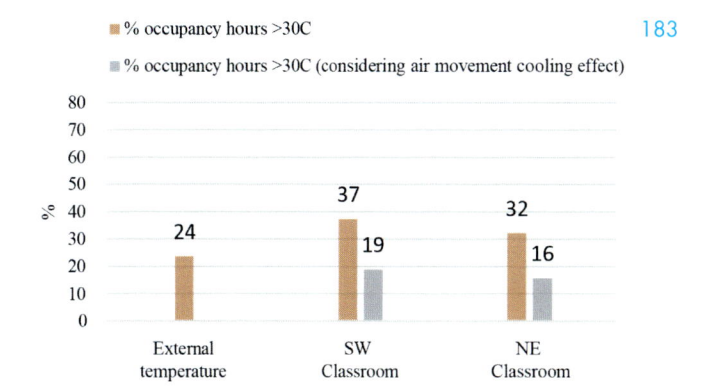

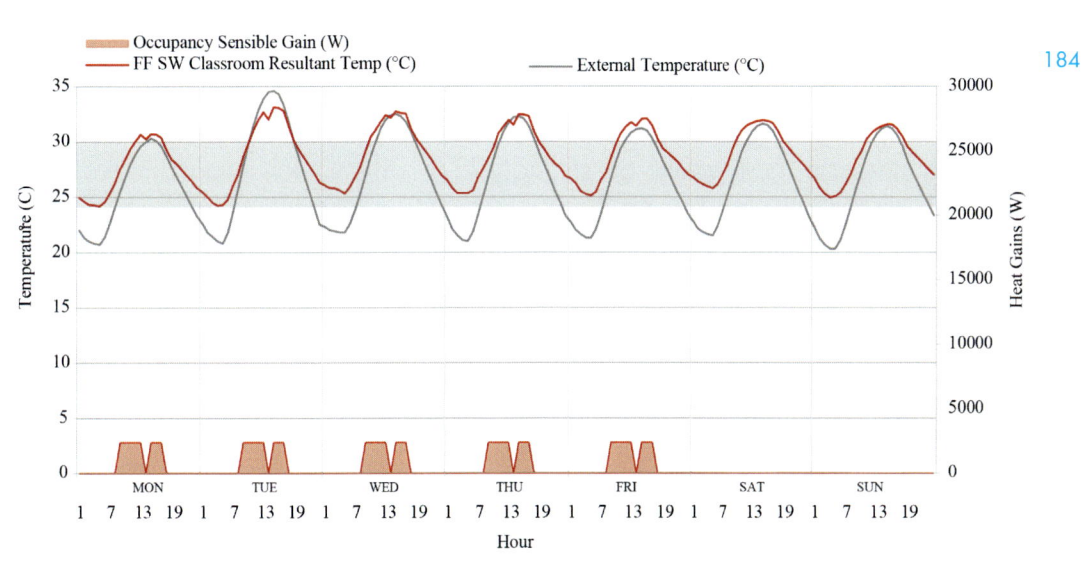

183. Frequency of predicted resultant temperatures during the warm season, Kuwait School.
184. Resultant temperatures & internal heat gains in SW classroom for a typical hot week in June, Kuwait School.

CHAPTER 6 | PERFORMANCE ANALYSIS

Further applications of thermal modelling allowed an overall evaluation of the building performance throughout the year. The impact of multiple passive strategies integrated in design was evaluated by comparing internal air temperatures against a thermal comfort criteria for classrooms during the warm season. Data was graphically represented on an hourly basis for a typical warm week and processed to identify the frequency of temperatures achieved. Results suggested that a comfortable learning environment could be achieved for more than 70% of the occupied time during the warm season. The flow rates achieved kept the resultant temperature in the classrooms very similar to the external dry bulb temperature (figs 183–184).

The above example demonstrated how dynamic simulations can provide valuable data that can be used to refine the natural ventilation or alternative passive strategies responding to the internal conditions and the dynamic response of the building. The integration of evaporative cooling systems in the thermal model is, however, limited, and the estimated cooling capacity or impact on building performance would be strongly influenced by the user inputs and the evaporative media used. The proportional relationship between the airstream properties and the evaporation rate require three-dimensional domains to simulate air stratification and distribution around the evaporative media. This can only be achieved with computational fluid dynamic (CFD) tools, commonly used in architecture to visualise air velocities and temperature stratification in the given space as explained in section 6.4.

MODELLING EVAPORATIVE COOLING

The modelling and simulation of the evaporative cooling process is complex and multi-phased and despite some progress by a selected number of researchers, there are still not many building simulation codes which can deal with the modelling of direct evaporative cooling or of evaporation through porous ceramic media. A historical review of tools dealing with evaporative cooling was undertaken by Martinez in his PhD thesis (*Martinez, 2000*) and a similar review was included in a subsequent paper by *Robinson et al. (2004)*. Some of the reviewed tools such as the America programs BLAST (*Osbaugh & Moore, 1998*) and DOE2 (*Peterson & Hunn, 1983*) include models of active evaporative cooling and others like the Dutch program ACCURACY can model evaporative cooling towers (*Niu & Van der Kooi, 1997*). On the passive evaporative cooling front, programs like the South African EASY (*Mathews et al., 1994*) and the Austrian CHEETHA (*Giabaklou & Ballinger, 1996*) have been developed but these present limitations on the zoning and use specific correlations. RSPT (RoofSol/PDEC Tool) developed by AICIA, University of Seville (*Alvarez et al., 2002*) is a calculation tool for radiative and evaporative systems. The

> THE PROPORTIONAL RELATIONSHIP BETWEEN THE AIRSTREAM PROPERTIES AND THE EVAPORATION RATE REQUIRE THREE-DIMENSIONAL DOMAINS TO SIMULATE AIR STRATIFICATION AND DISTRIBUTION AROUND THE EVAPORATIVE MEDIA

program can perform hourly simulations for the selected building and cooling technique as well as for a reference configuration, automatically generated. The simulation, however, is not multi-zonal and the output is given as energy savings or hours of thermal comfort for the whole building.

Since many of these codes are not widely available and quite specific in the type of output or limitations, a new trend has emerged by which instead of modelling and simulating the evaporative cooling phase changes, the output of the evaporative cooling process is emulated using the modelling capabilities of standard programs. This technique has been applied originally in the work of *Robinson et al. (2004)* for the performance assessment of a speculative office building project using passive downdraught evaporative cooling and modelled using the software ESP-r. Subsequently a similar emulation technique was employed by *Schiano-Phan (2010)* in the modelling of passive evaporative cooling via porous ceramic media, using the dynamic thermal modelling software EDSL TAS (*EDSL, 2018*).

TAS was particularly appropriate for the simulation of the proposed system since it can model several zones at the same time, giving the opportunity to appreciate the cooling loads of the various rooms in the building. Moreover, it has built in facilities for the modelling of a *'supply'* zone emulating the effect of the passive evaporative cooling system, which is subsequently supplied to the occupied zone. In fact, the evaporative cooling system (either direct or via

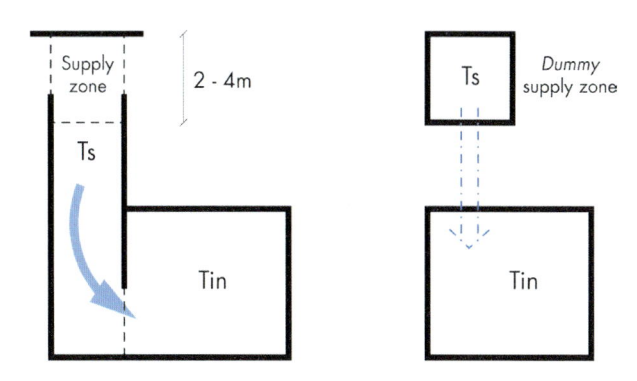

185. Schematic diagram of the application of evaporative cooling systems in dynamic thermal simulation tools.

porous media) can be emulated by creating a *'dummy'* supply zone cooled at a temperature equal to:

$$Ts = Tdb - 0.7 \times [Tdb-Twb] \quad \text{Equation 07}$$

$$Ts = Twb + 2°C \quad \text{Equation 08}$$

where *Ts* is the supply air temperature (°C), *Tdb* is the ambient dry bulb temperature (°C) and *Twb* the wet bulb temperature (°C).

Equation 07 characterises the supply temperature of a direct evaporative cooling system and equation 08 characterises the supply temperature of evaporative cooling using porous ceramic media (see Chapter 4). For the characterisation of the ventilation flow rate there are two possible modelling and emulation techniques. The original technique developed by Robinson implies a scheduled flow of air supplied from this zone to the occupied zone. However, with the improvement of the modelling software interface, airflow rates can be automatically computed as downdraught convective cooling. In practice this can be modelled in TAS according to the schematics in fig. 185. The whole emulation exercise is also aided by the possibility to specify hourly set points for the supply zone linked to the outdoor Tdb and Twb as per equation 07 and equation 08. However, the main limitation in this case is that the phase change and moisture generated in the process is completely neglected and the main assessment is based on the sensible cooling effect.

6.4. COMPUTATIONAL FLUID DYNAMICS (CFD)

When a deeper investigation of the influence of building geometry on convective and evaporative cooling is required, designers can opt for computational fluid dynamic (CFD) simulations. CFD tools allow the simulation of multi-physics (multiple physics coupled to mimic reality; i.e. heat transfer, phase change, chemical reactions, particle transfer, etc.) and multi-fluid interaction in a two-or-three-dimensional domain. It is for this reason that CFD have been applied to many fields involving prototyping design, such as mechanical engineering, electronics cooling, turbomachinery, etc. The nature of the software allows the user to practically solve any problem that could occur in reality, but the reliability of the results is strongly related to the skills of the user, their expertise and capacity to identify the physics involved and implement them in the software. In architecture, CFD has been used to visualise air movement in both indoor and outdoor environments. When applied to an urban scale model, designers can predict, for instance, air velocities and directions at pedestrian level, or accurately obtain wind pressure coefficients over the building envelopes for a detailed calculation of building infiltration or wind-driven natural ventilation airflow rates (fig. 186). If applied to indoor

> IN ARCHITECTURE, CFD HAS BEEN USED TO VISUALISE AIR MOVEMENT IN BOTH INDOOR AND OUTDOOR ENVIRONMENTS

environments, CFD allows the visualisation of air stratification and the identification of typical airflow paths driven, in most cases, by thermal forces. Illustrations like fig. 187 can be useful to identify poorly ventilated zones or overheated areas in large space volumes. CFD tools can also be used to simulate evaporation processes with high level of accuracy, and many authors have successfully validated CFD models for difference evaporative cooling systems and techniques, including cooling pads, cooling ponds, mist systems and water films *(Cook et al., 2000; Franco et al., 2010; Ramadan et al., 2014; Montazeri et al., 2015)*.

There are a wide range of CFD software tools available in the market, including Ansys Fluent *(Ansys, 2018)* and Comsol CFD *(Comsol, 2018)*, extensively used for their broad integration in industry, being extensively validated and offering an extensive suite of supporting tools; Autodesk CFD *(Autodesk, 2018)*, providing an intuitive interface and offering easy integration with other Autodesk products such as AutoCAD and Revit; and OpenFOAM *(OpenFOAM, 2018)* as the open source alternative, also extensively validated and recently integrated in parametric modelling tools like Grasshopper for Rhino *(Grasshopper, 2018)* using the plugin Butterfly *(Butterfly, 2018)*.

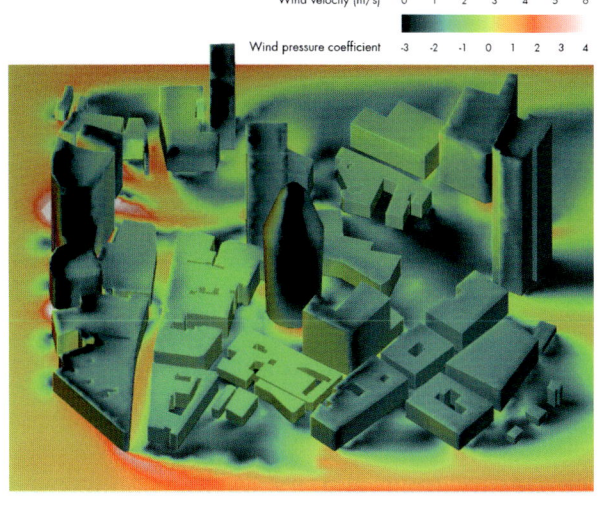

186. Example of outdoor CFD simulation to evaluate air velocities at pedestrian level and wind pressure on building envelopes.

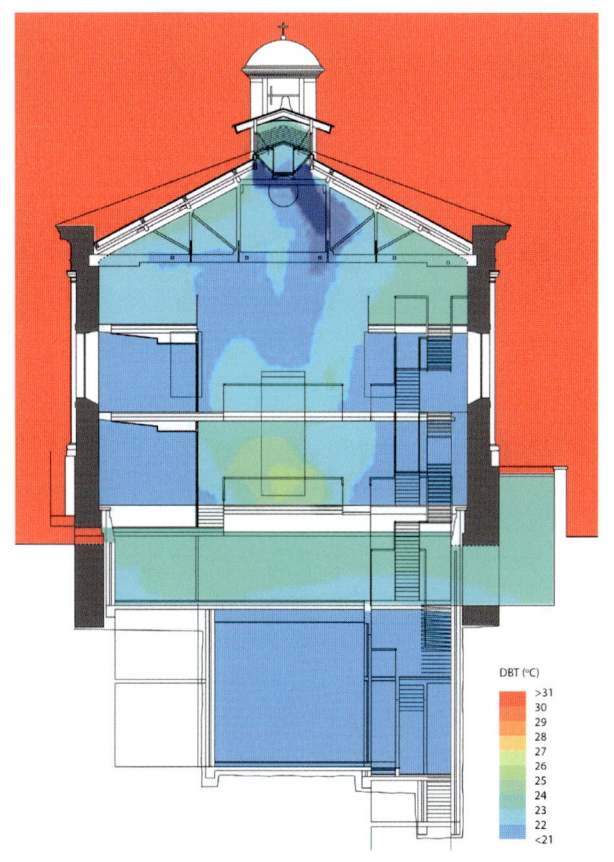

187. Indoor CFD temperature plot, Stock Exchange, Malta [Architecture Project].

CHAPTER 6 | PERFORMANCE ANALYSIS

The previous chapter has given general guidance and *'rules of thumb'* for preliminary sizing of the main elements of passive cooling systems. However, once basic feasibility has been established, and an outline design has been defined, a more detailed analysis is required in order to progress the design and give more confidence in predicting performance. A simple software tool may be considered for use, and give the design team and client the confidence to proceed to the next stage before CFD or more detailed dynamic thermal modelling is applied. In the field of passive cooling, OPTIVENT 2 (updated and re-issued in 2016) provides the opportunity to quickly evaluate feasibility and implications of both natural ventilation and evaporative cooling strategies. This integrated design approach is a step forward in environmental design in architecture, adding value to the project and maximising efficiency and performance while minimising effort and cost.

Detailed performance assessment provides the client and design team with the confidence to proceed with the agreed design strategy, but many details of the components which form the building may still have to be defined. For predicted performance to be delivered in practice the details of component design must support and reflect the assumptions built into the analysis. The next chapter considers the importance of these assumptions to the design of a number of important building components.

REFERENCES

- Alvarez, S. et al., (2002). *'RoofSol/PDEC Tool for Calculating Applicability Maps'*. AICIA Grupo de Termotecnica, Universidad de Sevilla.
- Ansys (2018). *'Ansys Fluent'* [Online]. http://www.ansys.com/en-gb/Products/Fluids/ANSYS-Fluent/. [Accessed 10/2018].
- ANSI/ASHRAE (2013). *'Standard 55-2013. Thermal Environmental Conditions for Human Occupancy'*. American Society of Heating, Refrigerating and Air-Conditioning Engineers, Atlanta.
- Autodesk (2018). *'Autodesk CFD'* [Online]. https://www.autodesk.com/products/cfd/overview/. [Accessed 10/2018].
- Butterfly (2018). *'Butterfly'* [Online]. http://www.ladybug.tools/butterfly.html/. [Accessed 10/2018].
- CIBSE (2005). *'Applications Manual 10: Natural Ventilation in Non-Domestic Buildings'*. London: Chartered Institution of Building Services Engineers.
- Comsol (2018). *'Comsol CFD'* [Online]. https://www.comsol.com/cfd-module/. [Accessed 10/2018].
- Cook, M. J., Robinson, D., Lomas, K. J., Bowman, N. T. & EPPEL, H. (2000). *'Passive Downdraught Evaporative Cooling: II. Airflow Modelling'*. Indoor and Built Environment: 9, 325–334.
- EDSL (2018). *'EDSL Tas'* [Online]. http://www.edsl.net/main/. [Accessed 10/2018].
- EN15251 (2007). *'Indoor Environmental Input Parameters for Design and Assessment of Energy Performance of Buildings Addressing Indoor Air Quality, Thermal Environment, Lighting and Acoustics'*. European Committee for Standardization, Brussels.
- EnergyPlus (2018). *'EnergyPlus'* [Online]. https://energyplus.net/. [Accessed 10/2018].
- Fanger, P. O. (1970). *'Thermal Comfort : Analysis and Applications in Environmental Engineering'*. Copenhagen, Danish Technical Press.
- Franco, A., Valera, D. L., Madueño, A. & Peña, A. (2010). *'Influence of Water and Air Flow on the Performance of Cellulose Evaporative Cooling Pads Used in Mediterranean Greenhouses'*. Transactions of the ASABE, 53: 565.
- Giabaklou & Ballinger, (1996). *'A Passive Evaporative Cooling System by Natural Ventilation'*. Building and Environment 6 (31): 503–507.
- Grasshopper (2018). *'Grasshopper 3D'* [Online]. http://www.grasshopper3d.com/. [Accessed 10/2018].
- Honeybee (2018). *'Honeybee'* [Online]. http://www.ladybug.tools/honeybee.html/. [Accessed 10/2018].
- Humphreys, M. (1976). *'Field Studies of Thermal Comfort Compared and Applied'*. Building Services Engineer, 44: 5–27.
- IES (20178). *'Integrated Environmental Solutions'* [Online]. https://www.iesve.com/. [Accessed 10/2018].
- Ladybug (2018). *'Ladybug Tools'* [Online]. http://www.ladybug.tools/. [Accessed 10/2018].
- Martinez, D. (2000). *'Thermal Simulation of Passive Downdraught Evaporative Cooling (PDEC) in Non-Domestic Buildings'*. PhD Thesis. De Montfort University.

- Mathews, E. H., Kleingeld, M. & Grobler, L. J. (1994). *'Integrated Simulation of Buildings and Evaporative Cooling Systems'*. Building and Environment 2 (29): 197–206.
- Montazeri, H., Blocken, B. & Hesen, J. L. M. (2015). *'Evaporative Cooling by Water Spray Systems: CFD Simulation, Experimental Validation and Sensitivity Analysis'*. Building and Environment, 83: 129–141.
- Niu, J. & Van der Kooi, J. (1997). *'Dynamic Simulation of Combination of Evaporative Cooling with Cooled Ceiling System for Office Room Cooling'*. Proc. of Building Simulation 1997. Fifth International Conference, Czech Republic, pp. 503–507.
- OpenFOAM (2018). *'OpenFOAM'* [Online]. https://www.openfoam.com/. [Accessed 10/2018].
- OpenStudio (2018). *'OpenStudio'* [Online]. https://www.openstudio.net/. [Accessed 10/2018].
- Osbaugh, D. & Moore, T. B. (1998). *'Applying Two Stage Evaporative Cooling'*. ASHRAE Journal 30, July: 26–30.
- Peterson, J. L. & Hunn, B. D. (1983). *'Evaporative Cooling Potential for Office Buildings'*. Proc. 2nd International Congress on Building Energy Management, American Mechanical Engineering Society, 1A pp. 3.31–6.40.
- Ramadan, A., Hasan, R. & Tudor J. (2014). *'Simulation of Flow and Heat Transfer of Humid Air in Spent Fuel Cooling Ponds'*. Proceedings of the World Congress on Engineering 2014 Vol II. London, UK.
- Robinson, D., Lomas, K. J., Cook, M. & Eppel, H. (2004). *'Passive Downdraught Evaporative Cooling: Thermal Modelling of an Office Building'*. Indoor Built Environment, (13): 205–221.
- Schiano-Phan, R. (2010). *'Environmental Retrofit: Building Integrated Passive Cooling in Housing'*. Architectural Research Quarterly, 14 (2): 139–151.

CHAPTER 7
COMPONENT DESIGN

Having developed a clear natural cooling strategy as early as possible in the design process, and subsequently predicted performance in detail (as described in the previous two chapters), careful consideration needs to be given to component design and integration, as part of the overall design development. Achieving successful convective & evaporative cooling in practice depends on the detailed design and specification of the different components and on the careful integration of the ventilators, actuators and controls in the building.

7.1. EFFECTIVE INTEGRATION

Effective integration requires close collaboration between architect and engineer. For example, it is important to agree who specifies what. An architect will often specify the window, and an engineer the actuator and control. Will they be compatible? How will they be put together? A factor often overlooked is that ventilators and actuators are joined by linkages. Experience suggests that in many installations, linkages and fixings are major points of weakness and sometimes they fail entirely.

When architect and engineer have agreed on the performance required of the windows and ventilator openings, there are many advantages if the window/louvre supplier is given responsibility for integration of vent, actuator and control. Where they are not, vent and actuator suppliers all recommend early dialogue, during which issues can be raised, for example providing suitable fixings for the actuator, suitable linkages, and the need for any reinforcement of the frames. The individual components are rarely designed to be part of an integrated system, for example: windows, ventilators and rooflights which initially may have been designed for manual control need to be matched with a suitable actuator to operate successfully in response to changing ambient conditions. There are fewer problems perhaps when the opening, actuator and controls are specified as a

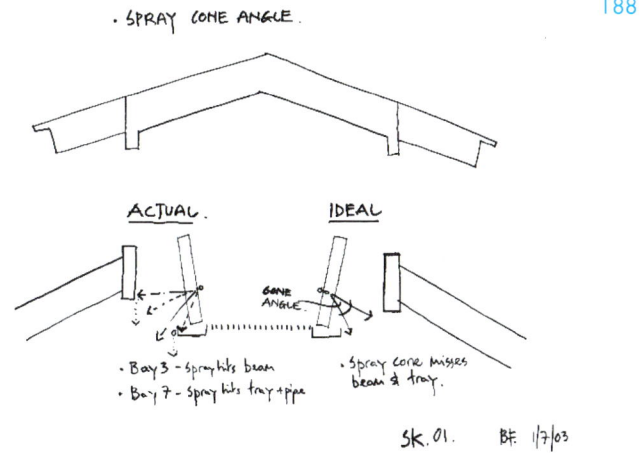

188. Sketch of relationship between ridge vents, rack & pinion actuators, cooling coils and misting nozzles.
189. Rack & pinion gear to glazed ridge vents + cooling coils on maintenance walkway, Malta Stock Exchange.

CHAPTER 7 | COMPONENT DESIGN

fully-integrated system, so that installation can be largely maintenance free.

With water misting systems, it is important that the effect of the *'cone angle'* of the spray nozzle or microniser is assessed in order to avoid wetting of adjacent building components or structure, and subsequent dripping into areas below (figs 188–189). The cone angle from the nozzles also needs to be set to avoid adjacent drip trays.

Where possible, it is desirable to combine automatically controlled ventilators (for example at high level or out of reach of occupants) with occupant controlled ventilators. Occupants will be much more tolerant of conditions in a building where they perceive that they have some influence on their surroundings *(Bordass et al., 1994)*. Equally, allowing manual override of automatically controlled vents can help, and still provide effective control (with control going back to a default position after a set period).

> WHERE POSSIBLE, IT IS DESIRABLE TO COMBINE AUTOMATICALLY CONTROLLED VENTILATORS (FOR EXAMPLE AT HIGH LEVEL OUT OF REACH OF OCCUPANTS) WITH OCCUPANT CONTROLLED VENTILATORS

7.2. VENTILATOR DESIGN

Having defined the cooling strategy, the choice of vent opening type can be made. The strategy is likely to have identified the need for different airflow rates in summer and winter. Decisions will also have been made about which ventilators are manually or automatically controlled. People like to be able to make alterations to their environment, so providing some manual control at low level is often a good idea. Conversely, high level vents are difficult to reach and their status (open/closed) is sometimes difficult to determine, so automatic control is advisable. This arrangement also means that night ventilation can be controlled automatically and securely.

Windows and rooflights can be made to shut tight relatively easily and, in comparison with louvres and dampers, they have a shorter crack length, and more effective seals. Window *'stays'* vary enormously in terms of their *'throw'*, geometry and robustness. Care must be taken to ensure that the effective free area will be achieved in practice.

Most patterns of window were originally designed for manual operation. Windows used for automatic control may require adaptation to accept motorised actuators and strengthening to accommodate the forces applied (fig. 190). Detailed guidance on window and ventilator design is provided by *CIBSE (2005)*.

Different window and louvre designs may be assessed under the following criteria:

- Ventilation Capacity. This is related to the way the ventilator opens and the surrounding head, sill and jamb details. Determining the effective area for a particular window type needs careful consideration. With top hung vents for example, the triangular opening each side of the open window is significant. In the event, an estimate is usually the basis for calculations. Fig. 191 illustrates the difference between structural opening, throw (travel), and effective opening. As a design develops, window sills, reveals, internal and external blinds have a major impact on the final effective area which is achieved. It is important to ensure that the strategy is carried through into detail design by providing continuity in the design team.
- Controllability. Where it is considered desirable and acceptable for occupants to control opening of windows, dampers or louvres, stays should be both adjustable and robust. With manual and automatic operation, the stay increases the angle of opening of the vent and the effective area will increase, but the relationship is not linear, and this will also vary according to the opening type *(ibid)*.
- Comfort. Generally, the use of opening vents results in high levels of user satisfaction, when local occupant control is part of the strategy. However, the use of the same opening for winter

190. Motorised glazed louvres with overlapping frames providing a tight seal when closed, ©naco.

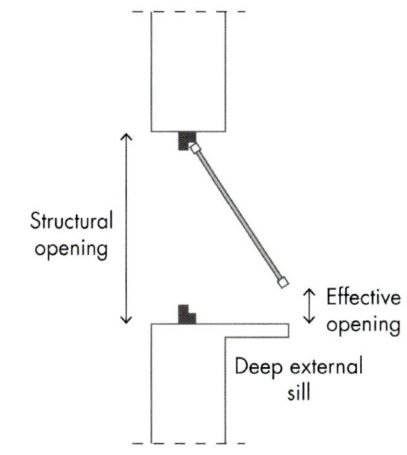

191. Section of window showing distinction between structural opening, effective opening area and travel distance.

CHAPTER 7 | COMPONENT DESIGN

and summer ventilation may be unsatisfactory in certain situations. This can be a reason for maintaining low level openings under manual operation, and high-level openings under automatic control. The provision of automatically controlled high-level window openings for night ventilation can incorporate manual override to provide occupants with a greater degree of daytime control.

- Security. It could be argued that open windows are only likely to be a security problem at ground or first floors of a building. Restricting the length of throw of stays or actuator arms may be sufficient in many situations.

> PEOPLE LIKE TO BE ABLE TO MAKE ALTERATIONS TO THEIR ENVIRONMENT, SO PROVIDING SOME MANUAL CONTROL AT LOW LEVEL IS OFTEN A GOOD IDEA

- Sealing. The sealing of windows is usually achieved by EPDM expanded rubber gaskets. In aluminium framed windows the tolerance achieved is usually very fine and good sealing is achieved. However, with some steel framed and timber windows, racking or twisting of the frame can occur, particularly with large windows, resulting in poor sealing. Motorised dampers and louvres do not shut as tight as most windows, owing to: longer crack lengths; difficulties with rotating seals; and problems with mechanical strength and closing forces. This may lead to excessive uncontrolled infiltration, and increased cooling (and heating) loads.

- Integration with Vent Actuators. The choice of window type and its integration with different actuator options and (where applicable) internal blinds requires careful consideration if the performance of one or other of these components is not to be compromised. Studies of a number of different installations *(ibid)* have revealed that one approach is to have one actuator per ventilator. If one actuator powers several ventilators then good initial alignment is vital or all the ventilators may not close tightly, leading to problems with air tightness and security. This can be aggravated by racking of the ventilator frames; by distortions from poor fixing; by flexing of the linkages, lever arms and brackets; or from shock loads from wind or at the ends of travel. On the other hand, rack and pinion actuators (linking many vents in series, as in commercial greenhouses) have been found to be robust. A common cause of poor performance with vent actuators in general can be the failure of limit switches, which are intended to switch off the drive to actuators at the end of their throw. This can cause damage to the actuator or vent frame and failure of control, and it is therefore vital that easy access for regular inspection and maintenance is provided, and that this is implemented by the building manager.

IMPROVEMENTS IN ACTUATOR DESIGN

Experience of naturally ventilated buildings over the last 25 years in different parts of the world has been mixed. There are some undoubted success stories, but also there are many where problems have arisen with building management and maintenance. This may partly relate to the need for improvements in training among facilities managers. However, where problems have occurred they have often related to the performance of actuators, their associated controls, and the linkages between them, as described in the previous section. Manufacturers and suppliers of vent actuators recognise some of the weaknesses which have been revealed (and have improved their products), but equally it is evident that the products specified must be appropriate to the situation and the control regime defined by the design team.

Many manufacturers will engage with designers at an early stage. Some, like the company 'Passivent' (*https://www.passivent.com*) also recognise that issues such as acoustic attenuation and/or humidity control often need to be part of a successful solution and produce bespoke products. There is also recognition of the need for seasonally adapted ventilation products, like automated trickle ventilators which are responsive to humidity (e.g. 'Glidevale'), and natural ventilators which can reduce particulate pollutants by means of electrostatic precipitators (ESPs), removing pollen and mould spores as well as fine dust particles. The market leaders of natural ventilators and vent actuator products have their own research departments and are involved in a continuous pursuit of improved resilience and performance.

192. Chain actuator for large, heavy windows, ©WindowMaster.

CHAPTER 7 | COMPONENT DESIGN

Aesthetic integration is also an important issue. *'WindowMaster'* (*https://www.windowmaster.com/*) supply automated natural ventilation products and solutions, some of which can be hidden within the framing to achieve a *'throw'* (or *'stroke'*) of between 50–1000mm, providing linked controls which are responsive to room temperature, CO_2 level and/or humidity (fig. 192).

The reference *CIBSE (2005)* provides a checklist for designers to review the specification of ventilators, actuators and their controls, to help meet expectations and avoid problems for the users and managers of the building.

> THERE IS A NEED FOR SEASONALLY ADAPTED VENTILATION PRODUCTS, LIKE AUTOMATED TRICKLE VENTILATORS RESPONSIVE TO HUMIDITY AND NATURAL VENTILATORS TO REDUCE PARTICULATE POLLUTANTS

7.3. SHAFTS, DUCTS AND COOL TOWERS

It is evident when exploring potential natural ventilation airflow paths through a building that vertical shafts or ducts will potentially generate much higher airflow rates when compared with single sided or cross-ventilated airflow paths. The inclusion of vertical shafts also enables the creation of deeper plans, on multiple levels, exhausting (or supplying) air at the highest level of the building. Such ducts and shafts can also help with the removal of smoke in the event of a fire, and incorporate acoustic insulation to reduce noise ingress or egress. They also provide the opportunity of promoting flow reversal, to achieve both downdraught as well as updraught ventilation. The design of these components (which typically may account for 5% of the total floor area), and in particular the design of the *'terminations'*, will therefore contribute significantly to a successful solution.

SUPPLY & EXHAUST AIR TERMINATION CONDITIONS

Updraught convective cooling requires exhaust at high level, preferably in a zone of negative pressure (suction). This can normally be achieved by locating exhaust vents about 2 metres above the general roof level, and the design of the termination needs to promote suction irrespective of the wind direction.

In designing for downdraught evaporative cooling hydraulic misting nozzles generally require a height of 2–4m for water droplets to be completely evaporated, depending on the ambient temperature and humidity. Therefore, the supply air termination must stand at least 3m above the ceiling level of the top floor of the building, if the top floor is to be served by the PDEC system (figs 193–194). Where pneumatic nozzles are used, and droplet sizes of less than 20 microns can be achieved, it may be possible to locate the nozzles close to the ceiling level of the top floor, and to avoid the need for a supply air tower. Such decisions clearly have significant design and cost implications.

It has already been established that the temperature difference between the supply air and the room air will be small (possibly only 2–3°C). In order to achieve substantial cooling, the supply and exhaust openings therefore need to be large to achieve the required air volume flow rate. Measures to baffle the wind, and to prevent the ingress of rain and possibly dust, must not be allowed to reduce the effective free area of inlet and outlet openings. The resistance to airflow of baffles, louvres and insect/bird mesh must be allowed for in the calculation of effective free area.

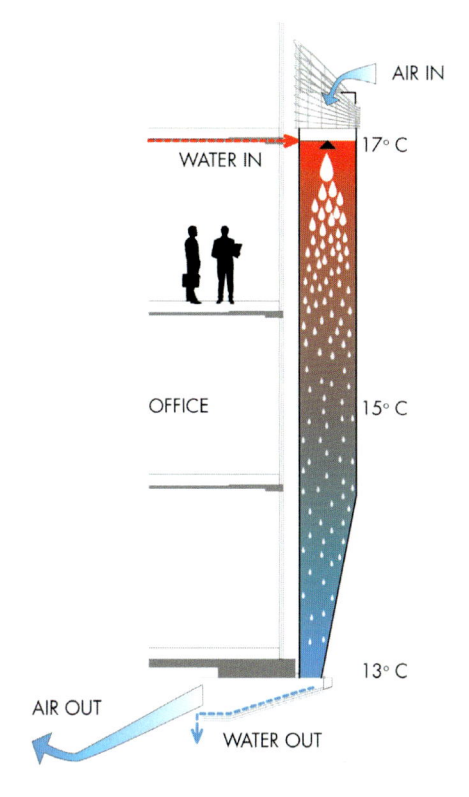

> THE INCLUSION OF VERTICAL SHAFTS ENABLES THE CREATION OF DEEPER PLANS AND PROVIDE THE OPPORTUNITY OF PROMOTING BOTH DOWNDRAUGHT AND UPDRAUGHT VENTILATION

193. Shower tower at CH2 Office Building, Melbourne [DesignInc].

194. External view of ventilation shaft of Queens Building, De Montfort University, Leicester UK [Short Ford Architects].

CHAPTER 7 | COMPONENT DESIGN

195. Ventilation shaft at the Habitat Centre, Namibia [Nina Maritz Architects].

WIND BAFFLES

As already stated, the termination of shafts or cool towers will normally be well above the roof level, in a zone of negative pressure (suction) in windy conditions. However, positive pressure at the supply entry point may be desirable to promote a daytime downdraught, and this has led to attempts to 'catch' the wind *(Pearlmutter et al., 1996)*. While there are numerous historic examples of wind-catchers in the extreme climates of the Middle East, catching the wind can also lead to over-pressure, uneven air distribution, dust and rain penetration. While traditional Middle Eastern wind catchers enhanced the air velocity (and hence convective cooling) within the building, this was highly variable and unlikely to be appropriate for a contemporary office environment. The contemporary application of downdraught cooling with misting nozzles can deliver consistent high airflow rates through buoyancy alone, and wind effects should be inhibited rather than encouraged.

196.

197.

196. Supply and exhaust air terminations at TRC, Ahmedabad, India [Abhikram Architects].
197. Half round pipe baffle detail at TRC, India.

CHAPTER 7 | COMPONENT DESIGN

An assessment of the microclimate of the site will confirm whether the wind and associated driving rain or dust penetration are likely to be a significant factor when developing the detailed design of downdraught cooling supply towers. Much will depend on the location and the building type (for example, a low rise building within an urban area is unlikely to require wind baffles). However, when wind baffles are required to deflect the wind and prevent it from counteracting the desired airflow path, experience suggests that louvres on their own may not be sufficient. The baffles specified for the Torrent Research Centre, Ahmedabad, were vertical half-round pipes, which help to turn the wind back on itself to further reduce the localised wind velocity. The half-round pipes were located outside fixed louvres and an inner layer of insect mesh, as shown in figs 195–197.

> WHEN WIND BAFFLES ARE REQUIRED TO DEFLECT THE WIND AND PREVENT IT FROM COUNTERACTING THE DESIRED AIRFLOW PATH, EXPERIENCE SUGGESTS THAT LOUVRES ON THEIR OWN MAY NOT BE SUFFICIENT

7.4. CONTROLS

Control systems, either manual or automatic, should be kept as simple as possible, consistent with the successful operation of the building. Training is required for the owner's representative or facilities manager, but too often insufficient time is given to commissioning of systems prior to occupancy of the building, often resulting in a frustrating and extended period during which all concerned learn how the building works.

The amount of air flowing through a given opening in a given position can also vary widely, as the driving forces (wind, natural buoyancy) change with indoor and outdoor conditions. In addition, the internal resistance to cross ventilation will vary as doors etc. are opened and closed. With the wide variations in the amounts of air needed and the associated driving forces, it can be difficult to achieve the full range of requirements using the same ventilation device. For example, the large openings required for downdraught cooling in summer may need to be closed in winter with only minimum fresh air supplied. Also, while people may welcome air movement for its additional cooling effect in summer, they will complain of draughts in winter unless the incoming air is either well-diffused or pre-heated.

While high level openings may need to be automated, low level openings are preferably left in the control of the occupants. Whether automated or manually operated, it is vital that the status of all ventilators can be easily confirmed.

FLOW RATES THROUGH VENTILATORS

The flow rate through a ventilator is a function of the open area, but often this is not directly proportional to the actuator movement. In particular, the thickness of the ventilator and the way it fits into the building fabric can greatly affect the relationship. With a centre pivot or hinged window, the lower opening can be much restricted by the frame, the window sill and the reveal, particularly for the first 100mm or so of travel. This can provide useful fine control, but it may restrict the required opening area for summer ventilation, particularly if safety considerations also restrict the permissible amount of window opening. For centre pivot or bottom-hung hopper windows, the upper opening may also be restricted by the reveal and by any internal blinds. External shading may also restrict flow rates and openings (it may also pre-heat the outside air, particularly if it is dark in colour and of a metallic finish).

COMBINED USER AND AUTOMATED CONTROL

Occupants need to be informed about how the building is designed to work, so that the goals of comfort, air quality and energy efficiency are all achieved (fig. 198). Ideally, the strategy should be clear, straightforward and as far as possible intuitive. The control requirement will often change with the context for the user, for example, occupants sitting beside a window are most affected by the draughts from the lower

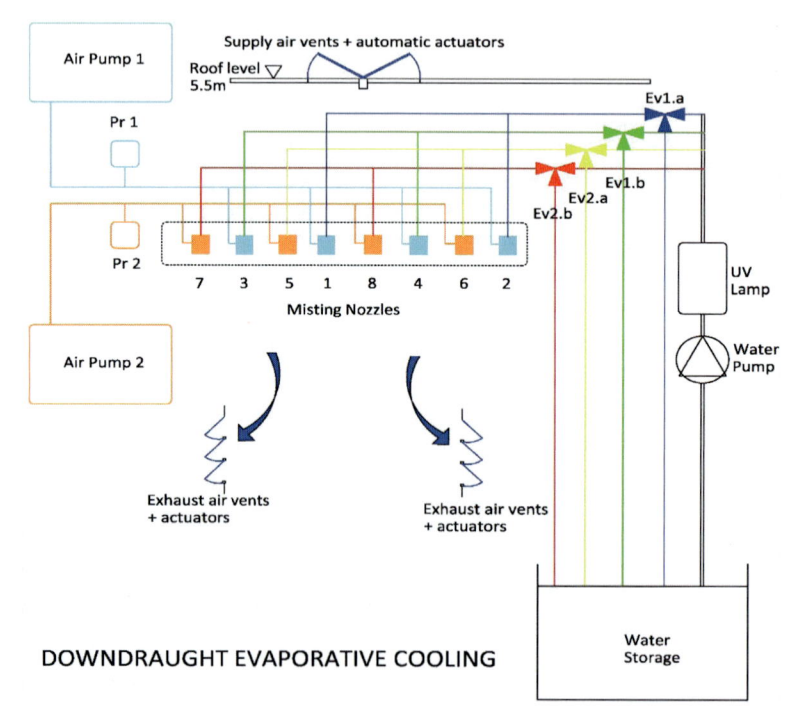

198. Water and air pipework schematic for Solar Decathlon Nottingham HOUSE in Madrid, ©Ingeniatrics/Frialia.

casements, while those in the centre of the room are more likely to want to adjust the high level vents.

Suitable combinations of automated and manual control will not only provide backup in the event of BMS malfunctions, but will also empower the occupants, improving their adaptive opportunities, and making variations in conditions more acceptable than when there are no over-ride facilities. Window-type ventilators are often used at high level, but even where they are visible, it is not always easy to tell whether they are tight shut or a little bit open. In such situations, status lamps or other forms of indication are helpful. Dampers or louvres are frequently hidden from view: status feedback then becomes particularly important. Doing without status feedback is a false economy.

> CARE MUST BE TAKEN TO AVOID MICRO-BIOLOGICAL CONTAMINATION, BUT AS LONG AS SIMPLE MEASURES ARE TAKEN THIS NEED WILL NOT BECOME A PROBLEM

Maintenance

Lack of familiarity with the systems and the cost of seasonal maintenance appears to be the reason for some (previously successful) installations to fall into disuse. Building owners and facilities managers should be made aware that since these cooling facilities operate on a seasonal basis, they may require attention at the start and end of the cooling season, so that it is built into the maintenance schedule for the building as a whole.

7.5. WATER SYSTEMS

A large part of this book is devoted to the design of direct evaporative cooling systems for buildings. The different techniques available to promote evaporative cooling vary in the quality of water required to deliver evaporative cooling and are characterised by different rates of water consumption. These issues, and the treatment required to ensure the risk of micro-biological contamination is minimised, are reviewed in the following section.

WATER CONSUMPTION IN EVAPORATIVE COOLING

Like energy, water is a resource whose use must be managed carefully. This is especially true in the environments most suited to application of evaporative cooling in hot dry climates. The amount of water used will vary substantially, depending on building use as well as climate. In addition, water consumption in an evaporative cooling installation depends on:
- The volume of air supplied to the space being cooled.
- The thermal properties of this air.
- The amount of water that must be evaporated to bring a unit volume of the air from its initial condition to the specified supply condition.

Analysis of the microclimate (as described in Chapter 2 and 3) will reveal the range of dry bulb temperature and humidity which is likely to prove acceptable within the building, within the context of adaptive thermal comfort theory. After internal and external heat gains have been estimated for a *'typical'* summer day condition then the air volume flow rate required to remove heat gains can be estimated (as described in Chapter 5). The amount of water required by misting nozzles to meet the cooling load can then be calculated by determining the difference in absolute humidity as described in the section on *'misting systems'* in Chapter 5.

WATER TREATMENT

The need for filtration and other water treatment measures obviously depends on local water quality, as well as the type of evaporative cooling system envisaged. One of the benefits of using the 'shower tower' principle is that water quality is not an issue, although scale build-up is likely to occur in hard water areas.

Care must be taken to avoid micro-biological contamination, but as long as simple measures are taken this need will not become a problem. *ANSI/ASHRAE (2018)*, establishes minimum legionellosis risk management requirements for building water systems. Following the provisions of this recent standard will mitigate the risk of microbiological contamination. Generally, making provision for systems to drain back when switched off and for water supply to include UV filters in the supply to the nozzles will need to be included. Water generally should be taken from the mains supply, where the temperature is unlikely to be conducive to growth of Legionella or other organisms (which require a temp of approx. 32°C). Where storage is necessary, then water must be kept well below this temperature. A regular programme of inspection and maintenance must be implemented by an appropriately trained building or facilities manager.

LOW AND HIGH PRESSURE MISTING SYSTEMS

Misting nozzles which rely on hydraulic pressure alone require a high pressure (25–50 bar) to create droplets of 30 microns or less. Hydraulic nozzles tend to be made of either brass or stainless steel, requiring appropriate pumps and pipework with compression fittings. Such a system can be relatively expensive to install, but can then run successfully for many years, requiring relatively little maintenance (as at the Torrent Research Centre in Ahmedabad).

The alternative of using pneumatic technology (compressed air blown through nozzles supplied with a low pressure water supply) can use relatively low cost plastic nozzles to create droplets of less than 15 microns in diameter. Pneumatic technology requires compressors (instead of pumps), but does have a number of advantages over hydraulic systems, including:

- Relatively low cost system components (namely pumps, pipes, connections and nozzles) when compared to those in the high pressure systems.
- Generally higher robustness of the system as a whole; on the one hand, the lower pressure puts the system into

much less stressful conditions; on the other hand, the wider orifice used in pneumatic nozzles reduces the risk of clogging.
- No final jet *'spitting'*; high pressure systems tend to *'spit'* a final short water jet when they are shut down as a result of the decrease in pressure. This issue makes the system unfit for many applications where these *'drips'* may fall from above onto people or equipment below.

A pneumatic system was used in the University of Nottingham's House for the 2010 Solar Decathlon in Madrid (described in Chapter 4). The system made use of a novel nozzle design (by Frialia Microclimas of Spain) that combines the motive energy from a pressurised water circuit with a compressed airflow (figs 199–200). This provides the desired level of atomisation of the water flow and avoids the creation of drips when the nozzles start up and stop. The system requires energy to drive a pump and small air compressors, to run an ultraviolet water treatment cell and to operate the window controls. Peak power requirement of the installed system (pumps, UV filters and vent controls) was approximately 700W, but this varies with the number of nozzles in operation. There were eight nozzles installed and peak water consumption per nozzle was 5 litres per hour if run continuously. In practice it was found that a maximum of six nozzles met the cooling requirement during occupied periods, and the system was run intermittently to avoid over-humidification.

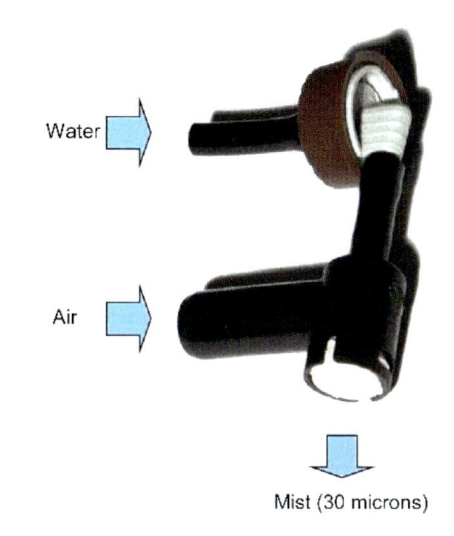

199. Example of pneumatic nozzle (plastic).
200. Pneumatic nozzles in operation below rooflight in Nottingham HOUSE, Madrid, for the Solar Decathlon Europe 2010.

CHAPTER 7 | COMPONENT DESIGN   173

MAINTENANCE OF NOZZLES

At Torrent Research Centre (TRC), where 316 stainless steel high pressure pipework and nozzles were installed, they soften and filter the water to reduce scale build up. The facilities manager at TRC (an engineer who has been with the project since its inception) has reported that blockage of nozzles at TRC has not been a problem (they are still using the same nozzles they installed 10 years ago). A monthly maintenance check includes removal and cleaning of any blocked nozzles and they have even developed a simple tool which enables them to unscrew nozzles from the walkways below to allow cleaning and replacement.

In most naturally ventilated and passively cooled buildings the building form (plan and section) and its construction will have a significant influence on the success of the cooling strategy. The previous chapters have emphasised the role the design team plays in defining a successful strategy and then implementing it through detailed design and construction. While much of a natural ventilation design is therefore 'bespoke' many components for natural ventilation and passive cooling are familiar products which are widely available and suitable for many different locations. This chapter has reviewed the design and specification of some of these components to highlight the importance of detailed design in contributing to the delivery of workable and robust solutions.

REFERENCES

- ANSI/ASHRAE (2018). 'Standard 188–2018. Legionellosis: Risk Management for Building Water Systems'. American Society of Heating, Refrigerating and Air-Conditioning Engineers, Atlanta.
- Bordass, W. et al. (1994). *'Control Strategies for Building Services: the Role of the User'*. In Buildings and the Environment Conference, Building Research Establishment, Garston, Watford.
- CIBSE (2005). *'Applications Manual 10: Natural Ventilation in Non-Domestic Buildings'*. London: Chartered Institution of Building Services Engineers.
- Pearlmutter, D. et al. (1996). *'Refining the Use of Evaporation in an Experimental Down-Draft Cool Tower'*. Energy and Buildings 23, 3: 191–197.

PART 2

INTRODUCTION

In Part 2 the performance and occupant perception of buildings employing various combinations of Passive Downdraught Evaporative Cooling (PDEC) and Active Downdraught Cooling (ADC) is presented. The feedback obtained on the selected case studies evidence the challenges and requirements associated with the design and operation of buildings which pioneer alternative forms of cooling and break free, to some extent, from the conventional assumption that non-domestic (mainly office) buildings will require air-conditioning almost irrespective of climate and location. The study, conducted in a specific moment in time, aims to derive lessons from the challenges and opportunities associated with the operation of a relatively innovative form of cooling, which is intrinsically linked to the architectural design and the building environmental design strategy. Hence the authors hope to highlight the fundamental link that exists between strategic design and building performance and which every design team should be aware of. Moreover, it is important to note that the metrics for such an evaluation can only be found in the ultimate recipient: the building user. For this purpose, a holistic and multi-criteria user satisfaction survey was used to evaluate the building performance and identify areas for improvement from which to derive lessons for future best practice.

Six case study buildings are presented in some detail and they include: one building in Europe, three in the USA, one in India and one in China. The buildings are presented according to their geographic location, climatic context and cooling typological classification. The location of the case study buildings is shown in fig. 01. Four of the six cases presented were part of a study that used Post Occupancy Evaluation (POE) surveys of occupant perceptions as part of an EU funded research project (Ford et al., 2010). The other two case studies, the University Building in Ningbo, China and the Torrent Research Centre in India, were reviewed based on POE studies conducted by others. The study of the three US cases entailed a combination of literature review, review of predicted performance and fieldwork. The fieldwork included on-site visits, spot measurements, and interviews with the building manager, correspondence with members of the design team, and an occupants' satisfaction survey. The study of the Malta Stock Exchange involved a review of design stage performance prediction, measurements made during commissioning and a post occupancy survey. Additional information such as energy consumption was obtained when possible from in-house monitoring and energy bills. Monitoring of environmental variables such as temperature and RH was not the main objective of the study, however, when possible either spot measurements were made or continuous monitoring was sourced from previous field studies.

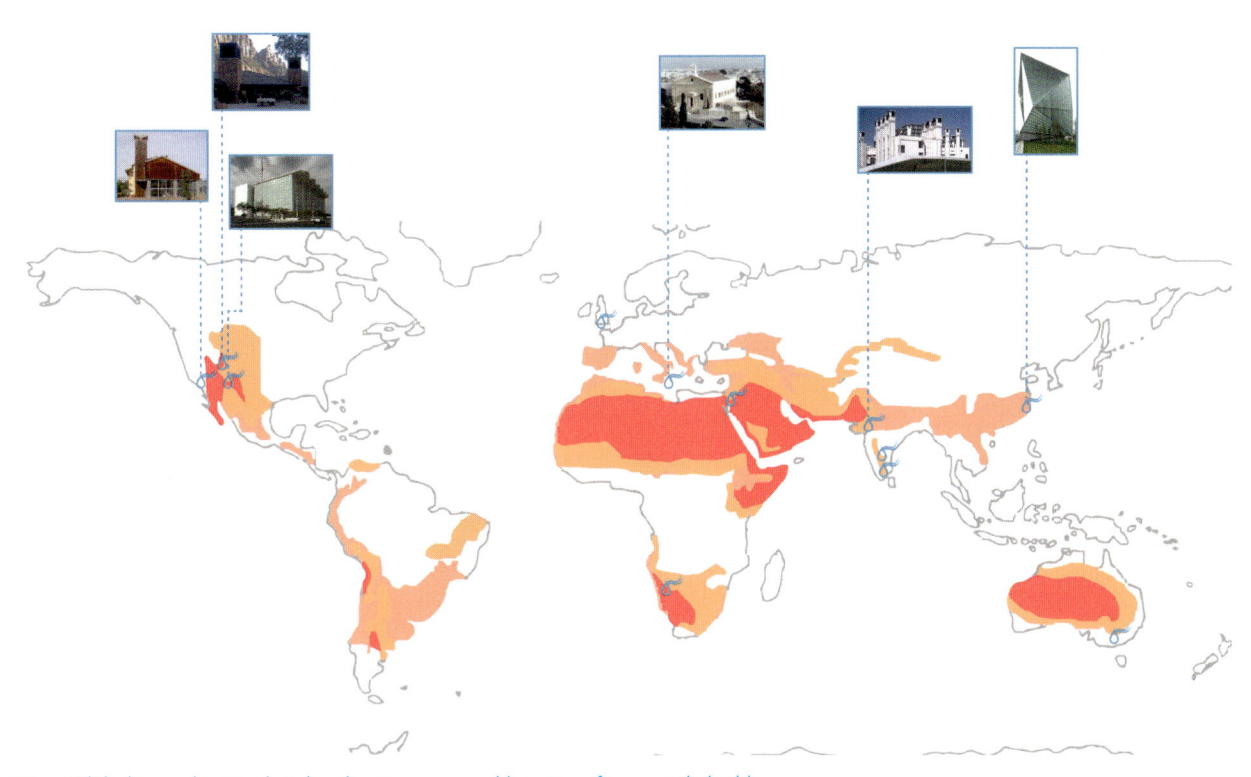

01. Global map showing hot dry climate zones and location of case study buildings.

INTRODUCTION   177

The occupants' surveys were undertaken using the Workplace Questionnaire developed and licensed by the Building Use Studies (BUS), UK *(Bordass et al., 2006)*. The BUS occupant questionnaire and methodology is described by *Leaman (1997)* and has been extensively employed since the PROBE (Post Occupancy Review of Buildings and their Environment) project between 1995 and 2002 *(Bordass, 1999)*. Over the years, as a result of the application of this survey tool and methodology, the BUS has collated a database of hundreds of surveyed buildings worldwide, including air-conditioned, mechanically ventilated and naturally ventilated buildings. This generated the dataset against which the POE results of each survey are compared. The questionnaire produces a comprehensive evaluation of the occupants' perceptions of the building, filtered through their 'historic' memory of the building environment over the duration of their occupancy, and does not relate to a specific set of contingent environmental conditions.

The questionnaire includes questions on environmental variables such as: temperature (in summer and winter), air movement and quality, lighting, noise, but also on other variables of building performance such as occupants' perception of design, perceived health, productivity and image to visitors. The survey's responses for each variable is summarised in a mean value which can be assessed in relation to the variable's specific scale or to the BUS database benchmark. The responses to the variable are expressed on a scale 1 to 7, where 1 is worst and 7 is best and graphically represented by shaped symbols (squares represent mean values significantly better than both benchmark and scale midpoint, diamonds are mean values significantly worse than both benchmark and scale midpoint and circles represent mean values no different from both benchmark and scale midpoint). The benchmarks are highlighted by the small rectangle on the top scale of each variable.

For this study, particular emphasis was placed on analysing responses related to the summer performance, and aspects such as perceived thermal comfort, air quality and control. During the visits, the building manager's views of the cooling system were recorded in relation to characteristics, performance and maintenance requirements.

REFERENCES

- Bordass, W. (1999). *'PROBE Strategic Review 1999: Final Report 4 – Strategic Conclusions'*. The PROBE Team 1999 [Online]. https://usablebuildings.co.uk/. [Accessed 01/2018].
- Bordass, W., Leaman, A. & Eley, J. (2006). *'A Guide to Feedback and Post-Occupancy Evaluation'*. The Usable Buildings Trust 2006 [online]. https://usablebuildings.co.uk/. [Accessed 01/2018].
- Ford, B., Schiano-Phan, R. & Francis, E. (2010). *'The Architecture & Engineering of Downdraught Cooling: a Design Sourcebook'*. PHDC Press.
- Leaman, A. (1997). 'PROBE 10: Occupancy Survey Analysis'. BSJ: 37–41 (May 1997).

CASE STUDY 1
TORRENT RESEARCH CENTRE, AHMEDABAD, INDIA

The Torrent Research Centre is a pharmaceutical industrial development in Ahmedabad, India. The client, Torrent Pharmaceuticals Ltd, commissioned the building from Abhikram Architects (Nimish Patel & Parul Zaveri) who worked initially with Short Ford Architects and subsequently Brian Ford & Associates (London) as environmental design consultants, and Dastur Engineering (New Delhi) for building services. This design team developed an integrated approach to the application of environmental strategies within the architectural design of a typical laboratory block, including a downdraught evaporative cooling strategy. With advice and support from Solar Agni International of Pondicherry, and with detailed design support from in-house engineer SB Namjoshi, this strategy was then applied to the majority of the first phase buildings which comprised five linked laboratory buildings on three floors connected to a central administrative complex, designed to connect with a future second phase. Two of the five laboratory buildings were conventionally air-conditioned for process reasons, but the other three (plus the admin block) are entirely dependent on naturally driven convective cooling. The first buildings were completed and occupied in 1998, while the whole of the first phase was completed in 2000. This project, inspired in part by the cool towers of the Seville World's Fair *(Alvarez et al., 1991)*, was the first large scale application of downdraught evaporative cooling in India *(Ford et al., 1998)*.

02. External view of the Torrent Research Centre, Ahmedabad, India [Abhikram Architects].

LOCAL CONTEXT

The Torrent Research Centre is located on the edge of Ahmedabad within a large 'science park' setting beside the Sabarmati river. Ahmedabad (Lat. 23°N, Long. 72°E) is the major commercial and industrial city in Gujarat, North-West India, and has historically been associated with the production of textiles. Le Corbusier's *'Mill Owners'* building is located beside the Sabarmati river a few miles away.

The rapidly expanding modern city continues to be a vibrant commercial centre with a global reach. International corporations have tended to locate at the periphery of the city, avoiding the congestion and pollution of the centre. In the early 1990s, and even more now, energy was a major issue, as demand outstripped supply, resulting in frequent power cuts, and the need for major businesses to own their own back-up diesel powered electricity generators (with related high capital and running costs). Although the energy supply situation is improving, the need to also improve the energy efficiency of buildings is now widely appreciated.

The local climate is characterised by extremely hot dry summers (March to mid-June) when peak temperatures can regularly reach 45°C, mean max temperatures reach 39.5°C and the daytime wet bulb temperature can go as low as 21°C (mean min in May). The hot summer is followed by a warm (32°C peak) and humid monsoon (late June to September), brought by the prevailing south-west winds (fig. 03). To cope

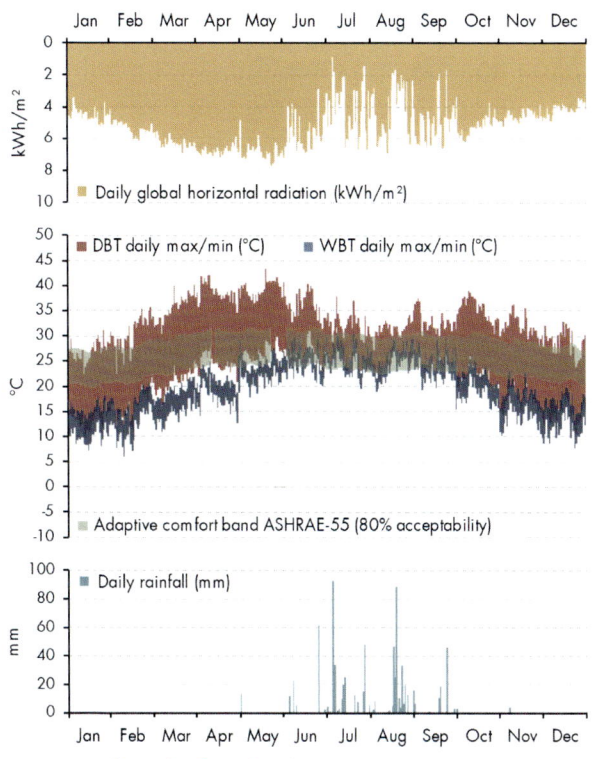

03. Typical weather for Ahmedabad, ®Meteonorm.

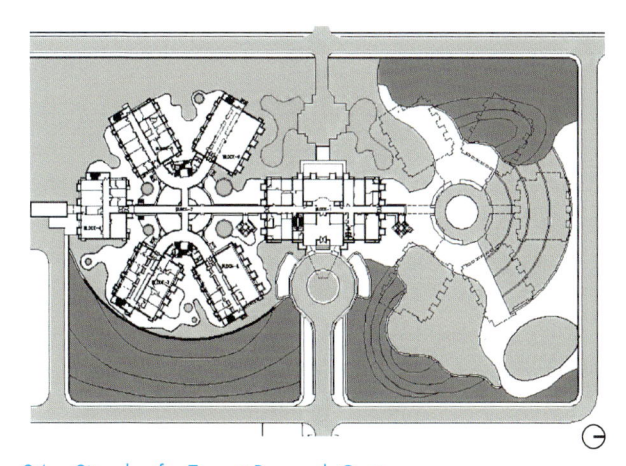

04. Site plan for Torrent Research Centre.

CASE STUDY 1 | TORRENT RESEARCH CENTRE, AHMEDABAD, INDIA

with these conditions, conventional air-conditioning has become the norm, although this places a huge burden of both capital and running costs on business. Alternatives to conventional air-conditioning therefore present a major challenge and opportunity.

THE BUILDING AND OVERALL STRATEGY

The brief required over 22,000m² of pharmaceutical research laboratories and offices, including 'unclean' areas like chemical synthesis laboratories, to 'very clean' areas like tissue culture, molecular biology and drug design areas (fig. 04). A conventional approach would suggest that over 50% of such a new laboratory building would require refrigerant based air-conditioning, in order to meet the environmental requirements within the laboratories. However, the client, aware of the compelling benefits of achieving improved energy efficiency in this project, sought to minimise dependence on conventional air-conditioning without compromising occupant comfort. In the final complex approximately 60% of the laboratories plus all adjacent administrative areas are cooled with PDEC.

> 60% OF THE LABORATORIES PLUS ALL ADJACENT ADMINISTRATIVE AREAS ARE COOLED WITH PDEC

Part of the energy strategy was to maximise the use of daylighting, while avoiding the risks of overheating through solar gain. In addition the architects used locally available materials wherever possible. The external masonry walls and roof incorporate the natural mineral 'vermiculite' to reduce thermal conduction, although by European standards the U-value is not very low. Half round ceramic pipes are used in the outer face of the inlet and exhaust shafts to reduce entry of the larger dust particles by creating local turbulence. Motorised louvres are located at roof level within the main shafts to seal the building in the event of a dust storm. The high-pressure water supply system (now run at approximately 40 bar) & installation of misting nozzles were designed by the resident engineer SB Namjoshi and Ramesh Borad, who has subsequently also been responsible for operation and maintenance.

ENVIRONMENTAL STRATEGY

Typically, laboratories and offices are arranged on three levels either side of an open concourse, which allows the circulation of people between spaces (fig. 05). This arrangement allows evaporatively cooled air to be introduced (via three 4m x 4m towers) to the occupied spaces at each level, and exhausted via perimeter stacks. The perimeter stacks are designed to exhaust air at both high and low level, depending on the existence (or absence) of wind. In this way, the working areas are both physically and thermally buffered from the external environment. The building envelope has also been designed to minimise solar heat gain while allowing diffuse and reflected daylight penetration.

Summer strategy: During the day from late February to mid-June the building is naturally ventilated (by stack or wind pressure differences) until the outside air temperature is several degrees above internal, when the misting nozzles are activated to induce a downdraught of cool air from the central towers into the concourse (fig. 06). The cool air drops either side of the circulation 'bridges' in the concourse and is drawn via 'hopper' vents in the concourse wall (fig. 08), through the laboratory spaces, to louvred vents in the stacks. At night the airflow path is reversed to flush out any residual daytime heat gains (fig. 07).

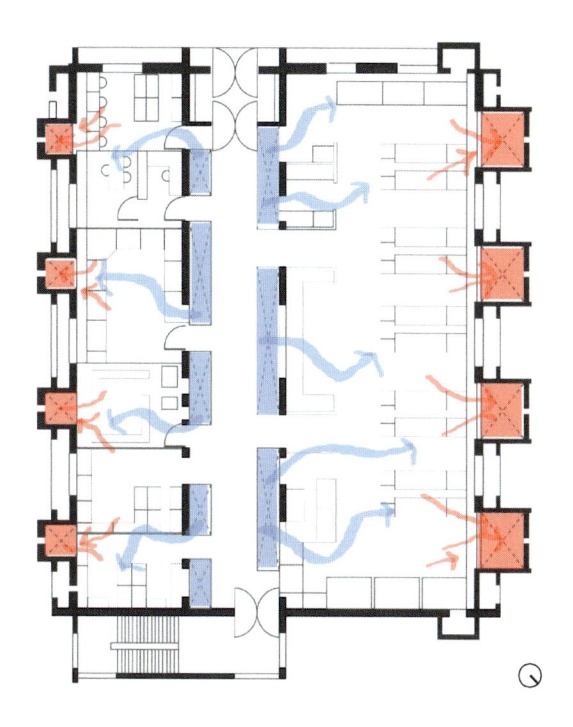

05. Plan of typical laboratory block of TRC, Ahmedabad, India.
06. Cross section of a typical TRC lab building showing summer daytime evaporative cooling strategy.
07. Cross section of a typical TRC lab building showing night-time convective cooling strategy.

CASE STUDY 1 | TORRENT RESEARCH CENTRE, AHMEDABAD, INDIA

08. Laboratory concourse with hopper vents to scoop cool air into the offices and labs.

Monsoon strategy: When the monsoon breaks in mid-June, the external temperature drops significantly, but the humidity rises. In these conditions, the client was happy initially to rely on air movement induced by natural ventilation alone. However, after the first year, following staff comments of *'stuffiness'*, it was decided to augment natural ventilation with the use of ceiling mounted low-speed fans.

Winter strategy: During the day, natural ventilation (from low to high level) is able to remove internal heat gains, while at night vents at the top of the building are closed to prevent excessive heat loss.

The definition of acceptable threshold conditions was the subject of extensive discussions between the consultants and the client, led by Dr C. Dutt, Torrent's Director of Research. Having discussed the reportedly more flexible attitude of the occupants of *'free-running'* buildings to variations in temperature and humidity *(Thomas & Baird, 2006)*, he agreed to the concept of designing for a threshold temperature of 28–28.5°C in still air conditions, which could be exceeded for a certain number of hours, rather than some absolute value. This was in line with an *'adaptive'* approach to thermal comfort. The occupants' own perceptions of the acceptability of the internal environmental conditions are discussed below.

COOLING SYSTEM AND BUILDING INTEGRATION

Each laboratory building has a similar 22m by 17m plan, with a 4m wide corridor flanked by 5m deep office spaces and 8m deep laboratory spaces (fig. 05). Two of the five laboratory buildings are air-conditioned, while the other three are equipped with the PDEC system. The larger main administrative building is located to the north of the laboratories, and a utilities building to the south, with a two level corridor spine linking all the buildings. The entire complex covers 22,600m² of floor space, of which only 3,200m² is mechanically air-conditioned. The central plant includes two oil fired steam boilers with a capacity of 4T/hr each, two 175cfm air compressors, two 725KVA diesel generator sets, and some 350 tons of refrigeration capacity.

Overall control of solar heat gains is achieved by careful design of the glazing and by external insulation of the building envelope. The fixed windows (controlled natural ventilation occurs via the perimeter shafts) are shaded by horizontal overhangs, and in the vertical plane by the air exhaust towers which project from the façade. The buildings are thermally massive: the reinforced concrete construction has plastered cavity brick infill walls and hollow concrete blocks filling the roof coffers plastered inside. Vermiculite is used as an insulating material on both roof and walls. External surfaces are white, while a china mosaic finish has been applied to the roof.

During the hot dry season mid-afternoon outside temperatures regularly reach above 40°C, while relative humidity is often below 20%. It is under these conditions that the PDEC system is designed to operate. Filtered water is pumped through nozzles at a pressure of 50Pa to produce a fine mist at the top of the three large (4m x 4m) air intake towers located above the central corridors of each laboratory building (fig. 09). Evaporation of the fine mist serves to cool the air which then descends slowly through the central corridor space via the openings on each side of the walkway (fig. 06). At each level, sets of hopper windows designed to catch the descending flow divert some of this cooled air into the adjacent space. Having passed through the space, the air may then exit via high level glass louvred openings which connect directly to the perimeter exhaust air towers. Exhaust air then either rises to the top of the stacks (in windy conditions) or falls out of the bottom of the stack (in still air conditions).

> **AT EACH LEVEL, SETS OF HOPPER WINDOWS DESIGNED TO CATCH THE DESCENDING FLOW, DIVERT SOME OF THE COOLED AIR INTO THE ADJACENT SPACE**

It has been noted *(Thomas & Baird, 2006)* that all of the five original laboratory buildings were originally designed for an occupancy of 25 scientists (approximately 15m²/p). With the expansion of activities, increase in staff and overlapping shifts in recent years, some of the buildings accommodate as many as 70–80 (approximately 5m²/p) people working at the same time.

CASE STUDY 1 | TORRENT RESEARCH CENTRE, AHMEDABAD, INDIA

09. Hydraulic nozzles and pipework at high level in central tower. Note 'khus' mats hung in opening – this was a temporary measure before half-round pipes were installed as a wind-shield.

INTERNAL MAXIMUM TEMPERATURES ARE MAINTAINED 12 TO14 DEGREES BELOW THE EXTERNAL PEAK

This will have significantly increased the internal heat gains, and therefore the cooling load. However, the building appears to have met this challenge without a significant rise in complaints from staff.

MEASURED PERFORMANCE

A number of studies have reported on the monitoring of thermal performance during the summer period. Monitored temperatures in 1997 and 1998 had indicated that internal maximum temperatures could be maintained 12 to14 degrees below the external peak and that internal temperatures were around 5 degrees lower than average external temperatures. Soon after the first labs were occupied, temperatures were reported of 27°C on the ground floor and 29°C on the first floor with outdoor temperatures at 38°C, and air change rates of 9 per hour on the ground floor and 6 per hour at first floor over the same period. The staff reported that during the summer (February to June) the laboratories are comfortable without fans and are not stuffy or smelly, as most chemistry labs are, even when air-conditioned. *Majumdar (2001)* reports temperatures of 29–30°C being achieved when outside temperatures reach 43–44°C. Majumdar also reported temperature fluctuations did not exceed a 4 degree range over any 24 hour period, when temperature fluctuations outdoor were as much as 14–17°C. One of the early issues noted was a tendency for air to by-pass the top floor. Water consumption for the PDEC system is reported to be approximately 5–6,000 litres/day (approximately 0.3l/m²·day).

After the first year of operation the completed building was reported to have used approximately 64% less electrical energy than the equivalent air-conditioned building (equivalent to approximately 66kWh/m²·year). Thomas & Baird report that in 2005, overall energy consumption for all four PDEC buildings plus two air-conditioned buildings was 54kWh/m². This can be compared with the typical energy consumption in Indian commercial buildings which has been reported to be in the range of 280–500kWh/m². This also compares very favourably to the target for fully air-conditioned office buildings under the recently introduced environmental rating scheme TERIGRIHA (140 kWh/m²·year for day use office building in a 'composite' climate). Thomas & Baird point out that *"not only is the Torrent building located in a 'hot dry' climate which is more demanding of energy for space conditioning than would be the case in a 'composite' climate, the building has equipment loads which are higher than typical offices and it is used over longer hours than typical office buildings"*.

OCCUPANT PERCEPTION

A survey of occupant perceptions was undertaken in 2004 (8 years after the building was opened. Questionnaires developed by the Building Use Studies were circulated to occupants of both PDEC lab buildings and air-conditioned lab buildings, which elicited 164 responses. Most of the respondents had worked in the building for longer than one year.

A total of 292 surveys were distributed and 164 responses returned, 64 from the AC blocks, 100 from the PDEC blocks. Thomas and Baird report a limited difference in terms of the demographics in the AC and PDEC buildings. Respondents in both groups were predominantly male and the majority were under the age of 30. Most of the respondents had worked in the building longer than one year, and roughly a quarter to a third of respondents reported that they were seated next to a window. The offices and labs were predominantly open plan with the exception of senior management.

Results of the survey (figs 10–11) indicate that both the PDEC buildings and the AC buildings returned mean scores that were *"significantly better or higher than both benchmark and scale mid-point for all of the categories"*. The survey points to *"consistently positive responses with respect to international benchmarks and scale mid-point"*. Thomas and Baird also point out that the co-location of passive downdraft evaporative cooling (PDEC) and air-conditioned (AC) blocks at Torrent offers a unique opportunity to compare performance.

Occupant perceptions of internal temperatures are clearly important in assessing the success of the PDEC system, and the overall results for Temperature Air and Comfort for PDEC in Thomas and Baird's survey corroborates an earlier report by Majumdar that *"comfort conditions have not been compromised".*

The summary table indicates that overall occupants have a high level of satisfaction in terms of Comfort, Air Quality and Temperature in the PDEC buildings. In all cases these are above the benchmarks, and compare favourably with the results for the air-conditioned buildings, the conditions of which are also favourably viewed by the occupants. These results for the PDEC buildings are particularly significant in the context of acceptable temperature ranges that are higher than those deemed acceptable in the air-conditioned buildings. These results would tend to confirm the results of field studies undertaken by *Nicol et al. (1999)* amongst office workers in Pakistan, where office workers were found to be comfortable at temperatures between 20 and 30°C with no cooling apart from fans.

Although the air-conditioned buildings produced marginally better results than buildings incorporating the passive downdraught evaporative cooling systems in the BUS survey, it is important to note that the BUS results for the PDEC buildings were also consistently better than international benchmarks and scale mid-points.

Thomas & Baird report that the occupants consistently rated temperatures on the colder side of neutral (mid point in a scale of too hot–too cold) in the air-conditioned buildings. Given that the controlled air-conditioned labs are maintained at temperatures

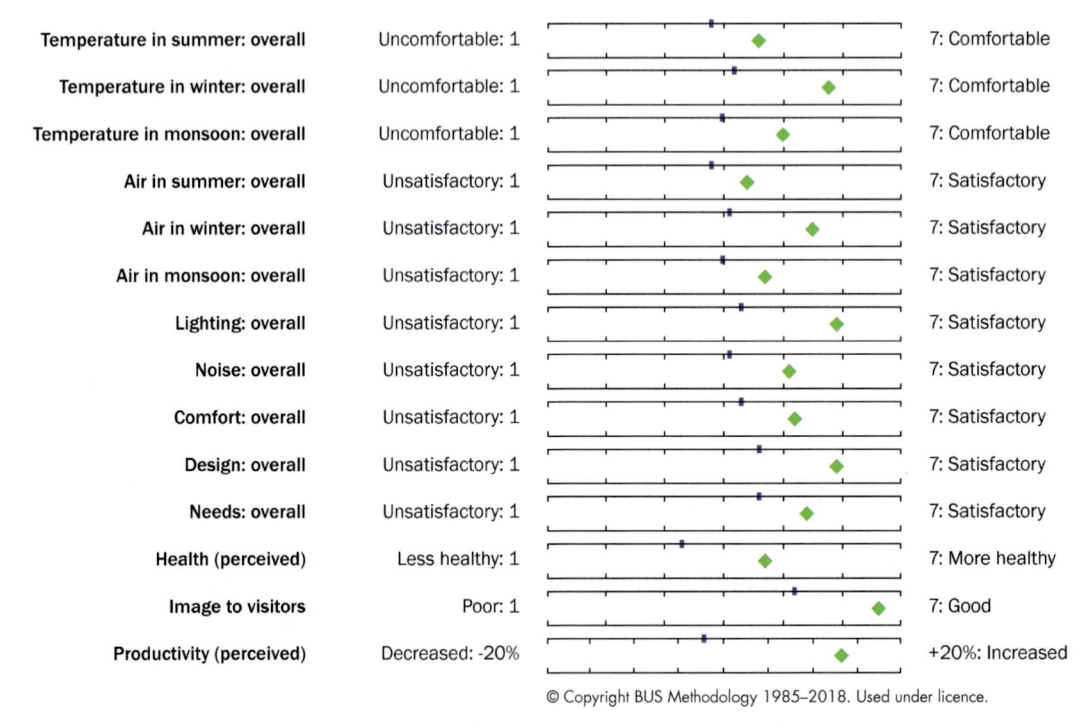

10. Occupant Perception Summary Chart for PDEC Buildings at TRC, ©Usable Buildings Trust.

around 22–24°C in comparison to generally 5 degrees below the outside mean temperatures in the PDEC buildings, the occupants' overall perception is that the PDEC buildings are only marginally less comfortable than the air-conditioned buildings.

Performance in the monsoon season was of particular interest in the PDEC buildings. As noted earlier, following the first year of occupancy ceiling fans were installed as a consequence of the experience of 'muggy conditions' in the building during the monsoon. Occupant responses for this season were generally positive, while occupants experienced some concern about air humidity, and moderate satisfaction with overall air conditions in monsoon. Mean scores for occupant votes regarding acceptable temperatures during the monsoon were close to neutral, indicating their view that, following the installation of the ceiling fans, temperatures were satisfactory.

Open ended comments regarding comfort and ventilation, arising from both PDEC and AC blocks, were predominantly positive. Comments from the PDEC surveys included *"everything in this building is well equipped for work and comfort"* (PDEC) *"satisfactory, well ventilated good infrastructure"* and *"good ventilation"*.

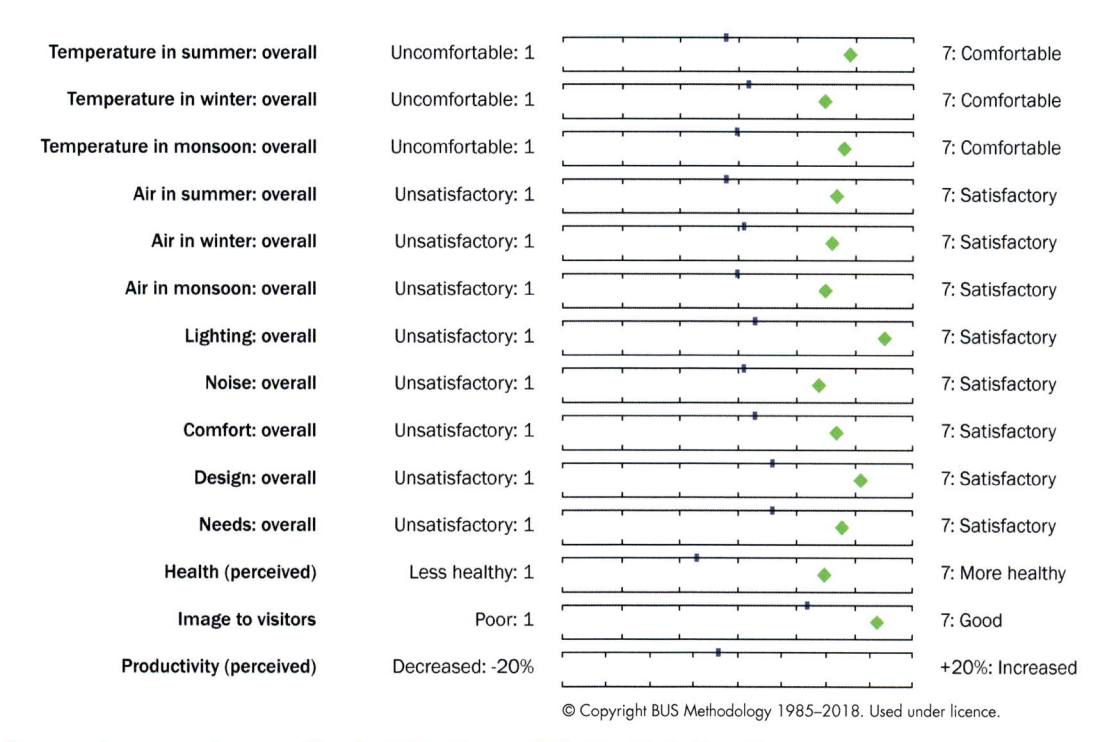

> **OVERALL OCCUPANTS HAVE A HIGH LEVEL OF SATISFACTION IN TERMS OF COMFORT, AIR QUALITY AND TEMPERATURE IN THE PDEC BUILDINGS**

Category	Scale Low	Scale High
Temperature in summer: overall	Uncomfortable: 1	7: Comfortable
Temperature in winter: overall	Uncomfortable: 1	7: Comfortable
Temperature in monsoon: overall	Uncomfortable: 1	7: Comfortable
Air in summer: overall	Unsatisfactory: 1	7: Satisfactory
Air in winter: overall	Unsatisfactory: 1	7: Satisfactory
Air in monsoon: overall	Unsatisfactory: 1	7: Satisfactory
Lighting: overall	Unsatisfactory: 1	7: Satisfactory
Noise: overall	Unsatisfactory: 1	7: Satisfactory
Comfort: overall	Unsatisfactory: 1	7: Satisfactory
Design: overall	Unsatisfactory: 1	7: Satisfactory
Needs: overall	Unsatisfactory: 1	7: Satisfactory
Health (perceived)	Less healthy: 1	7: More healthy
Image to visitors	Poor: 1	7: Good
Productivity (perceived)	Decreased: -20%	+20%: Increased

© Copyright BUS Methodology 1985–2018. Used under licence.

11. Occupant Perceptions Summary Chart for AC Buildings at TRC, ©Usable Buildings Trust.

CASE STUDY 1 | TORRENT RESEARCH CENTRE, AHMEDABAD, INDIA

LESSONS LEARNED

POSITIVE ASPECTS OF PROJECT IMPLEMENTATION

PDEC system works well

The passive downdraught evaporative cooling system has been working well since the first building was occupied in 1997. The results of both measured performance and occupant perceptions indicate that the building is providing acceptable conditions at the different times of the year in both the PDEC and conventionally conditioned buildings.

Occupant perceptions very positive

Independent assessments of occupant perceptions indicate that overall, occupants have a high level of satisfaction in terms of Comfort, Air Quality and Temperature in the PDEC buildings. In all cases these are above the benchmarks and compare favourably with the results for the air-conditioned buildings. These results for the PDEC buildings are particularly significant in the context of acceptable temperature ranges that are higher than those deemed acceptable in the air-conditioned buildings.

Substantial energy savings achieved

After the first year of operation the completed building was reported to have used approximately 64% less electrical energy than the equivalent air-conditioned building (equivalent to approximately 66kWh/m²·year). Thomas & Baird report that in 2005, overall energy consumption for all four PDEC buildings plus two air-conditioned buildings was 54kWh/m². This can be compared with the typical energy consumption in Indian commercial buildings which has been reported to be in the range of 280–500kWh/m²·year. This also compares very favourably with the target for fully air-conditioned office buildings under the recently introduced environmental rating scheme TERIGRIHA (140kWh/m²·year for day use office building in a *'composite'* climate).

Cost effective

Although the building cost 4% more than its air-conditioned equivalent, this was paid back in the first year of energy savings. Expenditure on mechanical and electrical plant was reduced by 36%, which has provided further reductions in maintenance costs

Resident engineer vital to success

The involvement of the resident engineer in the detailed design and maintenance of the pressurised water system has undoubtedly contributed to its success.

Robust performance following increase in occupancy

In spite of the significant increase in the density of occupation of the buildings (from 1 person/15m² to 1 person/5m²), overall occupant satisfaction with conditions in the building has been maintained.

PROBLEMS ENCOUNTERED AND OPPORTUNITIES FOR IMPROVEMENT

Over-spraying

Initial observation of water droplets not evaporating fully have been overcome by increasing the water pressure in the system. Although this was not reported as a problem and was quickly resolved, it could still occur when power is switched off. This issue can be completely avoided by adopting a low-pressure water and compressed air system (see Chapter 4).

Air movement on top floor

Air movement (and volume flow rate) on the top floor of the PDEC buildings is poorer than on lower floors, although this has not been commented on by the occupants. This could be addressed by balancing resistance to airflow on the different levels.

Increase in internal heat gains

The increase in internal heat gains and latent loads from increased occupancy of the PDEC buildings results in more incidences of internal conditions going above acceptable levels, particularly when temperatures peak outside in summer. This emphasises the role of management in understanding the natural limits of the PDEC buildings as originally designed and built.

12. Ventilation shafts on NW elevation for a laboratory block.
13. NE elevation of a laboratory block.

CASE STUDY 1 | TORRENT RESEARCH CENTRE, AHMEDABAD, INDIA

SUMMARY

The Torrent Research Centre is a successful example of integrated design which reflects close collaboration between the design team and the client. The building continues to operate successfully which also reflects regular and effective maintenance of the building and its systems. The continuity of the management personnel and their understanding of the system's maintenance requirement and operation has been vital for the successful implementation of the seasonal cooling strategies and the resulting energy savings and high level of satisfaction of the occupants. In drawing conclusions from their survey of occupant perceptions, Thomas and Baird state that: *"the overwhelmingly positive user satisfaction responses of the PDEC blocks coupled with their lower energy consumption validate the integration of alternative climate control systems such as evaporative cooling in contemporary buildings in India"*.

> **THE BUILDING CONTINUES TO OPERATE SUCCESSFULLY WHICH REFLECTS REGULAR AND EFFECTIVE MAINTENANCE OF THE BUILDING AND ITS SYSTEMS**

These positive attributes of the project, as well as the latest advances in the technology of nozzle design and system operation, demonstrate the feasibility of passive cooling strategies in composite climates within India. Such projects hold the potential to deliver healthier buildings, significant energy savings and high levels of occupant satisfaction. This building completed nearly 20 years ago *"continues to satisfy expectations for a contemporary workplace of high quality that is simultaneously energy efficient"*.

In brief:

- The PDEC system works well, delivering satisfactory cooling in summer.
- Occupant perceptions are very positive in comparison with international benchmarks.
- Substantial energy savings have been achieved (total measured electrical energy consumption 54kWh/m^2 in 2005, compared with the typical energy consumption in Indian commercial buildings of 280–500kWh/m^2·year).
- Integration of PDEC is Cost Effective (4% additional capital cost paid back in less than five years).
- The resident engineer/building manager has been vital to the successful operation of the PDEC system.
- The buildings and systems have been robust in their performance following an increase in occupant density.

REFERENCES

- Alvarez-Dominguez, S., Rodriguez-Garcia, E. A., Cejudo-Lopez, J. M., Velázquez-Vila, R. & Molina-Felix, J. L. (1991). *'Control Climatico en Espacios Abiertos. el Proyecto Expo'92'*. Secretaria General Tecnica del CIEMAT.
- Ford B., Patel, N., Zaveri, P. and Hewitt, M. (1998). 'Cooling Without Air Conditioning: The Torrent Research Centre, Ahmedabad, India'. Renewable Energy 15, 1–4: 177–182.
- Majumdar, M. (2001). *'Energy Efficient Buildings in India'*. Tata Energy Research Institute, TERI.
- Nicol, J. F., Raja, I. A., Allaudin, A. & Jamy, G. N. (1999). *'Climatic Variations in Comfortable Temperatures: the Pakistan Projects'*. Energy and Buildings 30: 261–279.
- Thomas, L. E. & Baird, G. (2006). *'Post-Occupancy Evaluation of Passive Downdraft Evaporative Cooling and Air-Conditioned Buildings at Torrent Research Centre, Ahmedabad, India'*. Proceedings of the 40th Annual Conference of the Architectural Science Association ANZAScA. Presented at the Challenges for Architectural Science in Changing Climates, Adelaide, Australia: The University of Adelaide and The Architectural Science Association ANZAScA, pp. 97–104, http://hdl.handle.net/10453/1370/.

CASE STUDY 2
CENTRE FOR SUSTAINABLE ENERGY TECHNOLOGIES, NINGBO, CHINA

The Centre for Sustainable Energy Technologies (CSET) building is located in the Chinese Campus of the University of Nottingham, in the south-eastern town of Ningbo. The CSET is a mixed-use office, laboratory and education building with a floor area of about 1,200m^2. It was commissioned by the University in 2005 as phase I of the Ko Lee Institute of Sustainable Environments. The centre was designed by Italian architectural practice Mario Cucinella Architects with mechanical and electrical engineering specialist, TiFS Engineers, and environmental design consultancy provided by the School of the Built Environment's modelling team, within the University of Nottingham.

This building was designed with the ambition of combining high quality architecture with being the first *Zero Carbon* building in China. Part of the design intention was that it would not require conventional heating and cooling systems, and that residual energy requirements would be met by renewable sources. This has been achieved and it provides a pioneering example of *Zero Carbon* design. It includes the combination of natural ventilation (during the inter-seasons) and active downdraught cooling in summer for the tower, and mechanical ventilation (via buried pipes) to the semi-basement lab space. In 2009 the building was awarded the MIPIM International Real Estate Green Building Award.

14. CSET Building, Ningbo, China [Mario Cucinella Architects], ©Daniele Domenicali.

LOCAL CONTEXT

Ningbo is located midway on the eastern coast of China, in the South of the Yangtze River Delta (Lat. 28°N, Long. 120°E). It has a subtropical climate, featuring mild temperatures, moderate to high humidity and distinct seasons. The annual average temperature is 16.5°C, with average ground temperatures at -2m reaching 8°C in January and 27°C in August, which implies that low grade heat exchanges with the ground can potentially be used to provide pre-heating in winter and marginal pre-cooling of ventilation air in summer. The hottest month is normally July, with mean max dry bulb temperature of 32°C and mean max wet bulb temperature of 26°C, indicating a low potential for direct evaporative cooling (fig. 15). Peak dry bulb temperature can easily reach 35°C, with a diurnal swing of about 7–8°C. The small diurnal swing during most of the summer implies that night ventilation may not be very effective in providing pre-cooling in summer, although it may be effective in the midseason.

THE BUILDING AND THE OVERALL STRATEGY

The building has been designed to respond to the diurnal and seasonal variation of the climate of Ningbo, to minimise heating requirement in winter and cooling in summer, and to promote natural ventilation in spring & autumn when ambient conditions allow. The building has one semi-basement floor for laboratory use and five floors for

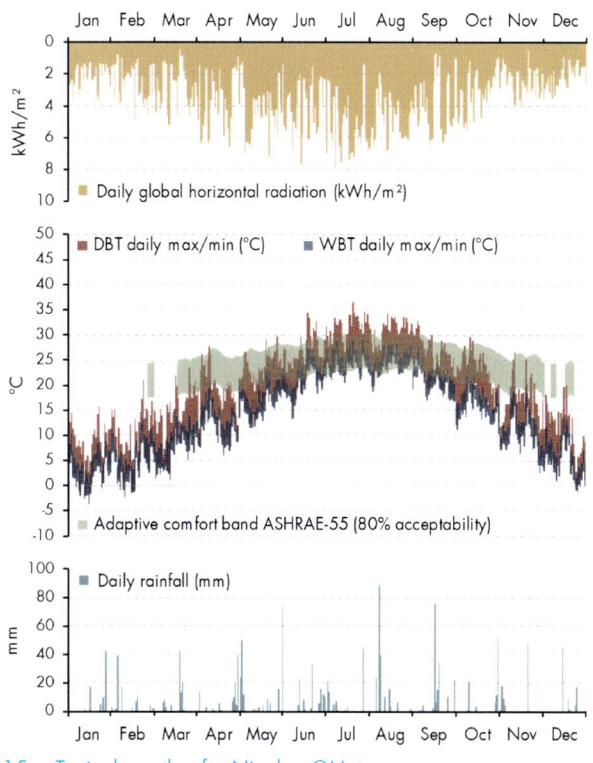

15. Typical weather for Ningbo, ®Meteonorm.

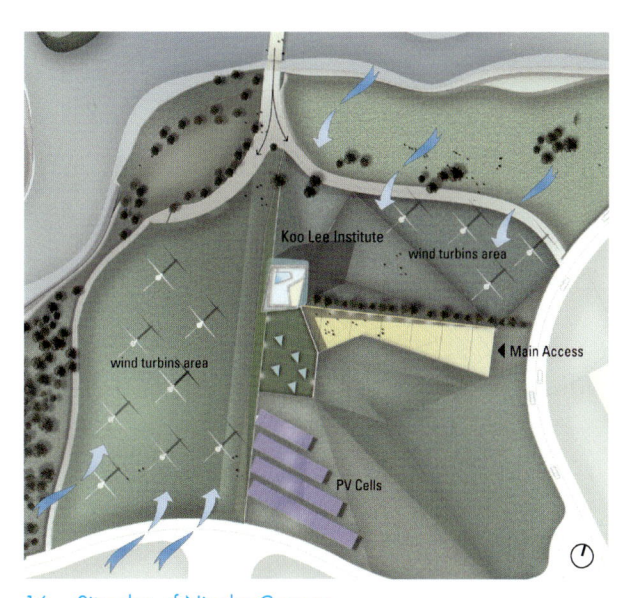

16. Site plan of Ningbo Campus.

CASE STUDY 2 | CENTRE FOR SUSTAINABLE ENERGY TECHNOLOGIES, NINGBO, CHINA

17

18

17. Typical 'tower' plan.
18. Semi-basement plan.

exhibition, teaching and offices arranged in a twisting 'tower' around a lightwell/ventilation shaft (fig. 19). This *'tower'* is surrounded by a glazed buffer space forming a clear double skin façade on the south side, and a diffusing glass *'jacket'* to the structural concrete walls on the west, and north sides. This buffer space also accommodates the escape stair.

ENVIRONMENTAL STRATEGY

The latitude of Ningbo (28°N) defines the sunpath – nearly overhead at midday on the summer solstice and at 40° from horizontal at midday on the winter solstice. Examination of the sunpath diagram suggests that large areas of glazing should be avoided on the east and west elevations (difficult to protect), while south elevations are not so vulnerable in the summer but will be the source of solar gain (and possibly glare) in winter, spring and autumn.

The diurnal and seasonal variation in air temperature, as described in fig. 15, suggests that the fresh air supply to the building will need to be pre-heated in winter and pre-cooled in summer to achieve thermal comfort inside, while in the midseasons outside air may not require pre-treatment to achieve satisfactory conditions.

The annual average air temperature in Ningbo of 16.2°C is strategically significant for this building. It implies that the ground temperature several metres below the surface will be close to this temperature (+/-2°C, as discussed in Chapter 2). This

CASE STUDY 2 | CENTRE FOR SUSTAINABLE ENERGY TECHNOLOGIES, NINGBO, CHINA

provides the basis for the direct pre-heating and pre-cooling of supply air to the lab areas. It also enables a ground source heat pump to pre-heat and pre-cool water within coils placed in the soffit of the intermediate floor slabs of the tower.

In the winter and summer, the pre-heating and pre-cooling of supply air to the building will require some mechanical assistance, although this is minimised by the exploitation of ambient heat sinks. In the summer, supply air needs to be de-humidified.

The building was designed to respond to the distinct climatic seasons of Ningbo, and is characterised by:

- A high performance (well insulated) envelope to the tower – to minimise fabric heat loss & heating demand.
- High thermal capacitance surfaces exposed internally – to minimise temperature fluctuations in summer & winter.
- A lightwell/ventilation shaft to promote day lighting without solar gain and glare, and to promote natural ventilation in winter, spring & summer.
- A glazed *'double-skin'* on the south side to provide side lighting without excessive solar gains, to act as a thermal buffer, to promote passive pre-heating, and to promote the exhaust of warm air in summer.
- A diffusing glass 'jacket' with top perforated openings to west, east and north walls – to provide a buffer zone with solar protection and to vent excess solar gains.

19. Interior view of office & lightwell.

- A half basement laboratory thermally coupled to the ground – to promote thermal stability & to exploit free cooling in summer and free heat in winter (via buried ventilation pipes).

The building has different ventilation strategies, for winter, midseason and summer peak conditions (figs 20–21).

In winter (typically December to March), when external air temperatures are well below internal comfort temperatures, external cold supply air enters at low level within the south facing double-skin façade, and is passively pre-heated before entering each floor of the tower. This supply air is then pre-heated further (as required) by finned tubes running within a perimeter floor duct, and air then passes across the space and rises by buoyancy through high level glazed vents into the lightwell, from where it is exhausted at roof level.

During the *'inter-seasons'* (typically April to May, and September to November) the airflow path is effectively the same as in the winter mode, except that the perimeter duct heating is switched off, and airflow rates may be raised (by increasing vent opening areas) to remove internal heat gains. Air temperature sensors at each floor level determine the extent of the opening of exhaust air vents.

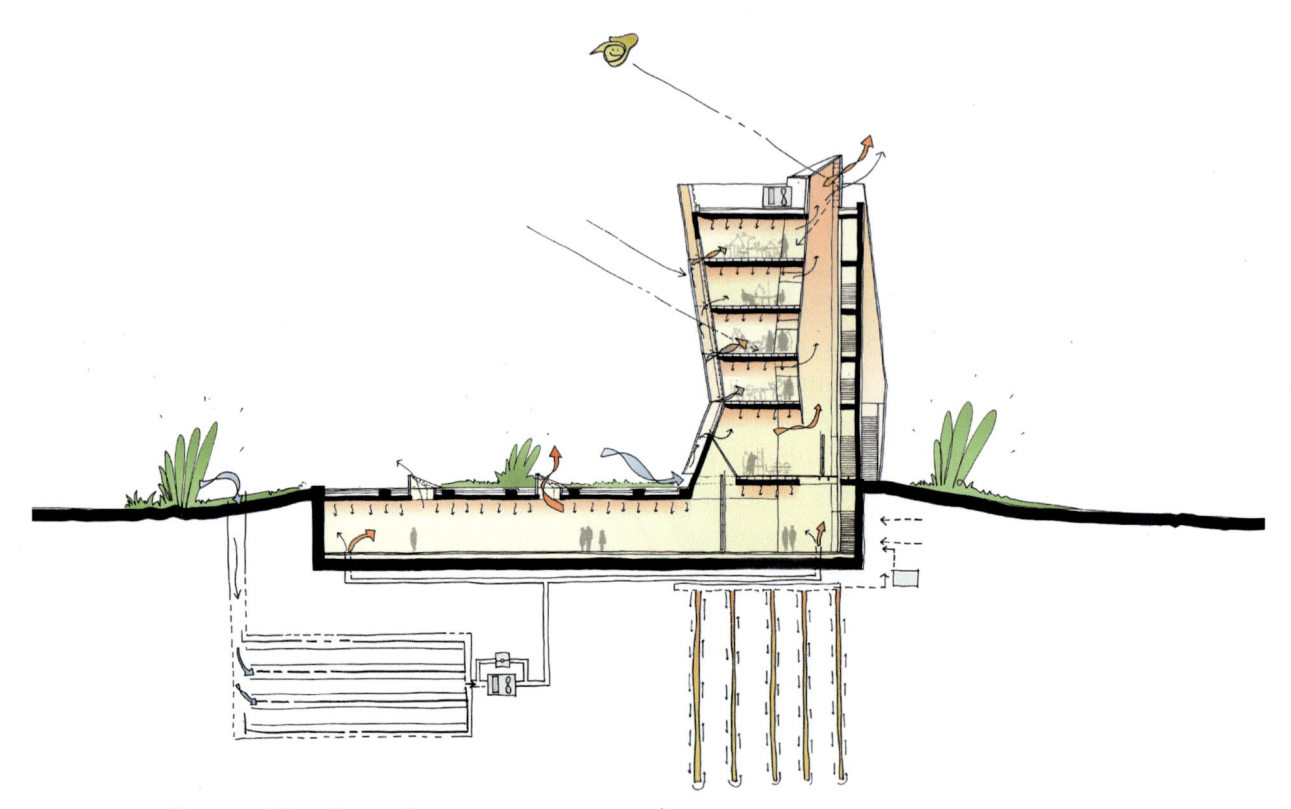

20. Section of basement lab and tower showing winter environmental strategy.

In summer (typically June to August), when daytime air temperatures and humidity preclude natural ventilation, supply air is cooled and dehumidified at roof level before being introduced to the lightwell, from where it falls to the hopper vents within the lightwell at each floor level. This cool air displaces warmer air within the occupied spaces, which is exhausted either through the high-level vents within the inner skin or at low level (depending on temperature differences within the airflow path). The cavity within the double skin façade will tend to be warmer than ambient, and with low and high-level outlets constantly open will help to drive air to exhaust from the parapet outlet.

The summer's prevailing wind will create an area of negative pressure which facilitates this exhaust air path. The cooling coil, de-humidifier and fan located on the roof provide cool humidified air at a rate of around 2000 m^3/h to the top of the lightwell.

The combination of the proposed strategies achieves a predicted total energy demand of approximately 60KWh/m^2.yr for cooling and heating and 30KWh/m^2.yr for de-humidification. Compared to the average annual electricity consumption for large scale public buildings (including educational) in the same climatic zone, which is about 286kWh/m^2.yr *(CABR,2011)*, the CSET building yields to potential energy savings

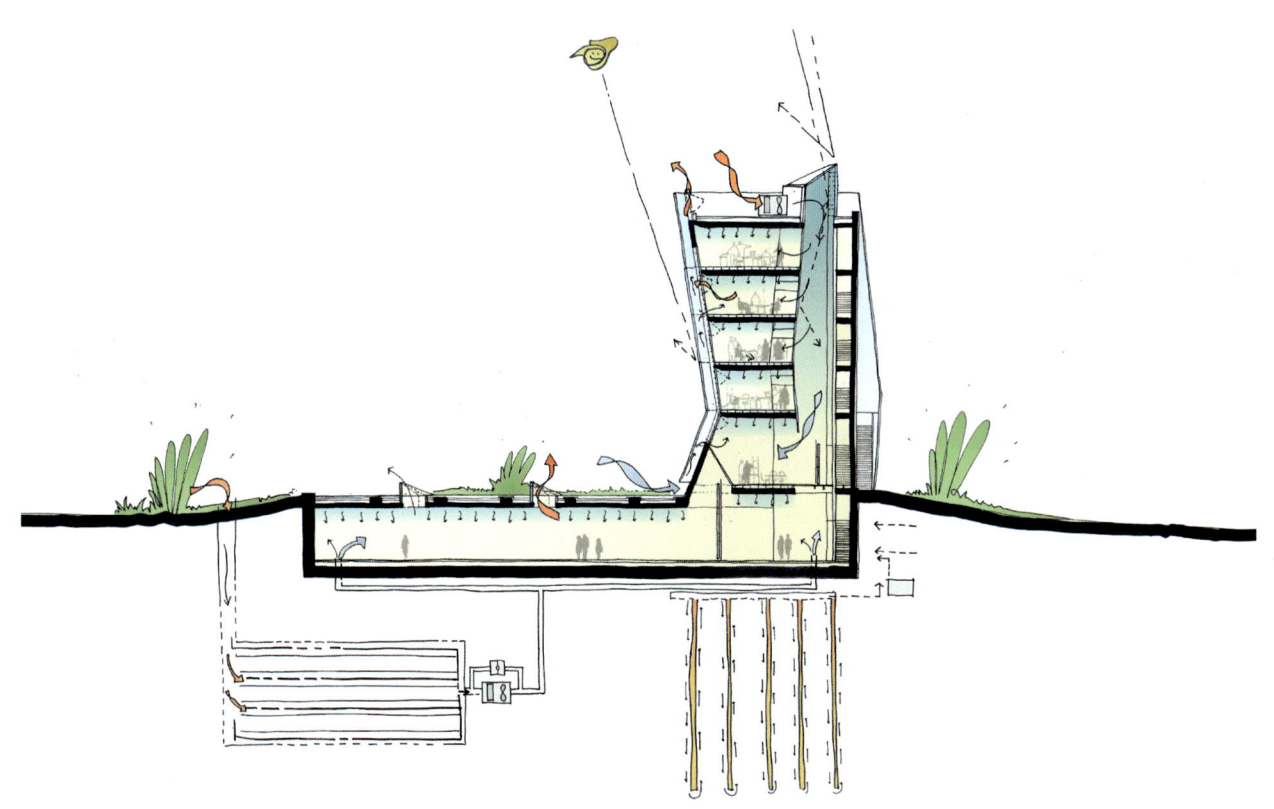

21. Section of basement lab and tower showing summer environmental strategy.

of about 226kWh/m²·year, i.e. of 80%. A linked array of photovoltaic panels provides the electrical energy required to meet this residual energy demand.

ENVIRONMENTAL SYSTEMS AND BUILDING INTEGRATION

The building exploits a number of ambient heat sinks and sources, and delivers heat and cool in a number of different ways. This was partly because the building is used for research into these systems, but also to provide a demonstration for students and professionals in practice. Chilled water is provided to both the rooftop dehumidifier and to the radiant cooling in the slabs from a ground source heat pump and/or solar panels with an absorption chiller. The rooftop dehumidifier drives cold air into the lightwell at a rate of 2000 m³/h, and reduces the level of humidity in the air to 60% (TiFS, 2006). The ceiling of each space incorporates a radiant cooling system. It was originally intended that night ventilation could be implemented coupled to the high thermal mass (350mm concrete wall with 150 external insulation and 400mm in-situ concrete for intermediate floors). However, because of security concerns, the window ventilators on the double skin façade are closed at night and night-time ventilation has not been implemented.

> THE VENTILATED SOUTH FAÇADE CONTRIBUTES TO THE PRE-HEATING OF FRESH AIR SUPPLY

CONTROL SYSTEM

In the proposed design, the building and both cooling systems are controlled by a Building Management System (BMS) using automatic motorised vent actuators, which are controlled in response to both temperature and relative humidity sensors at key locations inside and outside the building. However, at the time of the first post occupancy evaluation the building was still being commissioned and so during the fieldwork the BMS sensors were disabled, not collecting data and not controlling the cooling system. Initially, the control of the ventilation and cooling systems needed to be done manually through a computer program, based on the comfort perception of the building manager, occupants or based on schedules.

DESIGN STAGE PERFORMANCE PREDICTIONS

Design stage performance analysis indicated that the heating demand of the building (with its high level of insulation) was primarily related to pre-heating the fresh air supply. The ventilated south façade contributes to this pre-heating, and the residual demand for heating was found to be approximately 20kWh/m² per year. This is only slightly above the *'Passivhaus'* standard for heating. The analysis found that this heating demand was sensitive to the period of opening of the double façade, and the occupancy period.

With regard to the summertime cooling load, the analysis found that it was sensitive to the threshold temperature used to activate the night ventilation. Not surprisingly, raising the threshold temperature increased the duration of the night ventilation and reduced the cooling load.

It was further noted that, with the prevailing high humidity in summer, the predicted energy required for de-humidification was significant (approximately 30kWh/m^2.yr).

The evaluation of different options also illustrated the sensitivity of performance to the control regime, both in terms of energy consumption and thermal comfort, which further emphasises the importance of the commissioning of all the environmental control systems within the building.

The airflow analysis through CFD modelling demonstrated that the design was able to achieve useful rates of natural ventilation at different times of year, subject to the area of openings and the control regime imposed. The main design messages were:

- The Double Skin Façade requires a cavity depth of 800mm, with minimum structural opening areas of 15m^2 for the inlet and 12m^2 for the outlet.
- The façade should terminate in a parapet at roof level, at least 1/3 of the height of the previous floor, to allow the minimum required outlet surface area.
- The internal glazed façade must include inlets at the bottom and at outlets at the top level of each floor with a maximum effective opening area

22. Internal view of the double skin façade, ©Daniele Domenicali.
23. External view of SE/SW façades, ©Daniele Domenicali.

CASE STUDY 2 | CENTRE FOR SUSTAINABLE ENERGY TECHNOLOGIES, NINGBO, CHINA

of 0.128m² for inlets and minimum 2.38m² for outlets. This reflects the different opening areas required in winter (to pre-heat minimum supply air) and summer (to remove internal heat gains).
- The lightwell final outlet effective area required is a minimum 2.1m².

Of course the design advice arising from both thermal and airflow modelling is based on a set of assumptions regarding the occupancy and management of the building and its systems. The next section explores the results of (limited) post occupancy measurements.

POST OCCUPANCY MEASUREMENTS

The CSET building was the subject of two monitoring campaigns, one undertaken just under one year from occupation (Xuan, 2010) for a period of one month (May to June 2009) and another shorter monitoring period of five days in September 2011 (Sun, 2014). The second campaign also included the administration of a Post Occupancy Evaluation survey to the building's users.

A more targeted monitoring study on the performance of the double skin façade (figs 22–24) was undertaken in July 2012 and January 2013 for five days respectively, in order to characterise the performance of the façade during summer and winter conditions (Darkwa et al., 2014). The outcome of this third study, which compared analytic findings with monitored data, was that although the double skin façade is very effective in achieving pre-heating in winter with a differential temperature of up to 12°C from ambient, in summer it presents risks of overheating with temperatures reaching 41°C within the cavity when the outdoor temperature is 36°C. This demonstrates that although the operational strategy was to mix returned air from the offices (and cooled from the active downdraught cooling system) with the trapped air in the cavity of the double skin, the risk of overheating still persists.

The results of the first monitoring campaign (Xuan, 2010), which included smoke tests to verify airflow, spot measurements of air velocity and surface temperatures, and continuous monitoring of indoor dry bulb temperature and relative humidity, show that when the active downdraught cooling system is working, its performance is good and can provide effective cooling. Furthermore, for a large proportion of the time the ADC ventilation strategies are working as intended. However, the fieldwork also revealed some issues within the following areas.

AIRFLOW DISTURBANCE

The active downdraught cooling strategy relies on the effect of gravity to create a downdraught and to circulate air from the source of cooling to the occupied zone within the building, and on the stack effect to remove hot air. To ensure the intended air movement, the design included two main spaces: a lightwell/ventilation shaft promoting natural ventilation in winter, spring &

summer and a glazed *'double-skin'* on the south side of the building providing a thermal buffer and passive pre-heating in winter and the exhaust of undesired warm air in summer.

The smoke tests revealed a pattern of air movement, which is consistent with the air movement proposed at design stage, especially for the natural ventilation strategy and air movement in the façade. However, variances from the designed airflow path were also observed. Airflow reversal, in the form of ambient air coming in through the façade openings, was observed and smoke tests clearly showed the air exhausting via the *'inlets'*. The façade exhaust opening serves to 'catch' the wind when it comes from the north. Also, the obstruction by fluff and dirt found on the façade inlet and insect meshes blocking the openings may be the causes of this variation from the predicted airflow pattern (fig. 25).

The exhaust vents from the double-skin façade and from the lightwell were located on the leeward side of the prevailing wind. This is satisfactory when the wind is from this direction, but when the wind direction switches it can cause reversal of airflow as it is acting as a *'wind catcher'*. This emphasises the importance of ensuring that exhaust vents will work irrespective of wind direction.

Additional monitoring of absolute humidity recorded in September 2011 *(Sun, 2014)* showed higher values on the second-floor outlet than on the third floor, generating the likely hypothesis that during summer active

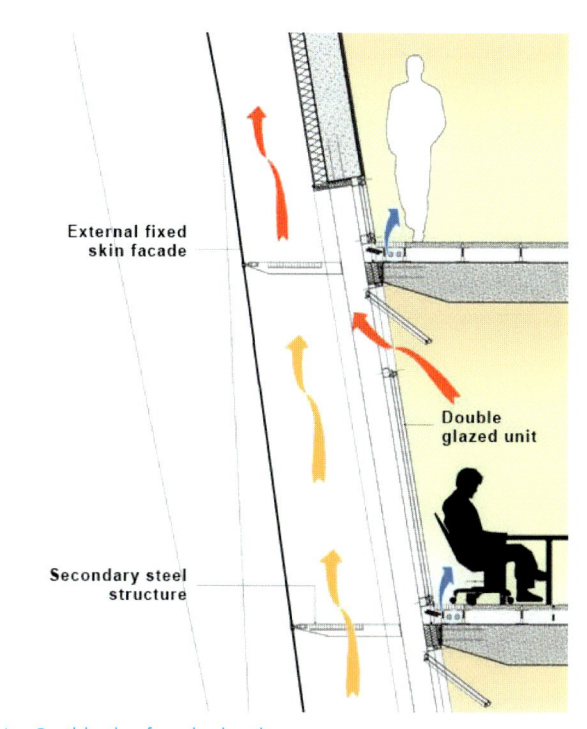

24. Double skin façade detail.

downdraught cooling mode the actual flow path in the double-skin façade is following a negative buoyancy path rather than a stack effect. This hypothesis is corroborated by the findings of *Darkwa & Chow (2014)* showing monitored mean air temperature profiles of the façade inlet lower than ambient and of every other floor's outlets during summer hot days (Darkwa & Chow, 2014, fig. 15, p. 138). This finding is significant and it correlates with performance observations from other case studies examined in this book. The direction of the exhaust airflow path is clearly significant if it negatively affects the total air volume flow rate. However, flow reversal in itself may not be a problem if adequately sized and strategically positioned openings are provided.

Subsequent spot measurements and observations of the airflow in September 2011 did not reveal the backflow from the north side of the tower termination into the façade but interestingly this *'improvement'* was due to a fault in the louvres of the north side of the double skin façade termination which, although reduced the overall outlet area, also reduced the positive pressure from the north wind causing the unwanted downdraught (Sun, 2014).

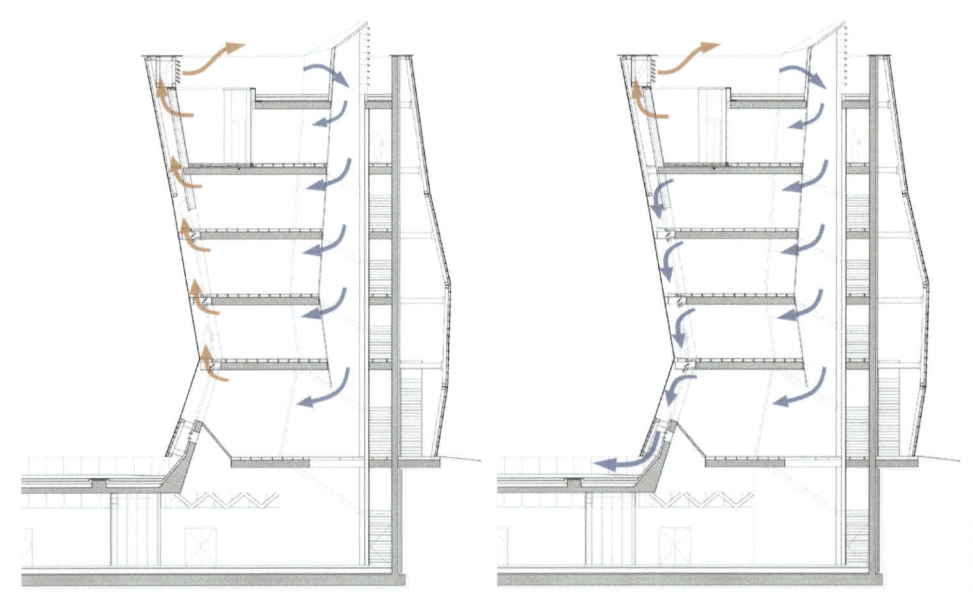

> IT IS IMPORTANT TO ENSURE THAT EXHAUST VENTS WILL WORK IRRESPECTIVE OF WIND DIRECTION

25. Comparison between designed airflow path and actual flow path.

204 CASE STUDY 2 | CENTRE FOR SUSTAINABLE ENERGY TECHNOLOGIES, NINGBO, CHINA

CONTROL SYSTEM

In order to promote a dynamic response to outside and inside conditions, automatic vent actuators (controlled in response to both temperature and relative humidity sensors inside and outside the building) were proposed at key locations in the original design. At the time of the first field study in 2009 the building was not fully commissioned and thus not all the systems proposed in the design were functioning. The automatic control system was disabled and it was controlled by a person through a computer. Settings were based on the working hours and the individual perception and judgement of the controlling person. The opening size and times were rather random not really following inside and outside conditions. The openings on the lightwell side were kept open all the time, with the same opening mode at the first floor and the third floor. This is in contradiction with the analysis' result and rule of thumb indicated by Francis, that the windows at the different floors of a PDEC tower should be opened with the size decreasing from top to bottom to achieve an equal cooling effect (Francis, 2000). The problem of consequently higher temperatures at the 3rd floor can be addressed by adjusting the openings according to the rule above. The two cooling systems were also controlled manually. However, these problems may be attributed in part to the fact that the building was occupied before the ADC system could be properly commissioned.

The problems emerging from this can be addressed by the full commissioning of the system and by applying the right settings. Another problem related to the control system was the dissatisfaction of occupants, who complained about lack of control, which was even less tolerable due to unsatisfactory air quality. Hence by addressing the problems associated with airflow and control, the problem of poor air quality can be addressed as well.

The original design of the building combined various passive strategies (solar shading, high thermal mass, natural ventilation, etc.) to maximise energy savings. To achieve comfortable conditions in the varied climate, besides the downdraught cooling system, fabric cooling was also installed. To achieve the desired results, the separate systems need to be tuned, and enhance each other's effects.

> **OPENINGS AT DIFFERENT LEVELS OF A DOWNDRAUGHT SHAFT SHOULD HAVE VARYING OPENING AREAS TO ACHIEVE A BALANCED COOLING EFFECT**

The climate analysis and Dynamic Thermal Simulation indicated that most of the time in the middle season and sometimes in early summer, the cooling load can be addressed by using natural convective cooling instead of fabric cooling and active downdraught cooling system. This can reduce the cooling system operation time and also suggests that these systems are not required throughout the summer.

Despite the good natural ventilation capacity revealed in the smoke tests and during the fieldwork period (June 2009), this passive cooling strategy was never used to cool the building. The employment of high thermal mass construction can be only truly beneficial in combination with night-time ventilation and, together with the shaded south façade and the ventilated jacket buffer skin on the other orientations, this should result in low cooling load of the building. However due to security issues, the lack of understanding of the active downdraught cooling system and the lack of connection of the control system with sensors, the windows on the double skin façade side were closed overnight, making the use of night ventilation cooling not possible inside the building.

Due to the disabled BMS, the two cooling systems were controlled manually, thus always used together. They were always switched on at the same time and closed together following the working hours rather than outside conditions. The collision of these two systems resulted in the disturbance of desired airflow patterns and in condensation and mould growth due to the wrong settings for the temperature of the radiant cooling system. Resetting of the control system to respond to the different outside and inside conditions could improve the effectiveness of the system.

If the outside conditions are suitable, free natural ventilation cooling can be used instead of the mechanical fan forced ventilation. In the peak summer conditions, the use of active cooling (radiant cooling and active downdraught cooling) is inevitable, but its operating time should be reduced in order to reduce the cooling energy consumption. The overuse of the active mechanical cooling system (long running time and simultaneous use of both systems) can result in unnecessary energy consumption and even in performance reduction.

Maintenance and setting

During the field monitoring several technical problems emerged and the settings of the systems were also changed. Faults in the ADC system at least twice during the month disturbed the operation of the system. Differences in cooling performance are clearly showed in the analysis of the data. In the first case the problem was discovered just by 'accident', due to some investigations done during the fieldwork. When discovered, the system was not operating correctly already for more than one month.

Another problem associated with the general building maintenance was the obstruction of airflow in the air inlet of the double-skin façade due to build-up of dust and spider nets, resulting in greater resistance and the reduction of the stack effect (fig. 26). The duct under the façade bottom opening requires regular cleaning to assure that adequate free area for the ventilation is available.

These observations show that the maintenance and management of this system is paramount to achieve the intended performance and to identify faults and troubleshoot, the building manager and maintenance personnel need to have a good understanding of the ADC cooling system. A system of this type will normally require a professional building energy manager who is in full commission of the system to ensure success. The current personnel need to receive training to have a full understanding.

OCCUPANTS' RESPONSE

A post occupancy survey on the level of occupants' satisfaction with environmental comfort provided by the building was undertaken in September 2011. Due to the small number of permanent staff working in the CSET building at the time of the survey (five researchers), the survey was extended to 60 students occupying two classrooms. Nevertheless, the results from the students' survey was compared to that of the permanent staff, considered as a benchmark due to their longer and more reliable experience of the building (fig. 27).

The feedback from the staff was positive (green) whereas students (blue) showed a relatively lower level of satisfaction in every aspects of the environmental performance which nevertheless is still higher than the baseline (dash line). Ventilation (rating 5.4/7) and thermal comfort (rating 5.2/7) appeared to be the two best performing factors in the six aspects investigated, whereas lighting and space efficiency were the two factors marginally passing and exceeding the baseline. This can be explained by referring to previous daylight analysis showing that the average daylight factor on the working area is 2% with a uniformity ratio

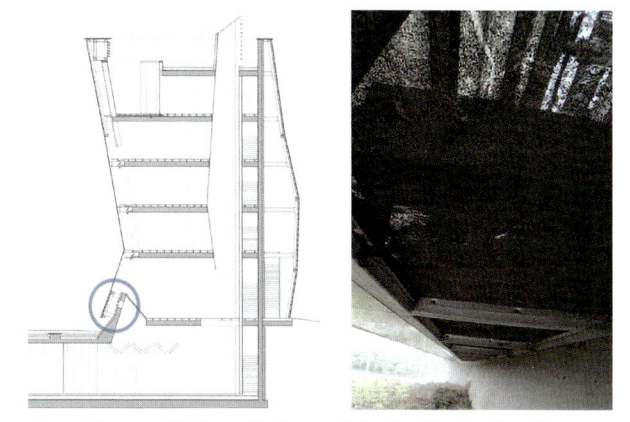

26. Obstructed inlet at the base of the double skin façade.

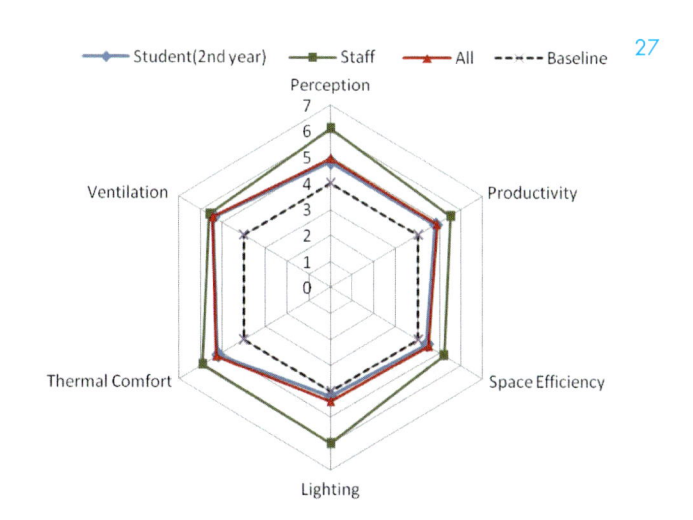

27. Occupants' perception. Data obtained from post occupancy survey (1: unsatisfactory to 7: satisfactory).
28. Furniture layout vs room geometry.

CASE STUDY 2 | CENTRE FOR SUSTAINABLE ENERGY TECHNOLOGIES, NINGBO, CHINA

at 0.24. In terms of the space efficiency, the impression of less efficiency was probably derived from the contradiction between the furniture layout and the irregular geometry of the spaces. In fact, the lack of bespoke furniture meant that it is difficult to accommodate generic office furniture in the small floorplan with leaning perimeter walls and atrium (fig. 28). As a result, even though the building externally self-shades itself and offers a neat and robust experience, the indoor space is perceived as with reduced functionality.

LESSONS LEARNED

POSITIVE ASPECTS OF PROJECT IMPLEMENTATION

Energy saved

The predicted total energy demand resulting from the combined low energy strategies for the entire building is 60.1kWh/m²·yr compared to 70–300kWh/m²·yr for a conventional building of similar typology. This notional performance could not be verified but if the building operation is optimised this could yield an energy savings of up to 240kWh/m²·yr or 288MWh/yr (80%).

Economic benefits

Projected energy savings represent an annual cost saving of 147,513.6RMB/yr based on 0.5122RMB/kWh.

Environmental benefits

The projected energy savings represent a reduction in CO_2 emissions of approximately 181,440kgCO_2/yr (151.2kgCO_2/m²·yr based on a carbon intensity of 0.63kgCO_2/kWh).

PROBLEMS ENCOUNTERED AND OPPORTUNITIES FOR IMPROVEMENT

Airflow pattern

Variances from the design stage airflow path were observed during the smoke test. This showed flow reversal, in the form of ambient air coming in through the double-skin façade openings during the summer cooling strategy. This problem was observed when the wind was from the north (opposite from the prevailing wind) but may also be influenced by the obstruction of the façade inlet (insect mesh not cleaned regularly) and poor control of the vents (not achieving the intended opening areas) are causes for the disturbance. Designing the tower termination to induce suction irrespective of the wind direction is therefore essential. However, in the absence of wind, flow reversal (i.e. downdraughting from the exhaust) is likely to occur and may not be a problem provided the designed volume flow rate can be achieved.

Control system

The automatic control system was disabled, and it was controlled remotely through a computer by the building manager. Settings were based on the working hours and the individual perception and judgement of the individual controlling it. The opening areas and times were rather random not really following inside and outside conditions. The openings on the lightwell side were kept open all the time, reducing effectiveness of the designed strategies due to lack of control.

SUMMARY

The cooling strategy of the CSET building showed much promise in innovative thinking and the pioneering of new systems and control strategies. However, the inconsistent control strategy and poor integration of the BMS as well as the infrequent and ineffective maintenance meant that the ventilation strategy was not implemented as intended in the design and that the total opening areas were not achieved. This resulted in issues of overheating and backflow, especially in the classrooms where a high occupancy was experienced.

Occupants expressed overall satisfaction with the building and generally exceeded the baseline score, but space efficiency, lighting and thermal comfort achieved the lowest score amongst the students. This difference between the groups can be attributed to higher density of occupation and lower flexibility in the use of space and free movement in the classroom. This was highlighted by comments on the lack of control by occupants in the classrooms. However, considering the issues highlighted by the fieldwork observation and monitoring, it is remarkable that the POE results are nevertheless very positive.

The actual cooling systems' performance, compared to the design stage predictions, is compromised by the lack of appropriate seasonal and diurnal control (either manual or remote), lack of maintenance of the inlet areas, and lack of implementation of the night-time cooling strategy during mid-season. These problems, however, can be easily overcome with the implementation of a comprehensive maintenance plan for the BMS system as well as the natural ventilation network (i.e. inlet and outlet openings, insect mesh, etc.) and the training of the building manager on the cooling and ventilation building requirements. The airflow disturbances during peak summer time can be minimised by: a) reducing the outlet opening at the façade termination in order to avoid backflow from northerly winds; b) differentiating the areas of the opening between the central atrium and the perimeter rooms in order to have larger areas on the top floors gradually diminishing towards the lower floors; c) allowing for the double façade to act as a downdraught exhaust and in summer peak conditions close or heavily minimise the top termination opening.

REFERENCES

- CABR (2011). *'The Development of Building Energy Performance Benchmarking Tools in China'*. Beijing: China Academy of Building Research.
- Darkwa, J. & Chow, Y. (2014). *'Heat Transfer and Air Movement Behaviour in a Double-Skin Façade'*. Sustainable Cities and Society 10: 130–139.
- Francis, E. (2000). *'The Application of Passive Downdraught Evaporative Cooling (PDEC) to Non-Domestic Buildings: Office Building Prototype Design in Catania Italy'*. Proceedings of the 21st Conference on Passive and Low Energy Architecture (PLEA). July 2000, Cambridge, UK, pp. 88–93.
- Sun, M. (2014). *'Office Design for Hot Humid Regions of China: An Evidence-Based Approach to Environmental Considerations'*. PhD Thesis, University of Nottingham, England.
- TiFS (2006). *'KLI M&E Services'*. TiFS Ingegneria Srl. Technical Report, October 2006.
- Xuan, H. (2010). *'The Application of Downdraught Cooling in China'*. PhD Thesis, University of Nottingham, England.

CASE STUDY 3
FEDERAL COURTHOUSE, PHOENIX, ARIZONA, USA

The Sandra Day O'Connor Federal Courthouse in Phoenix, California is an office and courthouse building with a floor area of about 46,500m² spread over two urban blocks and six floors. It was commissioned by the General Services Administration and it was designed by Richard Meier Architects with mechanical and electrical engineers Arup New York.

This building, completed in 2000, has since generated much debate for the extensive use of glass in a hot desert climate and for the complaints generated amongst some occupants. However, the building is commendable as one of the first examples of Passive Downdraught Evaporative Cooling (PDEC) applied to a very large and high-specification government building in the USA.

29. View of Sandra Day O'Connor Federal Courthouse [Richard Meier Architects], ©Scott Frances.

LOCAL CONTEXT

In the summer of 2011 Phoenix, capital city of Arizona, set a record of 31 days with temperatures above 43°C in a single year and more recently in 2017 it recorded nine consecutive days with a high of 44.4°C. Placed on the north margin of the Sonoran desert (Lat. 33° N, Long.112° W) and at an altitude of 330m above sea level, the town experiences very hot and dry summers (40°C mean maximum for August) with peak temperatures of 50°C and mild winters (7.5°C mean minimum in December) (fig. 30). Night-time temperatures drop considerably lower than the daytime, especially in winter, spring and autumn, but with the increased urban heat island phenomenon night-time temperatures in the dense city centre are increasing. The mean wet bulb temperature in summer is also very low, ranging between a min of 18.2°C and max of 22.6°C (August) which indicate a very high potential for direct evaporative cooling (fig. 30). Phoenix and its surrounding area benefit from 325 days of annual sunshine but receive only 158mm of rain per year, making this area ideal for the application of solar energy technologies but also vulnerable to drought. Water availability is, in fact, one of the major concerns in the south-western USA.

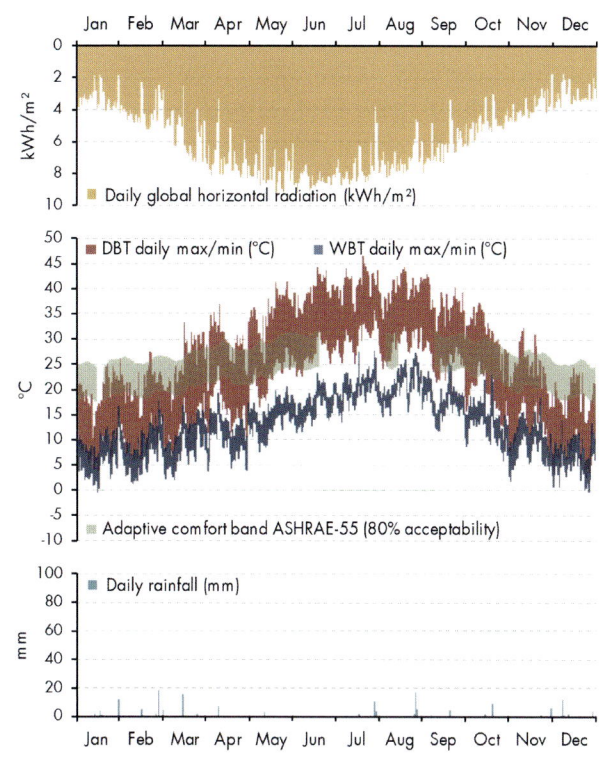

30. Typical weather for Phoenix, ®Meteonorm.

THE BUILDING AND THE OVERALL STRATEGY

Phoenix's urban form is similar to that of many new south-western cities in the USA.

The very low density suburbs give way to the high density centre downtown. The road infrastructure, however, stays the same and the multi-lane carriageways continue deep into the city centre. In this context, located on the west margin of downtown Phoenix, the 46,500m^2 six storey courthouse sits on two urban blocks. The building is oriented north south with the main entrance on the east side, preceded by a paved plaza (fig. 31). It consists of two volumes: a six storey office and courtroom block and a larger six storey atrium space.

Despite the large glazed area, this is an inward looking building concentrating most of the offices and courtrooms on the south side. These areas open into the vast 107x46m north atrium via the open access balconies (fig. 32). On the west side, the atrium houses the three storey cylindrical special proceedings courtroom, which is located at podium level. The atrium incorporates ventilation openings at low level on the east side and at high level on the north side and roof. The atrium's glazing is either shaded clear double glazing on the north elevation or white ceramic fritted low-e glass on the east and west sides.

DESIGN INTENTIONS AND ARCHITECT'S PHILOSOPHY

The design team's primary intention was to create a building that transposed an outdoor 'public' space where more commonly the car-park is located, into an internal transitional space between the extreme outdoor conditions and the controlled courtroom and office environment. The space acts as an imposing *'lobby'* and public plaza where the citizens of Phoenix can congregate in a way they would not be able to do outside. The atrium therefore shelters them from the harsh desert climate, while allowing diffuse daylight and sufficient air movement to mitigate the impact of the intense solar radiation and high temperatures.

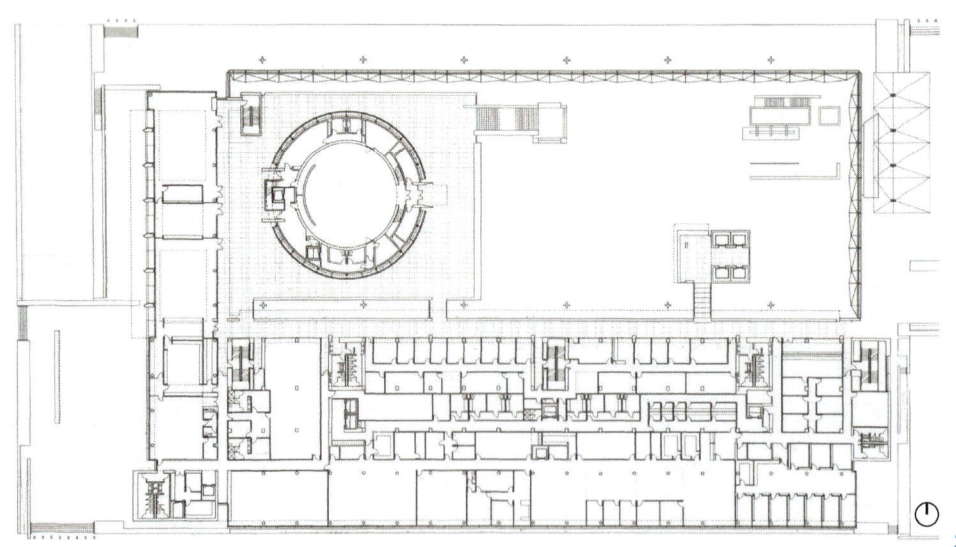

31. Second floor plan.

ENVIRONMENTAL STRATEGY

The building generally uses conventional (refrigerant based) air-conditioning systems to provide cooling in summer and heating in winter to the offices and courtrooms, but the atrium space, uses Passive Downdraught Evaporative Cooling (PDEC). This produces a 75% saving of the total air-conditioning costs for the atrium and 5M$ saving in the equivalent capital cost of air-conditioning. The system is operated via the Building Management System (BMS) and is controlled by temperature set-points.

One of the crucial aspects of the summer strategy is to reduce solar gains from the atrium's west and east elevations and the glazed roof, whilst maximising the daylighting. This is achieved with an extensive use of white horizontal external shading. Roof glazing is not shaded externally, but horizontal internal shading devices create a high level 'hot zone' which is vented at the gable ends of the roof (fig 33). Daylighting overall is sufficient in the atrium but the use of artificial lighting becomes unavoidable as we travel further away from the atrium corridors, creating a very typical deep plan office environment. Artificial lighting is operated via the BMS on a timed schedule.

> **PDEC PRODUCES A 75% SAVING OF THE TOTAL AIR-CONDITIONING COSTS FOR THE ATRIUM SPACE**

32. Federal Courthouse. Interior view, ©Scott Frances.

CASE STUDY 3 | FEDERAL COURTHOUSE, PHOENIX, ARIZONA, USA

33. North façade with detail of high level ventilators in atrium.

34. Security area in the atrium, ©Eric E. Johnson.

The energy efficiency of the building and its operation have been the focus of the building management team. The overall energy savings are tracked on a yearly basis but sub-metering has not been implemented so it is difficult to disaggregate energy use. Comparing the energy bills of the years 2006–07 with those of the previous years, it was possible to estimate an average annual saving of $109,636 equivalent to 1,916,400kWh/yr (41kWh/m^2·yr). Electrical energy consumption in 2007 was equivalent to approximately 144kWh/m^2 as opposed to 185kWh/m^2 in 2006. This compares favourably with the US national average electrical energy consumption for office buildings, which is 168kWh/m^2·yr.

Water consumption is an important issue. Average residential water usage in the USA amounts to over 600 litres/person·day (compared with 150 litres/person·day in the UK). Typical water consumption in offices for downdraught evaporative cooling is approximately 10–15 litres/person·day or approximately 1 litre/m^2·day (this actually varies significantly, for example, water consumption for PDEC at the Torrent Research Centre in India is approximately 0.3litres/m^2·day). Rainwater is not collected and stored by the building even though, theoretically, rainwater collected from the roof (approx. 1,000,000 litres) could satisfy the predicted water requirement of the system (6,000litres/day = 1.2litres/m^2/day) for about 6 months, which is roughly the amount of time when PDEC is operational.

COOLING SYSTEM AND BUILDING INTEGRATION

The Passive Downdraught Evaporative Cooling (PDEC) employed in the atrium was included after the schematic design stage, when Arup New York was involved as external consultant in order to overcome predicted risks of overheating in the atrium. The PDEC idea was inspired by the evaporative cooling system for green houses and second stage systems such as A/C with evaporative cooling. The intention was to cool the floor of the atrium and not the whole volume. The balcony corridors were originally part of the transitional spaces, but linear diffusers were added later to thermally separate them from the atrium. Although the designers were aware that stable conditions would be achieved in the atrium, the threshold set-point for the PDEC system were originally fixed at 23°C for the indoor dry bulb temperature and 60% for the RH. The statistical analysis undertaken by Arup showed that, although for 10% of the time the conditions would have been uncomfortable, that was deemed to be acceptable as the space is transitional. For the permanently occupied spaces in the atrium such as the guards' area (fig. 34), localised cooling was suggested but not implemented.

> THE INTENTION WAS TO COOL THE FLOOR OF THE ATRIUM AND NOT THE WHOLE VOLUME

Hydraulic misting nozzles were installed on the balustrade of the sixth floor walkway. Although the primary design intent of the engineers was to cool the floor of the atrium, a considerable secondary effect is created: a screen of cold air flowing down to the open corridors below, controlling the heat gains from the main atrium space and reducing the cooling demand of the office HVAC system (figs 35–36).

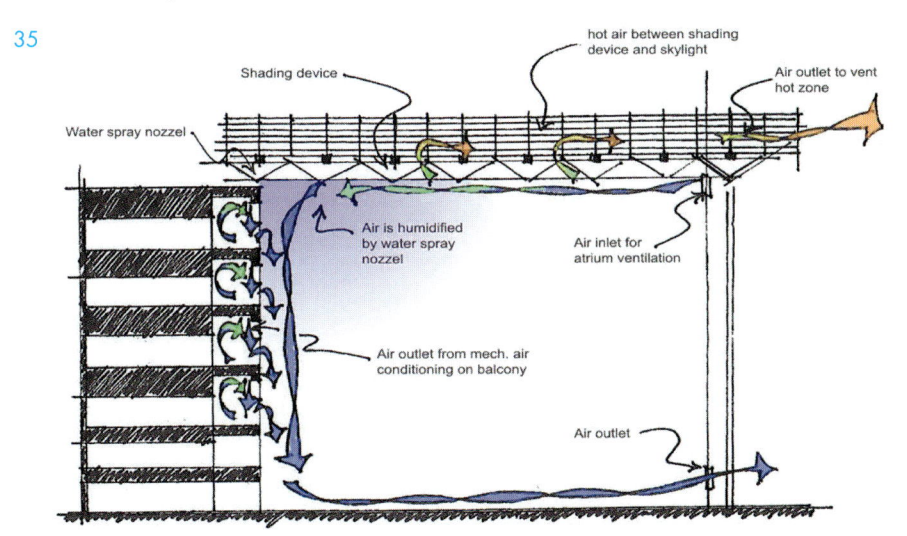

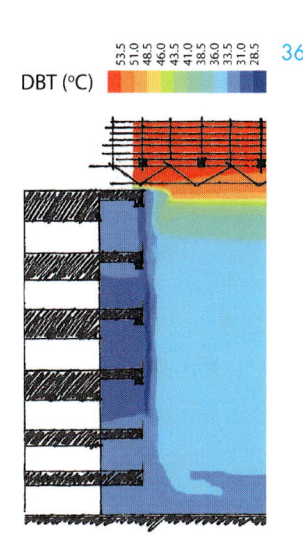

35. Main cooling and ventilation strategy, ©Daniel Griffin.
36. CFD sectional plot showing temperature stratification in the atrium. Federal Courthouse, Phoenix, Arizona, ©Daniel Griffin.

CASE STUDY 3 | FEDERAL COURTHOUSE, PHOENIX, ARIZONA, USA

Design-stage predictions suggested an average daytime temperature difference of 11°C would be achieved between the atrium floor and the outside. While this is undoubtedly achieved for much of the summer, significant variation is inevitable. Spot measurements in October 2007 revealed a difference of 7°C between the south end of the atrium and the outside. This difference decreased in proximity of the north glazing where it was only 5°C.

The Phoenix Courthouse's PDEC system is probably one of the largest in the world. It consists of nine water supply pipes attached to the 6th floor balustrade serving a total of approximately 1100 nozzles (figs 37–38). The nozzles are controlled by motorised valves linked to the building's BMS. The total water flow is 250 litres/hour with a pressure of 89 bar. At present there are no recorded figures of the daily water consumption but the measurement between 1 and 2 October 2007 show a figure of 6,317 litres which makes an average of 263 litres/hour (averaged over 24hrs). In a typical hot summer day, water consumptions in excess of 13,000 litres have been reported.

During the temperate months (October to March) when the outdoor temperature provides an acceptable level of comfort, natural ventilation via the east and north openings contributes to removing excess heat gains from the atrium. Fresh air is taken in from the three low level openings above the east entrance, which account for a total opening area of nearly 100m². The warmer exhaust air is drawn out by stack effect through the

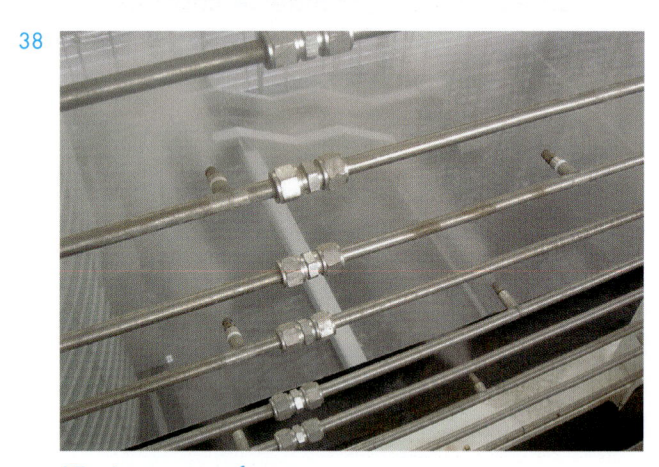

37. Interior view of atrium.
38. Detail of high pressure brass nozzles.

6x30m² triangular louvred openings at roof level.

OCCUPANTS' RESPONSE

A thermal comfort survey was undertaken with people occupying the ground level atrium space. This involved 19 subjects who answered the questionnaire in the course of one day (2 October 2007). The summary of results for the post occupancy study is illustrated in fig. 39. The subjects were located in the atrium and the surrounding perimeter offices on the ground floor where the impact of the PDEC system is likely to be most significant.

The results show that the majority of the interviewed subjects did not have a very favourable perception of the building, with aspects such as thermal comfort and air quality being the most unsatisfactory. However, care must be taken in interpreting these results, as the sample size was very small, and included many of the security staff members who are permanently located on the floor of the atrium. The interviewees have therefore been responding to their expectations of a *'working'* space rather than a *'transitional'* space.

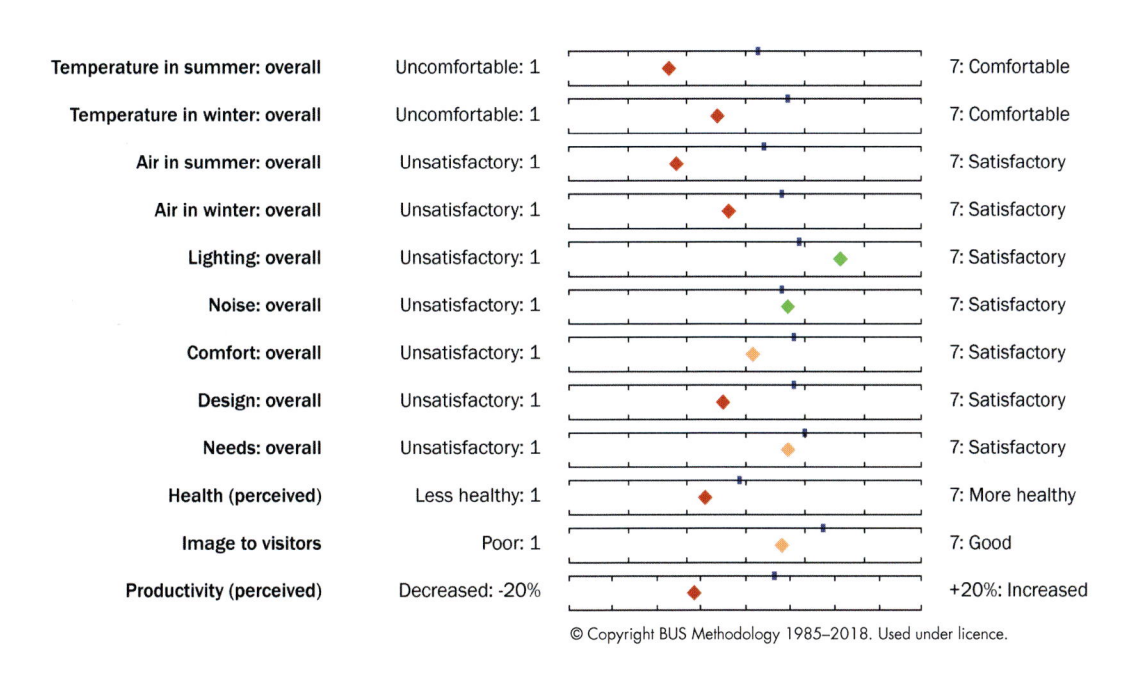

Temperature in summer: overall	Uncomfortable: 1		7: Comfortable
Temperature in winter: overall	Uncomfortable: 1		7: Comfortable
Air in summer: overall	Unsatisfactory: 1		7: Satisfactory
Air in winter: overall	Unsatisfactory: 1		7: Satisfactory
Lighting: overall	Unsatisfactory: 1		7: Satisfactory
Noise: overall	Unsatisfactory: 1		7: Satisfactory
Comfort: overall	Unsatisfactory: 1		7: Satisfactory
Design: overall	Unsatisfactory: 1		7: Satisfactory
Needs: overall	Unsatisfactory: 1		7: Satisfactory
Health (perceived)	Less healthy: 1		7: More healthy
Image to visitors	Poor: 1		7: Good
Productivity (perceived)	Decreased: -20%		+20%: Increased

© Copyright BUS Methodology 1985–2018. Used under licence.

39. Occupant Perceptions Summary Chart for Federal Courthouse, Phoenix, Arizona, USA, ©Usable Buildings Trust.

The detailed results on temperature, air quality in summer and control are discussed below.

Temperature in summer

74% think that it is uncomfortable with 88% perceiving it as too hot and 31% variable during the day.

Air quality in summer

For the majority of the staff air is still, and overall 68% of the sample considers the conditions unsatisfactory. Also, the air was perceived too stuffy by 44% whereas 39% considered it humid. For 83% it was either odourless or neutral.

Control

Between 74% and 79% of the occupants reported they had little or no control on heating, cooling and ventilation confirming the comments made during the informal interview.

The occupants' comments on overall comfort suggest offices and atrium had very different thermal environments: *"It is miserable during the summer outside the office in the atrium of the courthouse. Otherwise it is satisfactory in the office"*; *"It's warm in lobby area but it's cold in FCU building"*. Comments also related to the problem of *"Water on floor from misters"*.

The overall building strategy is questioned with several occupants commenting: *"It's a nice design but not for this climate"*; *"It's an amazing design, but not a good design for Arizona's summers. Too much wasted space in the atrium area"*; *"Stylish but not practical"*.

LESSONS LEARNED

POSITIVE ASPECTS OF PROJECT IMPLEMENTATION

Energy savings

Electrical energy consumption in 2007 was 144kWh/m^2. This compares favourably with the US national average electrical energy consumption for office buildings, which is 168kWh/m^2·yr *(EPA, 2008)*.

Cost savings

Passive Downdraught Evaporative Cooling produces a 5M$ saving in the equivalent capital cost of air-conditioning the atrium, and a 75% reduction in annual running costs associated with cooling the atrium.

PDEC system

Following *'fine tuning'* of the system, the predicted conditions of temperature and water consumption have now largely been met, and the system works well in tempering the outdoor conditions, providing a cooler environment in the atrium space. The environment on the floor of the atrium provides a successful transition space for most of the building's users.

PROBLEMS ENCOUNTERED AND OPPORTUNITIES FOR IMPROVEMENT

PDEC system and controls

The PDEC system experienced problems since the beginning of operation with dripping of nozzles due to blockage and pressure drops. Following early reports of failure, and although their appointment terminated during design development, the Arup engineers team went back to carry out a post-commissioning survey. Their survey highlighted the following problems *(Raman, 2009)*:

- None of the nine rows of pipes above the 6th floor worked on step control as specified.
- The quantity of water coming through the nozzles needed to be reduced.
- Linear diffusers were not working. The air speed of 1m/s could not be measured and more separation between the atrium and corridor spaces was required. 60cm drop down glass panels were installed near the diffusers (although 1m was originally recommended).
- The control thermostat was wrongly placed in the path of the downdraught.

Originally the 9.5mm diameter water pipes and nozzles were designed to withstand a pressure of 48 to 55 bar. By increasing the pressure to 89 bar and replacing the blocked Honeywell brass nozzles on a yearly basis, the external contractor, together with the General Service Administration maintenance engineers, report to have successfully addressed the problems of dripping and *'spitting'*. These practices are clearly effective but can result in increased maintenance costs, considering the relatively high cost of brass nebulisers, and in increased water flow rate, which could potentially increase water consumption if a tight control logic is not in place due to higher pressure. At the Torrent Research Centre in India, the resident engineer reported that during the cooling season any blocked nozzles would be removed, cleaned and reinstated on a monthly basis, and this was not regarded as an onerous task. Also, more recently, advances in nozzles technology using abs plastic and air pressure rather than pressurised water systems can potentially bring numerous advantages in avoiding nozzle blockage and reduce maintenance costs.

Another factor contributing to the teething problems of the system was the control logic. This is now regulated so that the nozzles operate only when the external relative humidity is lower than 45%, tightening the earlier control settings. The problem experienced by the occupants of the atrium corridors was dripping and/or condensing moisture on the lower balustrades and the surrounding areas. The perception that the atrium space is *'warm'* will be enhanced by the fact that the fully air-conditioned courtrooms and offices are kept at a considerably lower temperature (of approx. 21°C+/-1).

Water consumption and filtration

High pressure systems require high energy pumps and very high water quality. Currently, the water is purified by a reverse osmosis water filtration system (Nimbus ozone generation system). The filters are changed once a year and the tank is drained every other week in winter. The water is tested once a year but consumption and the system's operation is not automatically recorded by the BMS. The tank capacity is 300 gallons (1,135 litres) but this is an on-demand system and the water filtering happens by directly injecting the chemical treatment. The old system used twelve 300-gallon tanks storing filtered water but often this was not sufficient. Since the refurbishment of the system, a new drainage system was installed, which has individual drainage pipes, which are more effective when shutting off individual banks of nozzles.

> THE FILTERS ARE CHANGED ONCE A YEAR AND THE TANK IS DRAINED EVERY OTHER WEEK IN WINTER

SUMMARY

The cooling strategy for the atrium of the Federal Courthouse in Phoenix is a major demonstration of the potential of passive downdraught evaporative cooling which has provided significant savings in both capital and running costs. It is a very large volume, and PDEC on this scale was initially undoubtedly a challenge for the building managers. However, the predicted conditions of temperature and water consumption have now largely been met, and the system works well in tempering the outdoor conditions, providing a cooler environment in the atrium space in the proximity of the south side offices.

Water consumption is higher than other buildings (1.2litres/m^2/day compared with 0.3litres/m^2/day at the TRC building), but this is in line with the high level of water consumption in the USA (over four times the per capita consumption in Europe).

The views of the occupants (reflected in a small sample survey) identify a general dissatisfaction with temperatures in the floor of the atrium in the summer. However, the expectations of the occupants, in a city where shopping malls and public buildings are air-conditioned, are likely to have a significant influence on their response to more variable conditions. Not surprisingly perhaps, the rather negative perception of summer thermal comfort was particularly prevalent among the group (security guards) who work for long shifts in the middle of the atrium, with little opportunities for adaptation. It is possible that the inclusion of the security staff in the survey has skewed the results, and many of the staff working in offices may find the atrium satisfactory as a transition space.

On the other hand, most of the building's occupants may have expectations of a fully conditioned space, and thus feel 'dissatisfied' when they experience a wider range of conditionings inside a modern building.

This is particularly the case when the home environment, the journey to work and all public buildings are air-conditioned. Further surveys of occupant perceptions are required to obtain a better understanding of 'expectations' during the hot summers of south-western USA. However, in one of the fastest warming cities such as Phoenix, business as usual is no longer an option. A combination of adaptive comfort strategies, reduction of anthropogenic heat emitted by cars and air-conditioning compressors and increase of green, cool surfaces can help towards reduction of the urban heat island effect. This will in turn increase opportunities for night-time convective cooling strategies which can substantially improve indoor thermal conditions during daytime.

REFERENCES

- EPA (2008). *'National Action Plan for Energy Efficiency. Collaborative on Energy Efficiency'* [Online] http://www.epa.gov/RDEE/energy-programs/napee/meetings/sector.html/. [Accessed 10/2018].
- Raman, M. (2009). *'Notes from Teleconference between Dr. Rosa Schiano-Phan and Mr. Mahadev Raman at Arup'*. New York, on 31 May 2009.

CASE STUDY 4
STOCK EXCHANGE, VALLETTA, MALTA

The Stock Exchange in the city of Valletta, Malta is an office building created by the conversion and retrofit of a 19th century garrison church on the walls of the fortified military town. The building's conversion was completed in 2003 and it was designed by architects AP (Architecture Project) in collaboration with Environmental Design Consultants, Brian Ford and Associates, London, and MEP Engineers Mediterranean Technical Services (MTS). This building is the result of a close collaboration between the design team and the client, who fully supported the integrated design approach to environmental control strategies in the conversion of this heritage building. The strategies respond to diurnal and seasonal change, combining three different modes of convective cooling: stack driven natural ventilation, active downdraught cooling (ADC) and passive downdraught evaporative cooling (PDEC). For reasons that will be described later, the operation of the evaporative cooling system has been suspended, while natural stack driven ventilation and the active downdraught cooling systems continue to operate successfully, fifteen years after completion.

40. View of the Stock Exchange, Valletta, Malta [Architecture Project].

224   CASE STUDY 4 | STOCK EXCHANGE, VALLETTA, MALTA

LOCAL CONTEXT

The island of Malta (Latitude 36°N) lies in the southern Mediterranean, south of Sicily and 290km from the coast of North Africa. The entire island consists of beautiful honey coloured limestone, from which Neolithic builders constructed extraordinary temples, and much later Italian Baroque architects created some of the most handsome architecture in Europe. The tradition of high quality masonry construction has continued to the present day.

Malta experiences very hot summers (peak temperatures above 37°C in July and August), while winters are mild and sunny (rarely dropping below +5°C in January and February). Mean max dry bulb temperature is typically 30°C in July and the mean min wet bulb temperature 19.2°C, indicating a large temperature depression and hence a good potential for evaporative cooling (fig. 41). The sea also has a major influence on the climate, which can rapidly change the local microclimate and has a moderating effect. For this reason, summer days can also become warm and humid.

The site of the Stock Exchange is very exposed, perched on the south-western edge of the bastions of Valletta, overlooking the Grand Harbour 30 metres below. The elevated position exposes the building to the prevailing north-westerly winds, and also to the (less frequent) south-easterly winds in summer. A characteristic of the north-westerly winds is that they tend to blow during the day in summer, to leave the nights still

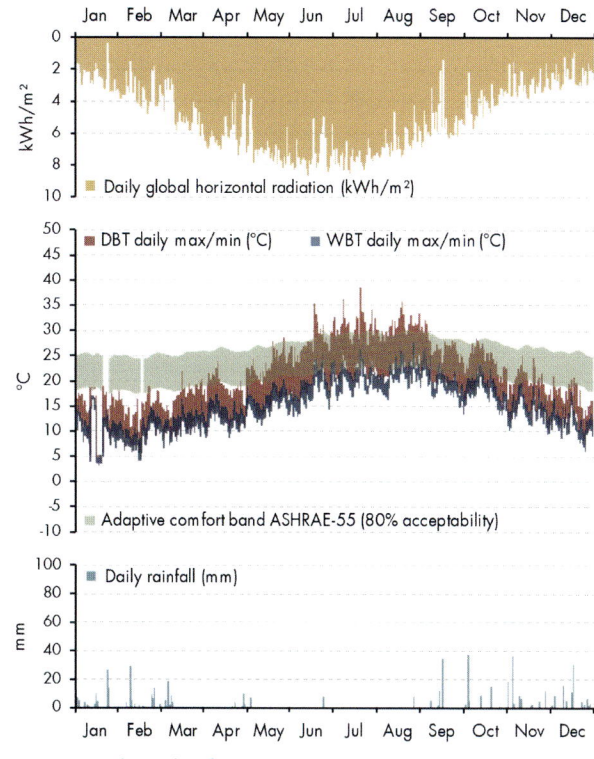

41. Typical weather for Valletta, Malta, ®Meteonorm.

and calm. The less frequent warm south-easterlies bring dust from the deserts of North Africa, and so ventilation openings must be capable of being closed.

Until recently, like many parts of Europe, the cost of energy for the conditioning of buildings has not been a major issue. However, with the more widespread recognition of the need to reduce carbon and other greenhouse gas emissions, and with Malta's accession to the European Union, buildings need to comply with EU Directives on energy performance. In the long term it is undoubtedly also in the interests of the Maltese, since (currently) nearly all energy on Malta is imported, and so improved building energy efficiency has become an important goal for everyone on the island.

> BUILDING ENERGY EFFICIENCY HAS BECOME AN IMPORTANT GOAL IN MALTA

THE BUILDING AND OVERALL STRATEGY

In converting the former garrison church of Valletta into the Malta Stock Exchange, the architects Architecture Project wanted to ensure that the mechanics of cooling (which could have involved extensive ductwork or numerous fan coil units) would not undermine the architectural qualities of the interior. They therefore sought a solution which would minimise the intrusion of plant on the interior.

The refurbishment of the existing listed building involved the incorporation of a five storey atrium surrounded by perimeter cellular offices (fig 42). This was created partly by digging several metres into the rock below the original floor, and removing the original flat ceiling soffit to reveal the magnificent trusses supporting the roof. The open

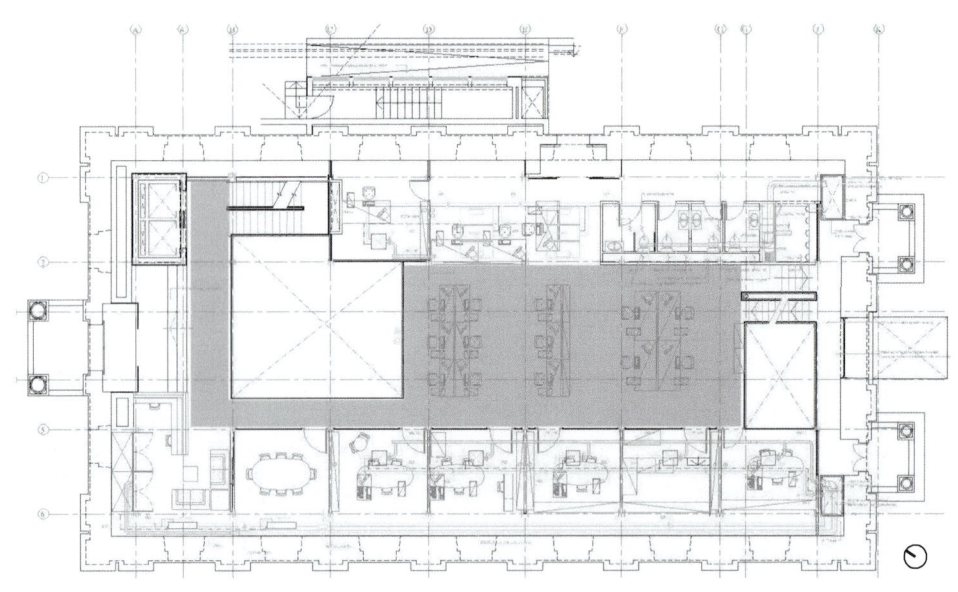

42. Plan of upper ground floor.

226 CASE STUDY 4 | STOCK EXCHANGE, VALLETTA, MALTA

atrium is also occupied as a work space, creating major challenges for environmental control in the central area of the building.

ENVIRONMENTAL STRATEGY

The architects were keen to avoid either very intrusive ductwork or large fan coil units within the central 14 metre high atrium. The cooling strategy adopted was therefore designed to minimise reliance on mechanical and electrical equipment through the application of passive and hybrid downdraught cooling. This avoids the visual intrusion of additional plant, and potentially reduces energy and maintenance costs. This same strategy does not apply to the cellular offices and the lower ground-floor meeting rooms, which are cooled with conventional 'cassette' units.

During summer days the design proposed to exploit downdraught cooling, which was induced in a column of air by either evaporative cooling (by means of hydraulic nozzles) when ambient conditions are dry, or direct cooling (by means of chilled water coils) when conditions are more humid (fig. 43). Previous research had suggested that a combination of these techniques was both technically and economically viable for many locations in southern Europe. The distribution of cool air with both techniques is from high level to low level and the pattern of resulting air temperatures and air velocities within the space is strongly influenced by the geometry of the major spaces.

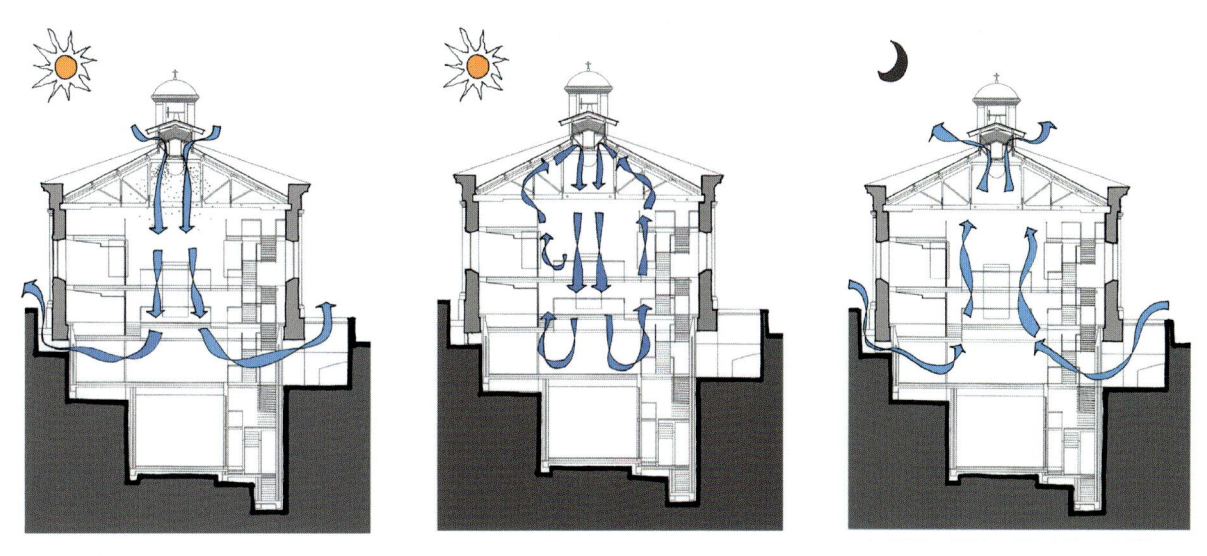

> **PASSIVE AND HYBRID DOWNDRAUGHT COOLING AVOIDS THE VISUAL INTRUSION OF ADDITIONAL PLANT**

43. From left to right: summer day cooling strategy using (i) PDEC, (ii) Cooling Coil and (iii) summer night ventilation strategy using convective night ventilation.

CASE STUDY 4 | STOCK EXCHANGE, VALLETTA, MALTA

44. Shutters in the SW façade.

Complementing the passive cooling strategy for this building, shutters were also placed on NE and SW façades to minimise solar heat gains while continuing to allow ventilation (fig. 44).

In addition, during summer nights (when there is little wind) the design strategy was to exploit buoyancy driven night ventilation (fig. 43) whereby air at low night-time temperatures removes heat built up during the day within the walls, floors and roof of the building, and is exhausted via ridge ventilators. The objective in combining these ventilation and cooling strategies was to provide a comfortable internal environment with minimum reliance on M&E equipment. As we shall see below under 'occupant perceptions', a number of factors have meant that this strategy has been modified to meet the needs of the client.

COOLING SYSTEM AND BUILDING INTEGRATION

The cooling and heating of the large central space was designed to be achieved by five complementary strategies:

Automated natural ventilation

The 18 high level vents (1.5x0.85m) are top hung windows at the ridge, with a maximum throw of 0.5m. These window vents are vulnerable to winds of the south-west (average 4mps), so each vent is operated by two steel toothed 'racks' or stays, rotated by pinions located on a tubular drive shaft operat-

228   CASE STUDY 4 | STOCK EXCHANGE, VALLETTA, MALTA

ed by three phase motors. The vents at lower ground level consist of motorised dampers mounted in the east, west and south walls. These ventilators have edge seals to minimise uncontrolled ventilation (although these seals have been found to be ineffective). The operation of the vents is controlled by the building management system, although this can be overridden (and often is) by the building manager (discussed below).

Chiller based downdraught cooling

Two chilled water circuits serve cooling coils from the header along each side of the walkway. The copper coils are inclined at approximately 70° to encourage airflow through the coils, and 300mm wide purpose made trays were installed to collect condensate. The cooling coils call for chilled water in response to variations of both internal temperature and internal relative humidity. The control strategy adopted assumes that the cooling coils are linked to the automatic dampers and vents and operate in conjunction with them. When the coils are switched on, the actuators at high level and the dampers at low level are both set to the minimum aperture.

Passive downdraught evaporative cooling (PDEC)

The PDEC system, as designed and installed, relies on 14 hydraulic nozzles at 1 pair/bay in four sets. Each set is controlled by motorised valves linked back to the Building Management System (BMS). The total water volume flow required is approximately 90 litres/hour with a pressure of 25 bar. The pipe work is attached to the balustrade and the nozzles have been located at the centre of the gap between the cooling coils. The nozzles also operate in conjunction with the motorised dampers. When the system is on, both banks of motorised dampers and high level vents are set to fully open. The aperture of openings may be overridden by the wind speed/direction sensor. The dampers are set to the minimum aperture when the wind speed is above 5m/s.

> BOTH EVAPORATIVE COOLING AND THE CHILLED WATER COILS RELY ON BUOYANCY TO DRIVE THE AIRFLOW THROUGH THE BUILDING

Both evaporative cooling and the chilled water coils (fig. 43) rely on buoyancy to drive the airflow through the building, and so fan power (which can sometimes be 30–40% of the electrical energy used in conventional air-conditioning) will be avoided, providing a potentially significant electrical energy saving.

Night-time convective cooling

It was anticipated that during the day in summer, internal temperatures would rise to 25°C or 26°C. Useful convective cooling is therefore promoted when the external temperature drops below 23°C, which occurs frequently at night during the summer period. The movement of air driven by either wind or thermal forces (fig. 43) reverses the daytime air movement pattern under downdraught cooling. Night cooling is controlled via the

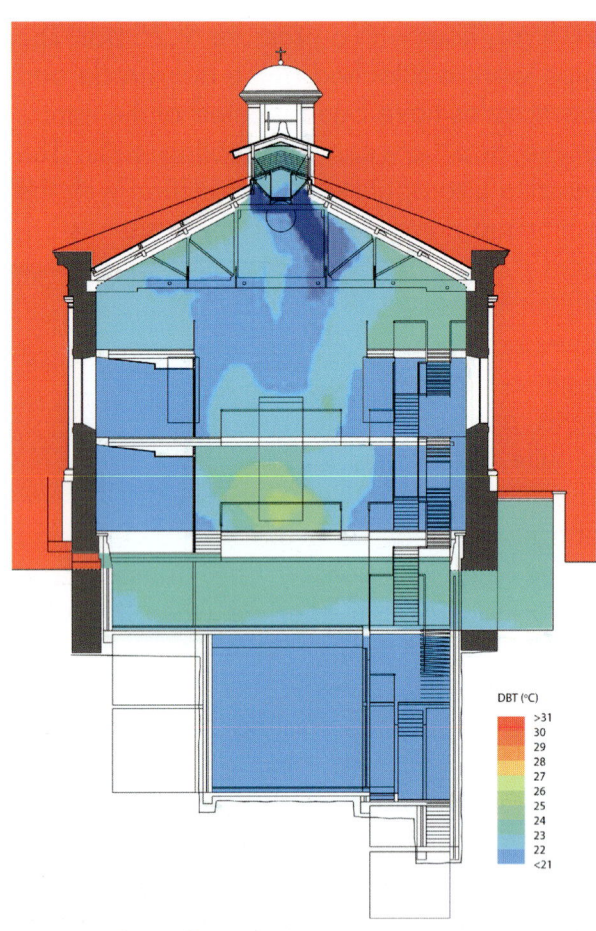

45. CFD plot overlay on section.

motorised dampers at lower ground level and the vent actuators at ridge level. Convective cooling relies on large air change rates, and so when night ventilation is initiated, all low and high level vents are fully opened. In spring and autumn, the control system is set to prevent over-cooling of the interior.

Heating

Fan coils have been used with the heat pump chillers for a maximum water temperature of 50°C. Simulations indicated a peak demand of 25–30kW to bring the central space up to 21°C from an ambient low of 8°C on a Monday morning in February. However, once the building is occupied the demand for heat is very low. It was important to make sure that the location of these fan coil units avoided the obstruction of the airflow path under natural ventilation.

DESIGN STAGE PERFORMANCE PREDICTION

Performance prediction at the design stage indicated that temperatures in the main occupied areas of the atrium would remain at an average of 24–25°C during summer daytime occupancy periods, but these vary according to localised internal gains across the plan. Warm air rises from the lower ground floor through the voids on the north side of the building. On the south side the downdraught effect from the cooling coils is more emphasised although with low velocities (between 0.3–0.4m/s). The general result indicated a good mixing of the air across and along the central space and a fairly even temperature distribution. With an outside temperature of 38°C, the downdraught cooling effect from the coils appeared to be able to maintain relatively low internal temperatures with no signs of overheating (fig. 45).

COMMISSIONING PERIOD MEASUREMENTS

The building was occupied (in August 2001) before the building's systems and controls had been fully commissioned. However, a limited set of measurements of the performance of the cooling coil system were made over a period of a few days in September 2001. The results indicated that although the cooling coil system was operating in line with predicted performance, during the period of measurements the relative humidity was too high for the evaporative cooling system to be activated. With the high level vents closed, an almost symmetrical pattern of air movement was generated down through the central space, returning up past the balustrades to the roof and through the coils at walkway level (figs 46–47).

The record of continuous measurements shows the impact of the cooling coils on temperatures in the atrium very clearly. Measurements for 21 September are closest to a *'controlled'* test in that there were no 'cooling gains' from the cellular offices and the ground floor door and gable vents were closed. They show the cooling coils bring the temperature at the first floor to just above

46. Internal view of the perforated plan by large openings at different levels to allow circulation of air within the building's interior [Architecture Project].
47. Internal view of the cooling coils at walkway level.

CASE STUDY 4 | STOCK EXCHANGE, VALLETTA, MALTA

24°C at an RH of 70–75%. On the top floor temperatures were 1–2°C higher, which are consistent with temperature predictions.

OCCUPANT PERCEPTIONS OF THE BUILDING IN USE

Most of the staff enjoy their working environment, describing the building as: *"modern and elegant"*, *"very welcoming and modern design"*. Unfortunately, the occupants reported problems with the evaporative cooling system from the first year of operation. The problems included droplets of water being discharged beyond the drip tray, falling on work stations and causing considerable annoyance. This was clearly not acceptable, and perhaps understandably the PDEC system was disabled.

However, these problems may be attributed in part to the fact that the building was occupied before the PDEC system could be properly commissioned. It was also discovered that poor alignment of the nozzles with the drip tray, or their close contact with adjacent water pipes and cooling coils, had also clearly been a cause of dripping. Some of these problems were identified during commissioning but subsequently not addressed.

Another matter arising from the fact that commissioning took place after the client had moved in, is that the control of the low level vents was not properly regulated, and staff on the lower floors reacted to unwelcome drafts by obstructing the openings. While appearing to solve the problem in winter, the reduced ventilation opening area has resulted in complaints of *'stuffiness'* in summer. It is also not clear whether the night ventilation regime is in operation. While the reaction of staff is understandable, full commissioning of the system could have resolved these problems. A survey of staff perceptions of the internal environment was conducted in May 2008.

DESCRIPTION & RESULTS OF THE SURVEY

A sample of 33 staff, representing 75% of the building's occupants, responded to the survey. Members of staff who responded were located both in the cellular offices and in the atrium open plan space and the majority of them answered the questionnaire in the course of one day. The results show that occupants do not have a very positive view of the building, with aspects such as noise and perceived health being the most unsatisfactory. The overall comfort in the building is rated as *'neutral'* (or amber in the BUS colour code) as the occupants' response is no lower than scale midpoint but lower than the benchmark. This means that the majority of the occupants gave a mid-scale vote (neither satisfactory nor unsatisfactory). However, this is lower than the benchmark derived from other buildings, where the overall score tends to be placed on the higher end of the scale (i.e. satisfactory). The same trend applied to 6 out of the 12 investigated aspects, with 5 rated as unsatisfactory and only one satisfactory (fig. 48).

Overall thermal comfort in summer was lower than both benchmark and mid-scale, suggesting that certain areas of the building tend to overheat (indeed just over half the respondents felt too hot in summer). A few years ago supplementary cooling was installed at level +2 of the atrium as this space was to be occupied (previously only used for storage). However, from occupants' comments it was apparent that there is a major problem with air quality and provision of natural ventilation both in summer and winter. Air quality in summer, however, was better perceived (amber) than air quality in winter (red). The main problem in summer seemed to be stuffiness (suggesting inadequate ventilation), whereas in winter it was draughts, suggesting that the seals on the inlet motorised dampers were unsatisfactory. It appears that the building manager frequently intervenes to over-ride the control system.

The detailed results on temperature, air quality in summer and control are discussed below:

Temperature in summer

From the survey it appears that 33% think that the temperature is comfortable against 42% perceiving it as uncomfortable. The remaining 25% expressed a *'neutral'* vote (i.e. not satisfactory nor unsatisfactory). The overall vote is significantly lower than both scale midpoint and benchmark. However, when prompted on hot/cold perception of

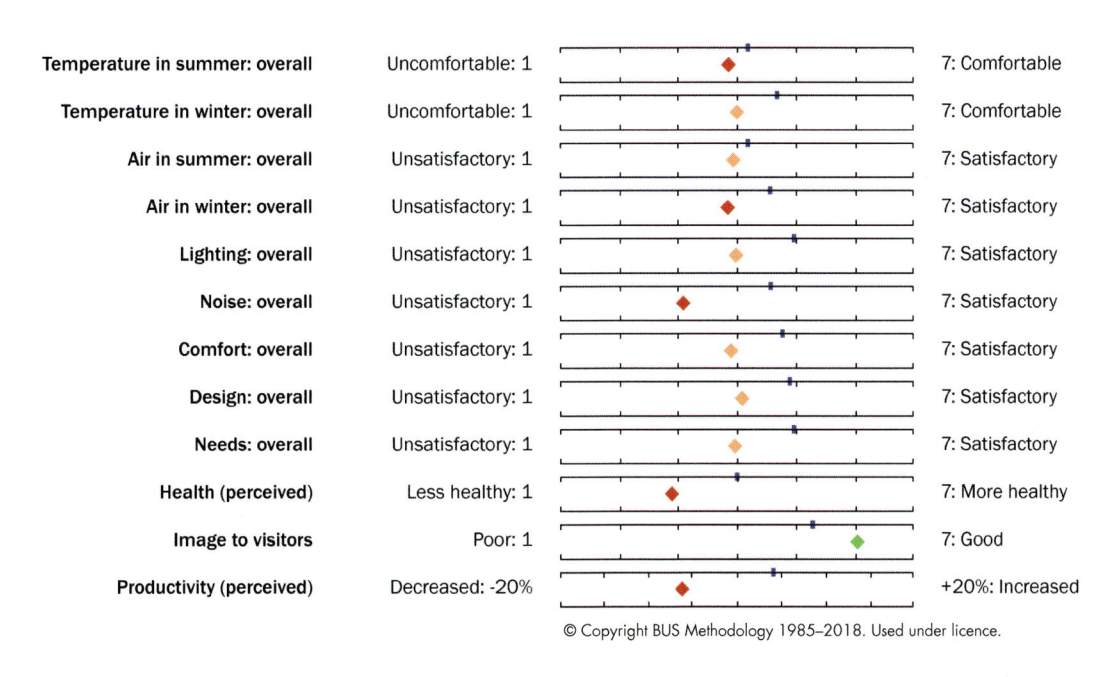

Category	Low	High
Temperature in summer: overall	Uncomfortable: 1	7: Comfortable
Temperature in winter: overall	Uncomfortable: 1	7: Comfortable
Air in summer: overall	Unsatisfactory: 1	7: Satisfactory
Air in winter: overall	Unsatisfactory: 1	7: Satisfactory
Lighting: overall	Unsatisfactory: 1	7: Satisfactory
Noise: overall	Unsatisfactory: 1	7: Satisfactory
Comfort: overall	Unsatisfactory: 1	7: Satisfactory
Design: overall	Unsatisfactory: 1	7: Satisfactory
Needs: overall	Unsatisfactory: 1	7: Satisfactory
Health (perceived)	Less healthy: 1	7: More healthy
Image to visitors	Poor: 1	7: Good
Productivity (perceived)	Decreased: -20%	+20%: Increased

© Copyright BUS Methodology 1985–2018. Used under licence.

48. Summary of post occupancy survey results for Stock Exchange, Valletta, Malta, ©Usable Buildings Trust.

the temperature 51% of occupants responded as too hot, which is lower than scale midpoint but not different from the benchmark. Not surprisingly, satisfaction with the temperature in summer depends on where the respondent works within the building.

Air quality in summer

For most of the staff (56%) the air is perceived to be still. It is interesting to note that with the evaporative cooling system disabled, 36% consider the air to be dry, whereas 70% consider it stuffy and 62% smelly (it appears that this latter issue relates to a drainage problem which has now been addressed). This confirms the comments gathered during the informal interviews and the reported problems with the controls of the ventilation system. Overall 45% consider the air quality to be unsatisfactory. This is lower than benchmark but not different from scale midpoint.

Control

Frustration with the environment in the building is apparent, since 80% of the occupants feel they have no control on heating, cooling, ventilation or lighting.

COMMENTS OF THE OCCUPANTS

Dissatisfaction with the internal environment is also expressed in comments from staff, particularly with regard to the ventilation: staff reported a *"lack of fresh air"*, and *"no fresh air"*. The interior is also described as *"poorly ventilated"*, *"stuffy air"*. Staff have also complained of: *"headaches due to stuffiness"*, and that *"more ventilation would be appreciated"*. There are also contradictory comments, some describing the air as *"still"* while others complained of draughts. This is clearly dependent on the location of the respondent in the building. Some employees on the lower ground floor, for instance, have obstructed part of the fan coil unit (positioned at high level on their backs) as it was causing draught and discomfort.

Not all the comments are negative. The building is described by staff as: *"modern and elegant"*, *"very welcoming and modern design"*, *"very nice design but in some factors not practical"*.

LESSONS LEARNED

One of the three elements of the summer cooling strategy for the Malta Stock Exchange has proved to be problematic. Direct evaporative cooling to open plan office areas within the atrium resulted in problems of over-spraying and dripping. This was due in large part to misalignment of nozzles. Such problems have not arisen at the Torrent Research Centre where direct evaporative cooling via high pressure spray nozzles was also applied, although at Torrent there are no people working directly below the nozzles. At the time the Stock Exchange was occupied, the evaporative cooling system could not be commissioned, and faults with the installation were not discovered until much later.

Night-time convective cooling is dependent on automated control of ventilators at high and low level. Motorised dampers at the lower level on the west side of the building have a poor seal and have been a source of unwanted draughts. Unfortunately due to this and other perceived problems, automatic control of cooling and ventilation was disabled.

However, the strategy of using cooling coils at high level has been successful in delivering cooling to the occupied spaces in the atrium as originally designed and forms a precedent for its application in large volume spaces elsewhere.

The premature occupation of new buildings prior to full commissioning of all the mechanical and electrical systems, is a common source of problems (PROBE study, 1995–2002). It is possible that the mechanical systems and controls were never properly commissioned. It is also likely that without a dedicated professional building manager, maintenance of the buildings systems and controls has not been properly coordinated.

Energy consumption figures for the building are not available at this time. However, the energy savings predicted at design stage will not have been achieved in full, although the success of buoyancy driven air movement from the cooling coils will have saved fan energy.

The smoke tests revealed a pattern of air movement, which is also consistent with the CFD analysis undertaken at design stage. Air velocities within the central occupied spaces are low (0.1 to 0.3 m/s), but the movement of large volumes of air at low velocities are desirable and part of the strategy (down draught 'plumes' of higher velocity, could become uncomfortable).

However, the combination of detailed air velocity measurement and smoke tests revealed that rising warm air passes over and through the coils where it gets cooled, resulting in a downdraught of cooled air as predicted. Substantial variation in flow over the coils is to be expected, but the overall effect is that the system is dealing with internal heat gains and providing stable, comfortable conditions generally.

THE HYBRID COOLING STRATEGY RESULTED IN A TOTAL ENERGY SAVING OF 48%

POSITIVE ASPECTS OF PROJECT IMPLEMENTATION

Energy saved

Buoyancy driven cooling from the cooling coils will provide an energy saving compared with fan coils or similar air-con units.

The general building improvements (shading, insulation, night ventilation) also benefit the cellular office accommodation, reducing the energy consumption of the cassette units by approximately 23% *(Ford & Diaz, 2003)*.

The total cooling energy consumption for the whole building using fan coils was estimated at approximately 103,924kWh (70kWh/m²), compared with 54,139kWh (36kWh/m²) for the built solution using cooling coils and PDEC, resulting in a total energy saving of 48%. With PDEC not operating, the likely energy savings attributable to the cooling coils in the atrium (i.e. not using fans) will be in the region of 30kWh/m²·yr.

Economic benefits

A comparison of the capital costs of the different options to ventilate and cool the building was not made. However, the adoption of a combination of active downdraught cooling and natural ventilation of the large central atrium space has avoided the need for fans and ductwork, providing a saving in capital and maintenance costs.

Environmental benefits

The estimated energy savings for the building as currently operated represent a reduction in CO_2 emissions of approximately 10,000kgCO_2 (6.5kgCO_2/m²). This is about one third of the savings predicted for the building as originally designed.

Reduced need for bulky plant

The high level cooling coils have been successful in delivering cooling to the atrium space, and this part of the system has worked as predicted. This demonstrates that the downdraught process drives the airflow pattern inside the building, avoiding the need for fans, ductwork and suspended ceilings. Downdraught cooling and night-time convective cooling strategies therefore have clear benefits.

PROBLEMS ENCOUNTERED AND OPPORTUNITIES FOR IMPROVEMENT

Energy saved

Energy use for building environment control and energy management has not been recorded, so savings cannot be verified. However, they are estimated to be about one third of those predicted (i.e. approximately 40kWh/m²), due to increased dependence on the cooling coils. This is based on the increased electrical load from the compressor.

PDEC system

'Over-spraying': Staff reported that when the evaporative cooling system was operated, droplets of water were discharged beyond the drip tray, and caused considerable annoyance. It was perhaps optimistic to locate misting nozzles above working areas, and not surprising that the high pressure nozzle system has been disabled. Subsequent investigation revealed that the dripping or *'spitting'* arose when the nozzles were switched off. It has also been found that misting nozzles which rely on compressed air linked to a low pressure water supply provide benefits of smaller droplets (quicker evaporation) and avoid *'spitting'* when switched off. These benefits were observed in the operation of the Nottingham Solar Decathlon House described in Chapter 4.

Commissioning

The building was occupied before it had been fully commissioned. As a result, the PDEC system was not commissioned and the control system has not been allowed to function as originally intended. It is widely acknowledged that full commissioning of all building energy systems is vital prior to building occupation but is still a frequent occurrence.

Building management

Staff have indicated their frustration with the system, and there was nobody who was able to fully comprehend and manipulate the building energy management system. A system of this type will normally require a professional building energy manager to ensure success.

SUMMARY

The design of the Malta Stock Exchange incorporated the ambition to architecturally integrate a hybrid cooling strategy. In the event, the implementation of this strategy and its operation in practice has only been partially successful. The application of chilled water cooling coils to provide active downdraught cooling (ADC) has been a success, both in terms of providing satisfactory cooling and in avoiding the installation of bulky and visually intrusive mechanical cooling systems. It is also the case that although the original evaporative cooling system has been disabled, technical developments suggest that this could now be implemented satisfactorily. Other aspects of the operation and management of the Stock Exchange building, which have resulted in localised discomfort and dissatisfaction, reflect common problems with regard to commissioning and the training required for building managers. However, a number of lessons have been learnt which will benefit designers and managers and will hopefully contribute to the greater success of the next generation of low carbon buildings.

REFERENCES

- Ford, B. & Diaz, C. (2003) *'Passive Downdraft Cooling: Hybrid Cooling in the Malta Stock Exchange'*. Rethinking Development: Are We Producing a People Oriented Habitat? Passive and Low Energy Architecture Conference, Santiago, Chile.

CASE STUDY 5
ZION NATIONAL PARK VISITOR CENTRE, SPRINGDALE, UTAH, USA

The Zion National Park Visitor Centre in Springdale, Utah is a mixed use office, commercial and exhibition space building with a floor area of about 800m². It was commissioned by the National Park Services (NPS) as a gateway to the park in order to improve access and transportation within the site. The centre was built in 2000 and it was designed internally by the NPS Denver Service Centre architects in collaboration with engineers from the Department of Energy's National Renewable Energy Lab (NREL). This building generated much interest for the application of several renewable energy forms and sustainable strategies and it still represents a successful example of the application of cool towers . In 2001 it was listed among the AIA Committee on the Environment Top Ten Projects award.

49. External view of Zion Visitor Centre [NPS Denver Service Centre Architects].

LOCAL CONTEXT

The stunning natural environment and the varied microclimate of the Zion National park (figs 49–51) are claimed to have inspired the designers of the visitor centre. The reverberating mass of the canyon surfaces, the contrast between the bright midday sun and the enclosed, sheltered feeling at dusk, all contribute to the experiential delight of the place. Moreover, the combination of the dramatic high peaks, the declining canyon river, the secluded waterfalls and pools, add to the creation of local microclimates and to the intense natural experience which makes this a very special site.

It is apparent, and so the designers claim, that this building tries to mimic and exploit both the potential of the surrounding microclimate and the local materiality. Thermal mass, day-lighting, solar energy, natural ventilation, evaporative cooling, all have been carefully applied within the building and resonate with the local context. The aesthetic result, however, is somehow conventional as the concerns with the local conservation laws and the mimetic intention have resulted in a building which possibly resembles a ranch farm more than anything else

The visitor centre is situated 40 miles north-east of the capital city of St George, in Springdale (Lat. 37°N, Long. 112°W), a small gateway village to the Zion canyon system. The visitor centre complex was designed as an answer to the air and noise pollution generated by traffic of the thou-

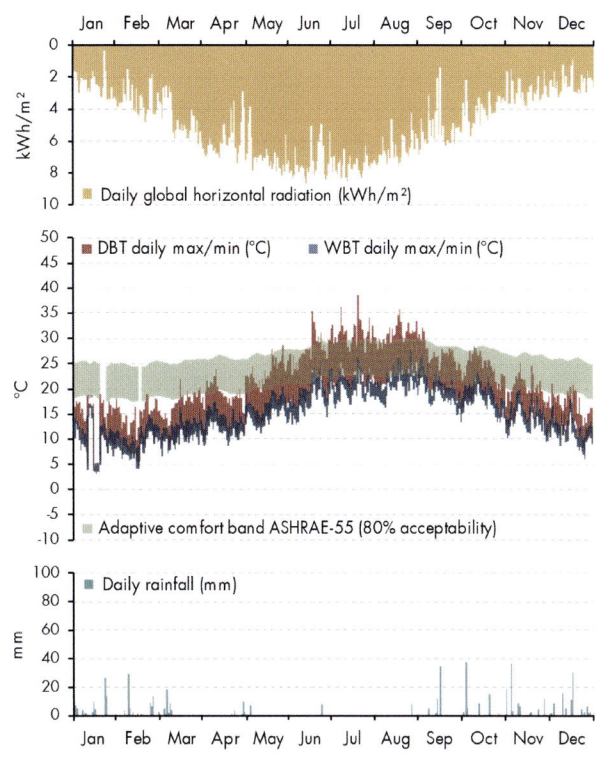

50. Typical weather for Zion National Park, ®Meteonorm.

51. Visitor Centre, view of indoors.

CASE STUDY 5 | ZION NATIONAL PARK VISITOR CENTRE, SPRINGDALE, UTAH, USA

sands visiting each year. The centre now operates on a *'park and ride'* basis reducing the adverse environmental impact of cars.

The local microclimate is dictated by a number of factors including: altitude, which varies from 1,200m at the lowest point to 2,660m at the highest; the exposed canyon mass (fig. 52), which absorbs heat during the day and releases it at night; and the proximity to the Virgin River, which has an evaporative cooling effect on the surrounding site. Winters are sunny and temperature are relatively milder during the day (10–15°C) but can reach below freezing at night *(NPS, 2017)*. Summers are very dry and

> THE EXPOSED CANYON MASS ABSORBS HEAT DURING THE DAY AND RELEASES IT AT NIGHT

hot with only 375mm of precipitation a year *(ibid)* and peak temperatures exceeding 38°C *(Torcellini et al., 2005)*. In the summer of 2006 a heat wave was experienced with temperatures reaching 47°C to 48°C. In 2007 the peak was 42°C outdoor, whilst indoor 27°C was recorded in the visitor centre. The relative humidity is generally very low, often below 10% creating great discomfort to visitors and residents. Also, the mean max dry bulb temperature in July is 38.3°C. This with a corresponding mean max wet bulb temperature of 18.1°C (i.e. high wet bulb depression) in July gives a measure of the high applicability of evaporative cooling in this location (fig. 50).

The prevailing south-westerly winds, whose pattern is influenced by the canyons'

52. Zion National Park, view of canyon peaks.

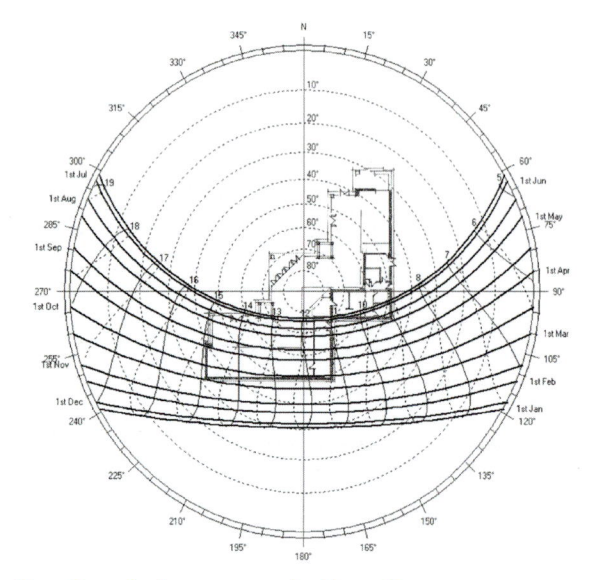

53. Sunpath diagram over plan Visitor Centre.

walls narrow up-river and wider down-river, create a diurnal chimney effect with winds going up-canyon between noon to midnight and down-canyon from midnight to noon (*Torcellini et al., 2005*). There is a wet season when it rains very frequently from mid-July to the end of September and where flash floods are often experienced. Springdale's average annual temperature is approximately 14°C, which could make the option of a ground source heat pump system theoretically viable to provide pre-cooled and pre-heated air to the building in summer and winter. Also, the abundance of clear skies during summer nights (with typically low temperatures) could also be exploited for radiant cooling as in the example of the Global Ecology Research Centre in California.

THE BUILDING AND THE OVERALL STRATEGY

PHYSICAL DESCRIPTION OF THE BUILDING, MATERIALS AND SYSTEMS

The Visitor Centre complex (fig. 54), designed by the architects of the US National Park Services, comprises of an outdoor exhibition area and three buildings: the main visitor centre building (818m²), the restrooms (256m²) and the fee station (15.8m²). The main building (fig. 55), which is the focus of this investigation, houses a book shop and exhibition space, the main reception desk and offices at the back on the east side.

The consultancy over the cooling strategy and the design of the cool towers was

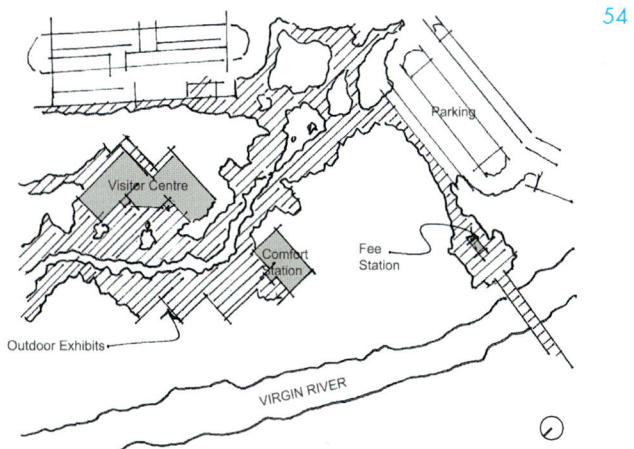

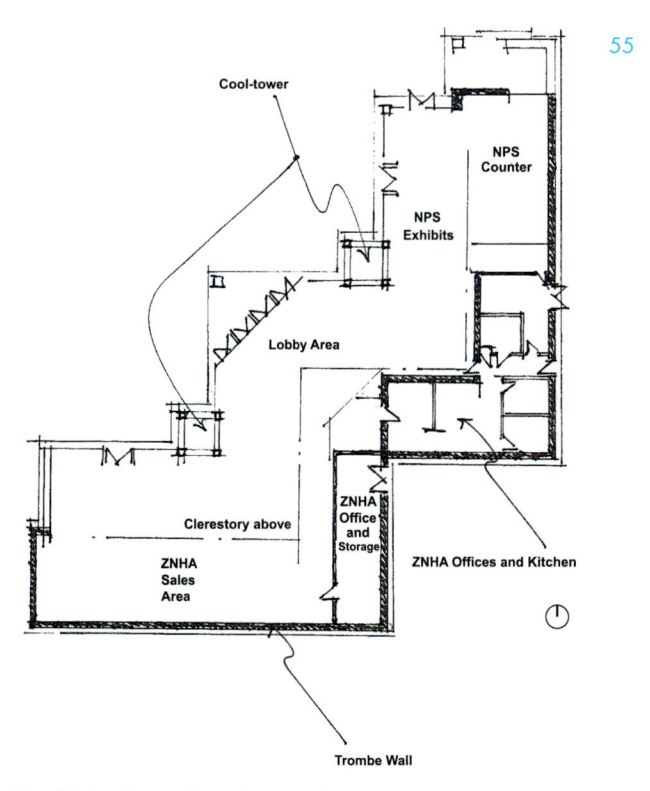

54. Visitor Centre Complex, site plan.
55. Floor plan of Visitor Centre.

CASE STUDY 5 | ZION NATIONAL PARK VISITOR CENTRE, SPRINGDALE, UTAH, USA

undertaken by the University of Arizona at the very early stages of the design. The low-energy strategy and the environmental and energy analysis during design and at the early stages of occupation, however, came from Mr Paul Torcellini, an energy engineer from the National Renewable Energy Laboratories (NREL). Mr Torcellini undertook the energy analysis of the various building design options and specified the original instrumentation for the building's automated control systems for lighting, cooling, heating and hot water.

ENVIRONMENTAL STRATEGY

The clear low-energy agenda set by the National Park Services and the resulting brief required a 70% saving compared to a conventional equivalent. This was achieved by a combination of energy demand reduction strategies and on-site renewable energy generation. The overall environmental strategy included the employment of day-lighting, low-energy lighting, passive solar heating, natural ventilation, and passive evaporative downdraught cooling.

> THE VERY HIGH DIURNAL VARIATION OF 17°C IN SUMMERTIME ALLOWS FOR NIGHT-TIME CONVECTIVE COOLING

The cooling and heating strategies clearly respond to the seasonal requirements. However, due to the local microclimate, cooling might be required not only in summer but also in spring and early autumn, when the peak daytime temperature can reach highs of 32°C. The very high diurnal variation of 17°C during midseason and summer allows for the additional exploitation of night-time convective cooling through perimeter openings.

Natural ventilation is employed to provide fresh air throughout the year and comfort cooling during periods of high occupancy. When natural ventilation is insufficient, comfort ventilation is achieved mechanically via ceiling fans.

Heating demand is predominant between December and March with sporadic smaller demand in the mornings of the midseason. This demand is met by passive solar heating through south facing windows and a Trombe wall. Residual loads are met by electric radiant ceiling panels.

Cooling demand is comparatively smaller than the heating one and it is mitigated by solar control via the roof overhangs and exterior vegetation; and by natural ventilation and night-time pre-cooling. Residual cooling requirements during dry conditions are met by a passive downdraught evaporative cooling system delivered by cool towers using wet cellulose mats. Summer conditions are mostly hot and dry but during wet conditions, stack ventilation is achieved between the low-level inlet windows and high-level motorised clerestory windows.

The building envelope is well sealed and insulated. It is characterised by a high thermal mass 1.2m concrete floor slab insulated with 250mm of rigid insulation board to the ground. The walls are 152mm steel-stud with expanding blown-in-place foam insulation. The exterior sheathing, covered by cladding, is 180mm extruded foam achieving a total U-value of 0.34W/m²·K. The timber roof has a U-value of 0.18W/m²·K and is made of structural insulated panels (oriented strand board with rigid foam insulation).

The 7.2kW grid-connected photovoltaic panels were integrated on the south roof (figs 56–57) and at the time of the visit in 2007 provided 10% of the power for the majority of the uses. The energy demand reduction and the contribution of the renewable on-site generation produced a considerable saving in the total electricity bill (including back-up electric heating, hot water, internal and external lights) which was $400 a month (compared to the $4,000 a month of

56. South side with top hung clerestory windows, roof mounted PVs and Trombe wall.

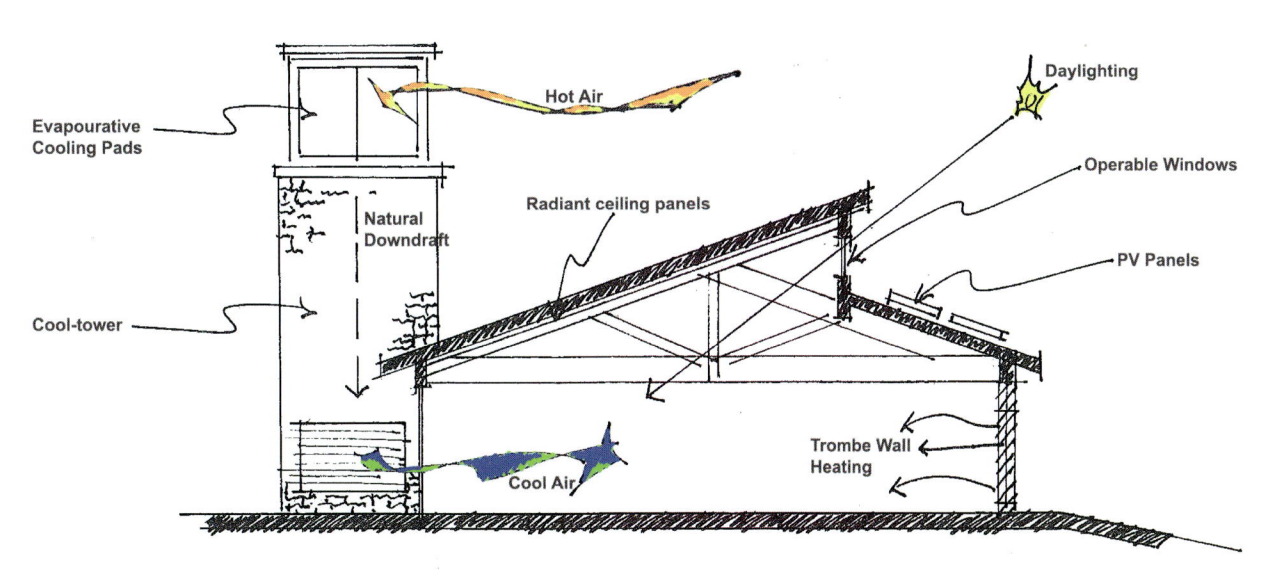

57. Section showing Cool Tower operation and other strategies.

CASE STUDY 5 | ZION NATIONAL PARK VISITOR CENTRE, SPRINGDALE, UTAH, USA

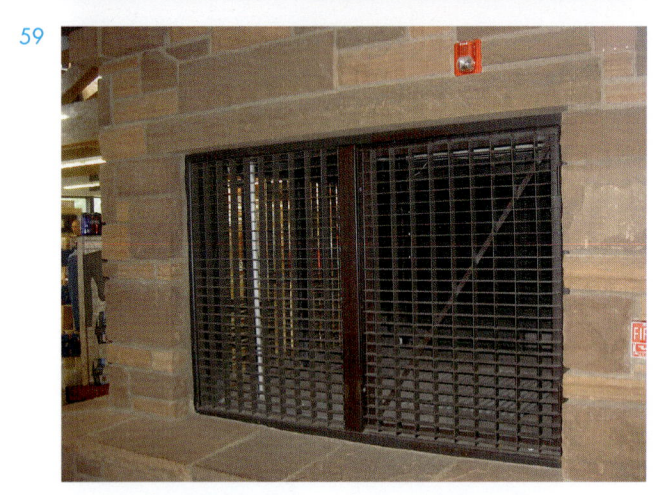

58. Cool Tower termination with wet pads screens.
59. Cool Tower outlet.

the pre-existing 1960 office building). Earlier post occupancy monitoring by the energy consultant, conducted soon after inception, revealed that the energy consumption for the operation of the cool towers' pumps and ceiling fans was in direct correlation with the seasonal weather conditions and ranging between 4.5kWh/m² and 6.6kWh/m² in 2001 and 2002 respectively. However, subsequent studies showed that the average cooling energy consumption was 4.5kWh/m²·yr as opposed to 17.3kWh/m²·yr of a conventionally cooled building in this location.

COOLING SYSTEM AND BUILDING INTEGRATION

The passive cooling strategy is deployed in summer and during periods of overheating. Depending on the relative outdoor-indoor conditions, cooling is achieved through natural stack ventilation via the low-level inlet windows and high-level outlet clerestory vents. Additional cooling is provided by ceiling fans and during peak hot-dry conditions passive downdraught evaporative cooling is delivered by two cool towers using wet cellulose pads at the top termination. The towers' operation is controlled by the building management system, which also controls the size of the opening at the bottom of the tower directing the cool air into the building visitor hall, into the outdoor patio or both. Night-time pre-cooling is often employed to reduce daytime temperatures during expected peak hot days but this is only implemented through the perimeter high and low level vents.

The cool towers use a wet cellulose pad system called CELdeck *(Munters, 2008)*. Each pad is 100mm thick, 1.8m tall and 450mm wide (fig. 58). The pads are wetted by an irrigation pipe which sprinkles mains water on the cellulose matrix. The un-evaporated excess water is collected at the bottom in a sump pit (fig. 59). From the water tank a re-circulation pump (120V, 249W for each tower) feeds back to the water pipes onto the pads.

The passive cooling strategy and the operation of the cool tower follow a specific sequence which is remotely controlled by the Building Management System (BMS). The typical sequence is as follows:
 1. Open clerestory windows.
 2. Open cool tower doors.
 3. Turn on pumps linked to water tank.

If this is not enough to maintain comfortable conditions, then:
 4. Turn on ceiling fans.
 5. Turn on floor fans.

The control system is maintained in house, which means that the engineer can change the sequence of operation and adjust the opening set points as and when necessary. The set points regulating operation of perimeter windows and cool tower doors and water pumps, for natural ventilation and cooling respectively, vary seasonally and are adjusted every two weeks. For example, for days when a peak outdoor dry bulb temperature of 26.6°C is expected then a set point will vary between 21–26°C; if the expected outside temperature is 43°C then the set point will be 17–18°C for morning pre-cooling, and this is achievable due to high diurnal variations with early morning and night-time temperatures dropping below 20°C.

Water efficiency is a very important issue due to the high water requirement and the relatively dry climate. However, rainwater harvesting has not been implemented even though the yearly rainfall of 375mm could yield about 337,500 litres of water on the main roof alone. The water usage is typically 1,800 litres a day (one 227 litres water tank per tower is flushed every 6 hours) and the typical cooling season runs from May to October (180 days on average), making water harvesting a viable option for satisfying most of the water requirement. There is a metering device on mains water; the drained water is collected and used to water the garden. The Cool Towers' water consumption metered in the year 2002 showed an average yearly water consumption of 421,000 litres (514 litres/m^2), which is considerably greater than the water consumption of the Phoenix courthouse's consumption of 216 litres/m^2·yr). This figure, however, was inflated by the wastage produced by the malfunctioning plumbing and sump pit overflow.

OCCUPANTS' PERCEPTION

The Post Occupancy Evaluation survey involved six subjects who answered the questionnaire in the course of one day (16 October 2007). The small sample was dictated by the small number of permanent staff.

The results (fig. 60) show that occupants have a fairly positive perception of the building, with aspects such as image, design and needs being the most satisfactory. Thermal comfort in summer was neutral with a slightly higher mark than the scale midpoint but not much different from the benchmark. The same applied for air quality and lighting. Noise in such a multi-functional open space seems to be problematic with 83% perceiving the acoustic conditions as unsatisfactory.

The detailed results on temperature, air quality in summer and control are discussed below:

Temperature

60% think that it is too hot and that the temperature varies during the day. This confirms some of the outcomes of the post occupancy monitoring done by the Mechanical and Electrical Engineers *(Torcellini et al., 2005)*. However, the fact that the temperature seems unstable for some can be due to the fluctuations of occupancy, with waves of visitors increasing the thermal loads instantaneously. The combination of the open plan layout and the lack of an effective lobby is thermally significant, especially for a public building such as this one where the proportion of permanent occupants is very small

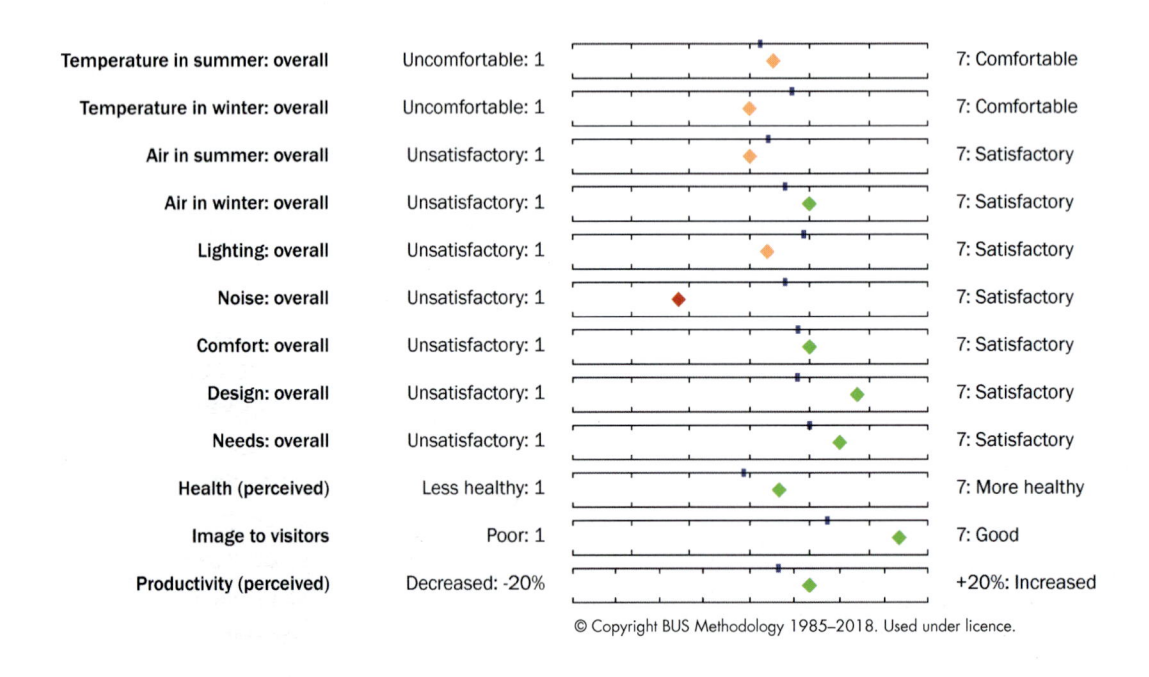

Category	Scale Low	Scale High
Temperature in summer: overall	Uncomfortable: 1	7: Comfortable
Temperature in winter: overall	Uncomfortable: 1	7: Comfortable
Air in summer: overall	Unsatisfactory: 1	7: Satisfactory
Air in winter: overall	Unsatisfactory: 1	7: Satisfactory
Lighting: overall	Unsatisfactory: 1	7: Satisfactory
Noise: overall	Unsatisfactory: 1	7: Satisfactory
Comfort: overall	Unsatisfactory: 1	7: Satisfactory
Design: overall	Unsatisfactory: 1	7: Satisfactory
Needs: overall	Unsatisfactory: 1	7: Satisfactory
Health (perceived)	Less healthy: 1	7: More healthy
Image to visitors	Poor: 1	7: Good
Productivity (perceived)	Decreased: -20%	+20%: Increased

© Copyright BUS Methodology 1985–2018. Used under licence.

60. Summary of post occupancy survey results for Zion National Park Visitor Centre, Springlade, Utah, ©Usable Buildings Trust.

CASE STUDY 5 | ZION NATIONAL PARK VISITOR CENTRE, SPRINGDALE, UTAH, USA

compared to the large number of visitors that enter the building intermittently. The presence of a lobby or of a 'decanting' space for the large number of visitors to transition before entering the space would have been beneficial in reducing the prolonged coupling of the indoor space with the warmer outdoor conditions, taking place at the time of the visitors entering. Also, the thermal mass is mainly placed on the floor and some of the walls, which perhaps it is not as effective as it could have been if spread more evenly on the roof and the south walls.

Air quality in summer

For most of the staff air is still, fresh and dry but overall half of the sample considers the conditions satisfactory and the other half unsatisfactory. In these cases and with such a small sample it is difficult to surmise on the occupants' response. However, measured airflows in each tower were approximately 7.55m³/s during operation in the summer 2002 *(Torcellini et al., 2005)*. This approximate value was obtained by measuring the air velocity at the cold air supply grill.

Control

83% of the occupants said they had little or no control on heating and cooling and noise confirming the comments made during the informal interviews.

Occupants' comments collected during the interviews mainly gravitated around the issues of daylight and discomfort from glare: *"Clerestory permits sunlight at desk area, making it difficult to see and read maps. – Insufficient lighting in cash register area. Sun rays during winter months blinding for cashier. – Most mornings for approx one hour sunlight creates glare on top desk, visitor eyes making services to them a bit tricky. – Sunlight from high windows can be very annoying to visitors approaching the desk. At other times it can be very annoying to the employee answering questions."*

Other comments refer to the problem of noise: *"Beautiful, nice floor plan. Hard to hear visitors from echo, etc. – Sound echoes through building – loud sometimes."* And temperatures: *"Temperature is fairly constant (no large fluctuations), air is fresh as a rule."*

According to the maintenance engineer, Mr James Lutterman, some of the most frequent complaints are related to the control system (lighting, thermostats, etc.). Some members of staff complain of too little heating in winter and the current policy is to address all complaints by compromising between the occupants' requirements and the low energy strategy. There are minimum complaints regarding the temperatures in summer in the main reception space but in the back offices mechanical ventilation is used to promote air movement and minimise overheating. Complaints about noise are received mainly when the place is almost empty and the concrete floor reverberates.

> **FOR MOST OF THE STAFF AIR IS STILL, FRESH AND DRY**

CASE STUDY 5 | ZION NATIONAL PARK VISITOR CENTRE, SPRINGDALE, UTAH, USA

LESSONS LEARNED

The building strategy overall and the cooling strategy were carefully thought through in a design process that was recorded and evaluated (Torcellini et al., 2005), delivering a coherent design and a design team satisfied with the level of communication and management.

Some faults, however, did emerge during the operation of the building and are possibly related to some of the strategies chosen as well as the planning of the space. The choice of Trombe walls on the south façade, for example, has contributed to overheating problems not only in summer but also in winter. This problem is particularly exacerbated in the relatively small and enclosed office space, which also lacks adequate air movement. Electric fans were subsequently installed to address this problem, with the additional burden of energy consumption and noise. Placing one or both towers deeper into the building rather than on the perimeter could perhaps have allowed their use as chimney stacks in winter and midseason, increasing the airflow in the offices and the reception area. Also, a more central location could have proved more effective in cooling the interior in the summer. The main reason for the towers' current location is the opportunity to open their outlets on the outdoor side towards the exhibition area. This, however, can hardly be considered effective (Yoklic & Thompson, 2004) as it can easily lead to a wastage of the cold air to the outdoor. Perhaps, the original design, which would have included three towers instead of the two currently built, was a more conservative yet realistic approach in dealing with the peak cooling loads experienced in summer.

One of the main problems found at the beginning of the cool towers' operation was the splashing of water on the tower floor occurring in windy conditions. Another important factor in reducing cooling performance was the pads' exposure to direct solar radiation, which caused them to dry out on the outer face much quicker than anticipated and to consume much more water than necessary. Moreover, the ABS plastic water collectors were not holding properly to the pads, causing leakages.

A small company called RCF came up with an interesting solution of creating metal frames all around the cellulose pad (figs 61–62), which solved the function of locating the pipes distributing the water at the top and at the bottom created a gutter gathering the excess water to the tank. The problem of the water spillage under wind conditions was addressed by increasing the thickness of the CELdek panel from 100 to 150mm whereas the exposure to solar radiation was reduced by creating a 25mm deep plastic honeycomb mesh, which acts as a shading device avoiding drying up and curving (bowing) of the pad. The CELdek manufacturers advise to replace the pads every five years but at the Zion National Park Visitor Centre they have been kept in good conditions for seven years until the time of the survey.

POSITIVE ASPECTS OF PROJECT IMPLEMENTATION

Energy saved

The average cooling energy consumption in the period 2001–2006 was 4.5 kWh/m²·yr as opposed to 17.3kWh/m²·yr of a conventionally cooled building of the same type.

Economic benefits

Projected energy savings represent an annual cost saving of $1,047/yr based on $0.10/kWh. Additionally, the 7.2kW grid-connected photovoltaic panels integrated on the south roof provide 10% of the power for the majority of the uses. The electricity bill for all the equipment (including back-up electric heating, hot water, internal and external lights) was in 2007 $400 a month (compared to the $4,000 a month of the existing 1960 office building) after the contribution of the PVs.

> PROJECTED REDUCTION IN CO_2 EMISSIONS ARE APPROXIMATELY 7KGCO_2/M²·YR

Environmental benefits

The projected energy savings for cooling represent a reduction in CO_2 emissions of approximately 5,674kgCO_2/yr (6.93kgCO_2/m²·yr).

61. Detail of wet pad with metal frame and water supply pipe.
62. Wet pads and drainage connections.

Building management

The presence of an on-site dedicated building manager made sure that the initial operating problems were addressed in a timely manner. The proposed solutions included the installation of metal frames around the cellulose pad, which solved the water distribution and guttering functions. Also, the issue of the water spillage under wind conditions was addressed by increasing the thickness of the CELdek panel from 100 to 150mm whereas the exposure to solar radiation was reduced by creating a 25mm deep plastic honeycomb mesh, which shaded the pad avoiding fast drying up and curving (bowing).

PROBLEMS ENCOUNTERED AND OPPORTUNITIES FOR IMPROVEMENT

Natural ventilation strategy and cool towers

The choice of Trombe walls on the south façade has contributed to overheating problems not only in summer but also in winter. This problem is particularly exacerbated in the relatively small and enclosed office space, which also lacks adequate air movement. Electric fans were subsequently installed to avoid this problem, with the additional burden of energy consumption and noise.

One of the main problems found at the beginning of the towers' operation was the splashing of water on the tower floor occurring in windy conditions. Another important factor in reducing cooling performance was the pads' exposure to direct solar radiation, which caused them to dry out on the outer face much quicker than anticipated and to consume much more water than necessary. Moreover, the ABS plastic water collectors were not holding properly to the pads, causing leakages.

Occupant perceptions

The most frequent complaints are related to the control system (lighting, thermostats, etc.). Some members of staff complain of too little heating in winter and the current policy is to address all complaints by compromising between the occupants' requirements and the low energy strategy.

SUMMARY

The Zion Visitor Centre is a successful example of low-energy building entirely designed and delivered by the US national administration, where great efforts were made to follow a best practice approach which reduced the building energy loads and employed renewable forms of energy. More specifically, in relation to the building's cooling, the designers opted for a climatically responsive strategy which exploited the potential for evaporative cooling. However, they implemented the well-known cool tower typology, which is widely used in the United States. The system of wet cellulose pads and the cool tower termination, however, presented various limitations and reduced the range of strategic opportunities which could have been achieved with alternative solutions within the building. Namely the opportunity for a night-time ventilation strategy involving the towers as stack up-draught ventilation chimneys was missed due to the high resistance of the cellulose pads. Also, given the strong diurnal variations experienced for most of the summer and midseason the combination of night-time ventilation and a greater proportion of exposed thermal mass could have been more successful in mitigating peak temperatures and reduce the cooling loads. Some of these options were included in the original design proposal, which also featured a larger building with greater number of towers, but this was not implemented due to cost reasons. The final built option did present a number of problems mostly related to the wet pad operation and the associated water distribution and drainage, but these have been resolved during operation. This was possible due to the presence of a dedicated on-site maintenance team. Moreover, the engineer, who originally designed the system, took a particular interest in the troubleshooting process, providing advice and support. Preventative maintenance is routinely provided and the unavoidable complaints addressed trying to compromise between occupants' satisfaction and energy efficiency.

The performance of the system seems to have been improved by technical adjustments such as thicker cellulose pads, shading, improved water supply and drainage system. Such adjustments address problems which are typical and recurrent in the operation of many cool towers and could be effectively implemented in the standard design of cool towers. However, depending on the desired overall strategies, it is important to select the most appropriate cooling system typology in light of water availability and opportunity for seasonal interchange.

REFERENCES

- Munters (2008). *'CELdek'* [Online]. https://www.munters.com/. [Accessed 10/2018]
- NPS, 2017. *'Zion National Park Map and Guide'*. Washington, DC: National Park Service. US Department of Interior [Online]. https://www.nps.gov/zion/planyourvisit/upload/Zion-Summer-2017-web.pdf/. [Accessed 10/2018]
- Torcellini, P., Long, N., Pless, S. & Judkoff, R. (2005). *'Evaluation of the Low-Energy Design and Energy Performance of the Zion National Park Visitors Centre'*. Golden, Colorado: National Renewable Energy Laboratory [Online]. https://www.osti.gov/biblio/15015115-evaluation-low-energy-design-energy-performance-zion-national-park-visitor-center [Accessed 11/2018].
- Yoklic, M. & Thompson, T. L. (2004). *'Cooltowers. Passive Cooling and the Case for Integrated Design'*. Proceedings 2004 American Solar Energy Society meeting, Portland, Oregon.

CASE STUDY 6
GLOBAL ECOLOGY RESEARCH CENTRE, PALO ALTO, CALIFORNIA, USA

The Global Ecology Research Centre in Stanford University, Palo Alto, California is a mixed-use office and laboratory building with a floor area of about 1,000m². It was commissioned by the Carnegie Institution of Washington in 2000 on a site previously owned by them within the Stanford University Campus. The centre was built in 2004 and it was designed by San Franciscan architectural practice EHDD with mechanical and electrical specialist Rumsey Engineers. This building has attracted much attention in the early 2000s as a pioneering example of sustainable architecture and for the employment of alternative and low-energy forms of cooling such as radiant cooling and passive downdraught evaporative cooling (PDEC). In 2007 it was listed among the AIA Committee on the Environment's Top Ten Projects Award.

63. External view of Global Ecology Research Centre, Palo Alto, CA [EHDD].

LOCAL CONTEXT

Thirty-seven miles south-east of San Francisco, CA, the University of Stanford hosts the Department of Global Ecology of the Carnegie Institute of Science. Stanford (Lat. 37°N, Long. 122°W) is located near the town of Palo Alto in the heart of Silicon Valley. Palo Alto experiences a Mediterranean climate with moderately hot and dry summers (24.7°C mean maximum for August) and mild winters (6.5°C mean minimum in January). The summer season is generally from the end of May to the end of September. Peak temperatures can occasionally reach 38 to 40°C and temperature variations between night and day tend to be moderate in summer with a difference that can reach maximum 10–11°C. The average wet bulb temperatures in summer afternoons is low and varies between 16 and 17°C (fig. 64). Palo Alto benefits from more than 300 days of clear sky days per year and the sky temperature between May and September at 5am varies between 0 and 3°C, indicating a very good potential for radiant cooling. It receives 388mm of rain yearly, mainly occurring in the winter season.

Palo Alto is one of the most expensive and wealthiest cities in the US. The proximity to Stanford University and Silicon Valley makes this area a very exclusive place where the majority of the global internet start-ups have flourished in the late 1990s. Following California's trends, environmental issues are quite high on the local government's agenda and there is great sensitivity to these issues amongst the general public.

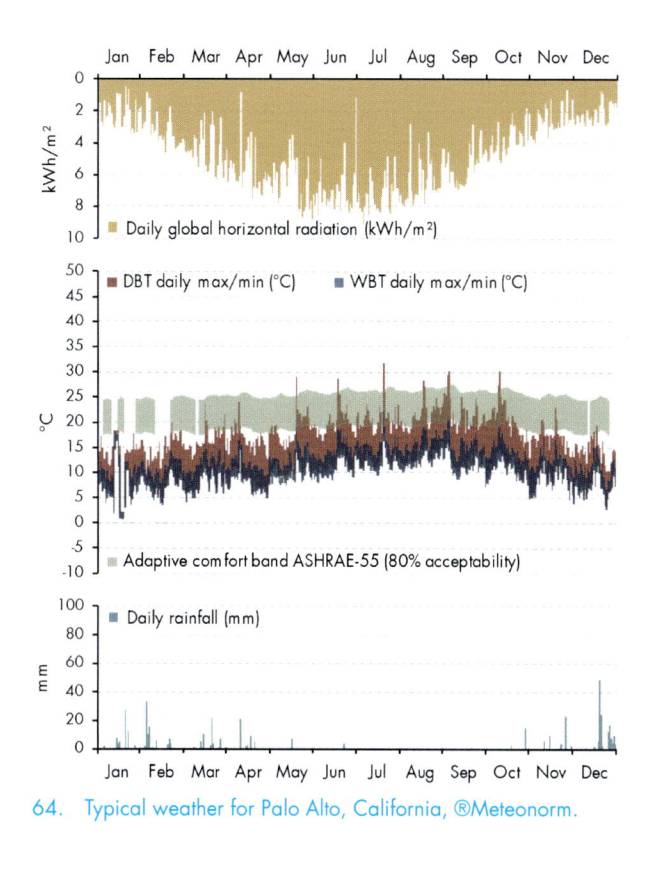

64. Typical weather for Palo Alto, California, ®Meteonorm.

THE BUILDING AND THE OVERALL STRATEGY

The two storey steel framed building housing the Department of Global Ecology is located in the Stanford University Campus. The site is fairly open with surrounding low-rise buildings which do not obstruct the sky view and sunlight (fig. 65). With the east facing main entrance and its north-south orientation, this 1,000m² building is arranged in such a way to maximise northerly daylight in the ground floor open lab area and first floor open plan office, relegating the cellular labs and offices on the south side (figs 66–70). This has clear benefit during the winter in maximising solar gains in the perimeter offices but can cause overheating in summer and the midseason, if sufficient areas for natural ventilation are not provided. The west elevation is almost completely obscured, housing emergency staircase and exit, and this is beneficial in reducing the afternoon solar heat gains in summer which are also more difficult to control. The entrance hall on the east side is characterised by the PDEC tower, located in the south corner, and a fully glazed double height space, which is designed for full visual and thermal permeability during the temperate and warm months of the year.

In designing the Global Ecology's building San Franciscan Architects, EHDD, took on board the environmental and energy efficient brief in a very comprehensive way and, with the aid of a highly motivated design team and sympathetic client, delivered

65. Aerial view of building site.
66. View of East and North elevations.
67. View of South elevation.
68. Entrance lobby.

CASE STUDY 6 | GLOBAL ECOLOGY RESEARCH CENTRE, PALO ALTO, CALIFORNIA, USA a building which could be deemed successful not just for the effective implementation of the low-energy agenda but, also, for its management and the effective participation of the occupants in its operation.

ENVIRONMENTAL STRATEGY

The building design strategy involved careful master planning and landscaping, including the choice of vegetation and optimum orientation. The environmental design strategy included *(EHDD, 2007)*: an energy demand reduction strategy aimed at minimising the heating and cooling requirements of the building; improvement of visual comfort and lighting energy efficiency; passive and low energy heating and cooling, employing natural heat sinks where possible; on-site renewable energy generation and water reuse. The building materiality was carefully considered, and low-embodied carbon materials were usually favoured. This meant that the building envelope, although very high performing in terms of high insulation specification, lacked thermal mass.

> AN UNDERSTANDING OF THE MEDITERRANEAN CLIMATE HELPED TO IDENTIFY A *HYBRID* LOW CARBON SOLUTION TO ENVIRONMENTAL CONTROL

The Mediterranean climate dictated three building strategy modes divided in winter (November to March), summer (May to September) and a short midseason outside these. Although the winters are mild, heating is still required and similarly in summer during peak conditions temperatures can reach

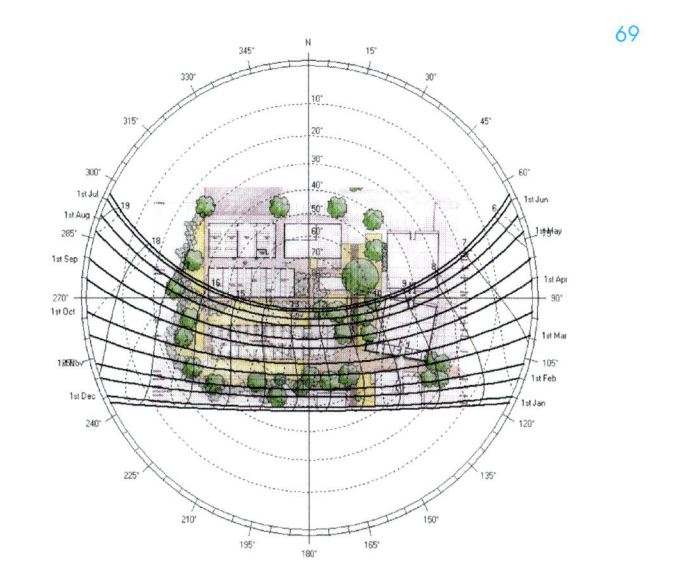

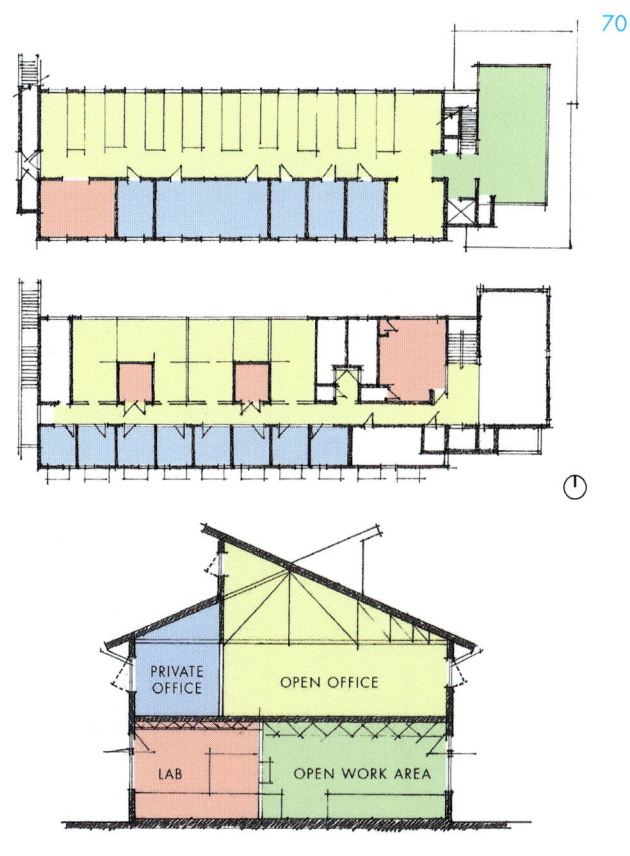

69. Site plan and sunpath diagram.
70. Ground floor, first floor plan and cross section.

CASE STUDY 6 | GLOBAL ECOLOGY RESEARCH CENTRE, PALO ALTO, CALIFORNIA, USA 255 the high thirties and risk of overheating may occur during the afternoon when the conditions are also drier.

In winter, passive solar heating is promoted in the cellular offices through the orientation of the building following an east-west long axis. Minimisation of the heating loads is achieved with a high-performance envelope, spectrally selective glazing and effective dimensioning of the solar shading which allow solar ingress in winter and shading in summer. Additional heating is provided through an underfloor radiant system, which is also used in summer for cooling.

According to the original design strategy in summer and in the midseason, natural cross and stack ventilation can provide convective cooling when required through perimeter windows and high-level vents (fig. 71). During summer peak conditions, radiant ceiling panels provide additional cooling in the first-floor open plan offices, whereas in the ground floor laboratory a combination of chilled water ceiling panels and a dedicated mechanical ventilation with heat recovery system is used. The night spray radiant cooling uses reclaimed water to recirculate on the roof at night in order to be cooled by evaporation and radiation to the night sky. This system still works well and provides the majority of the cold water required.

The combination of these strategies achieves a predicted total energy consumption of 97kWh/m²·yr compared to 194kWh/m²·yr for a conventional building, yielding energy savings of 50% (EHDD, 2007). Recent energy figures are not available but presumably there has been a slight

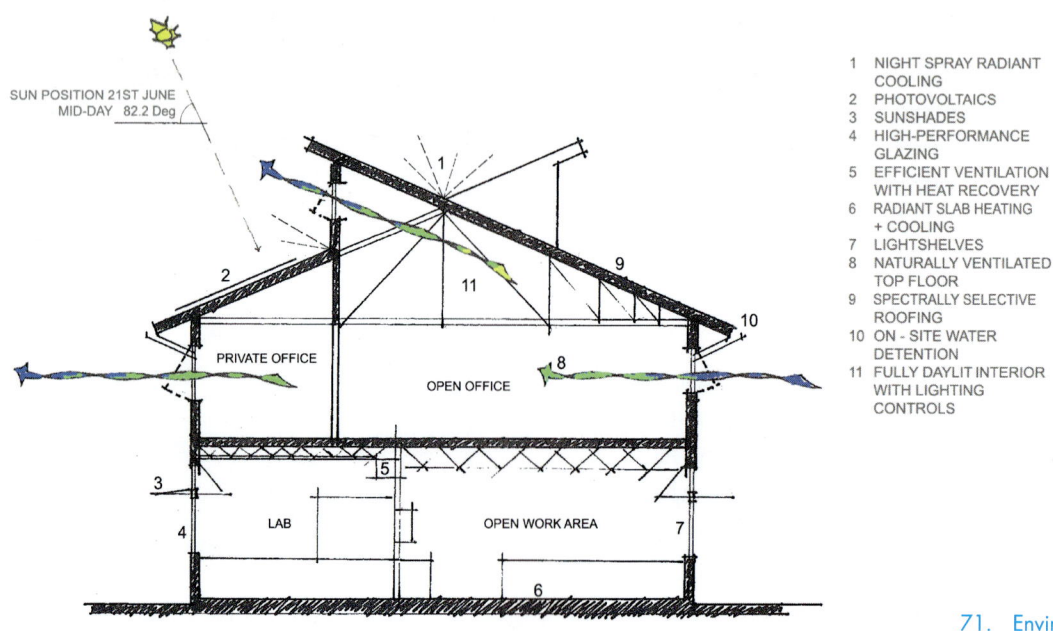

71. Environmental strategy.

increase due to the addition of individual fan coil units in the south facing perimeter offices on the first floor.

More recently (September 2018) due to increased summer temperatures and overheating in the upper floor, the building managers have slightly altered the cooling strategy by adding additional cooling provisions to both the open plan and the south facing perimeter offices. In peak hot conditions the open plan offices are exclusively mechanically ventilated receiving fresh air from the modified fan coil unit of the adjacent server room. In the past these were usually naturally ventilated via perimeter windows and high level vents, which were manually controlled by wall mounted winding mechanism linked to vent actuators (figs 72–73).

Due to the natural ventilation system being perceived to underperform, the server room's fan coil unit has been adapted with an *'economiser'* that mixes outside air with recirculated air. When outside air is cool enough but not too cold (typically overnight and most of the day) the recirculation damper is closed and only the fresh outside air is circulated, saving on refrigerants. This adaptation was possible due to the fact that apparently the fan coil unit of the server room was originally oversized and further developments in server systems and relocation of some equipment elsewhere has further contributed to the minimisation of the cooling loads of the server room, hence making

RADIANT CEILING PANELS PROVIDE ADDITIONAL COOLING

72. Manual control for high level vents on first floor.
73. High level vents on first floor.

additional cooling capacity available to the cooling of the open plan office. This strategy was devised by the building managers in order to have closer control on temperatures as they perceived that this would be more difficult with manually operated windows and felt that the mechanical system could achieve better *'air-filtration'* and advantages over noisy and dusty surroundings due to outdoor generators and construction sites.

Considering the potential of the climate and the diurnal variations of 10 to 11°C typically experienced in summer together with low humidity levels during hot afternoons, it could be said that alternative passive options for cooling strategies could be given by the employment of materials with higher thermal capacity and their associated thermal mass together with night-time ventilation. High density internal lining materials such as gypsum fireboards, which incorporates recycled paper, for example, has been successfully used in other projects to stabilise internal temperatures and retain the beneficial effect of night-time ventilation. Evaporative cooling could also have been more extensively used to cool a larger proportion of the building, but this should have been done only following a thorough review of the building design, internal layout and airflow patterns. If considering water consumption, however, the evaporative cooling option would obviously cause greater consumption than the night sky radiant cooling system with water reuse, which would be more water efficient but potentially more energy intensive in re-circulating water in the system.

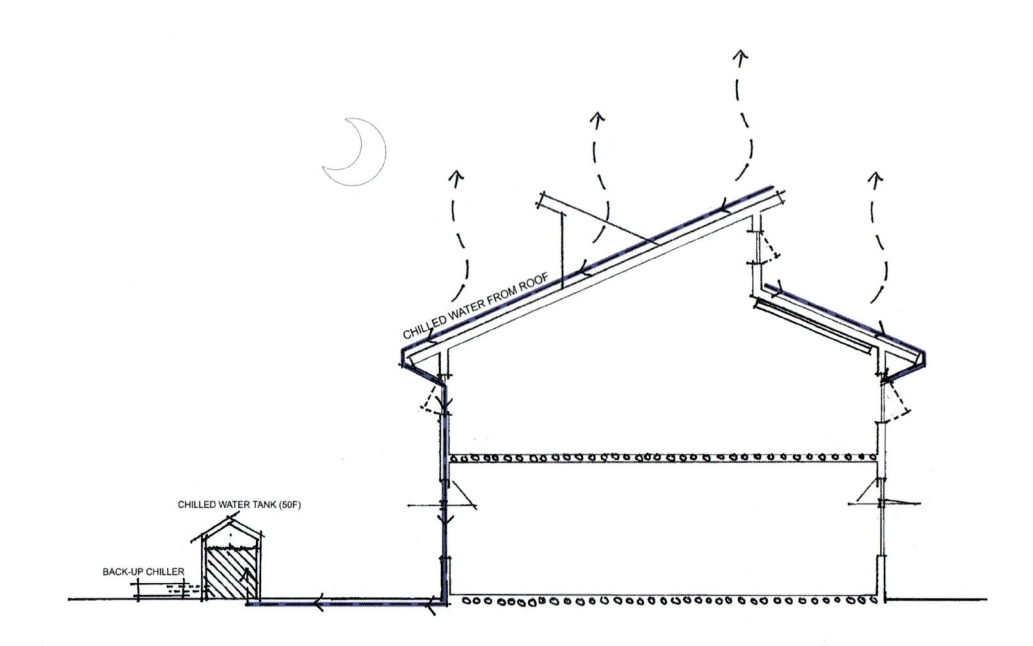

74. *'Night-spray'* radiant cooling strategy, night time.

COOLING SYSTEM AND BUILDING INTEGRATION

Similar to Lowara office building by Renzo Piano (described in Chapter 2.2), the main source of cooling in the building is provided by a radiant cooling system (branded *'night-spray radiant cooling'*). The water is sprayed on the roof and cooled by evaporation/radiation to the night sky; it is then collected in the gutters and drained into an outside tank (figs 74–75). Its temperature is kept at 10°C in the tank, the size of which is limited by the size of the roof and cannot meet the peak cooling requirements. To top-up the chilled water an existing 20 ton air-cooled chiller has been reused, which was anticipated to operate less than 200 hours a year (as opposed to a conventional 30–40 ton system that operates for 2,500–3,500 hours per year) *(EHDD, 2007)*. During the day the chilled water is re-circulated back into the building to supply three sub-systems:

1. Radiant cooling floor slabs on the ground and first floor (with fresh air supplied through infiltration in the initial strategy and through mechanical ventilation afterwards).
2. Radiant cooling ceiling panels on the first-floor open plan office (with ventilation as above).
3. Air handling unit for the mechanical ventilation and cooling of cellular labs.

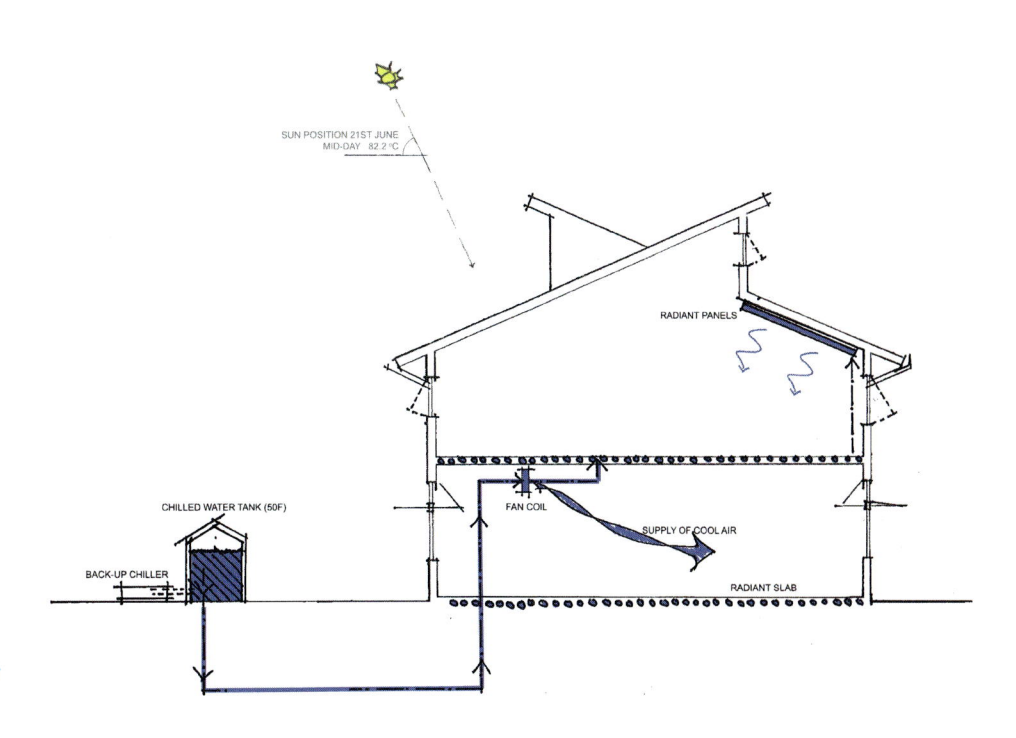

75. *'Night-spray'* radiant cooling strategy, daytime.

CASE STUDY 6 | GLOBAL ECOLOGY RESEARCH CENTRE, PALO ALTO, CALIFORNIA, USA

Recent reports on the performance of the night-spray radiant cooling strategy indicate that the system works very well and that the need for refrigerant based chiller usage is generally low. Records from the last three weeks of September 2018 indicate that the chiller ran for only 1 hour and 40 minutes despite eleven days with dry bulb temperature above 26.6°C and one above 32°C (fig. 76). These recordings from the Building Management System show that during the entire three-week period in September the cooling demand was mostly met by the natural heat sink generated by the cold water produced by the night-spray radiant cooling system and that the cumulative chiller time was less than the daily average for August (approx. 5.5hrs/day). This was partially due to clear, cold, nights which allowed the roof spray to cool the tank. Moreover, comparison of similar days in September and August shows that this was also due to the additional measures such as increase of setpoint temperature (from 23 to 24°C) and improvements in the fan coil unit circulating outside air when this is below setpoint temperature. These findings suggest that given the appropriate supply system (i.e. larger opening areas) outdoor air can effectively provide cooling for extended periods of time.

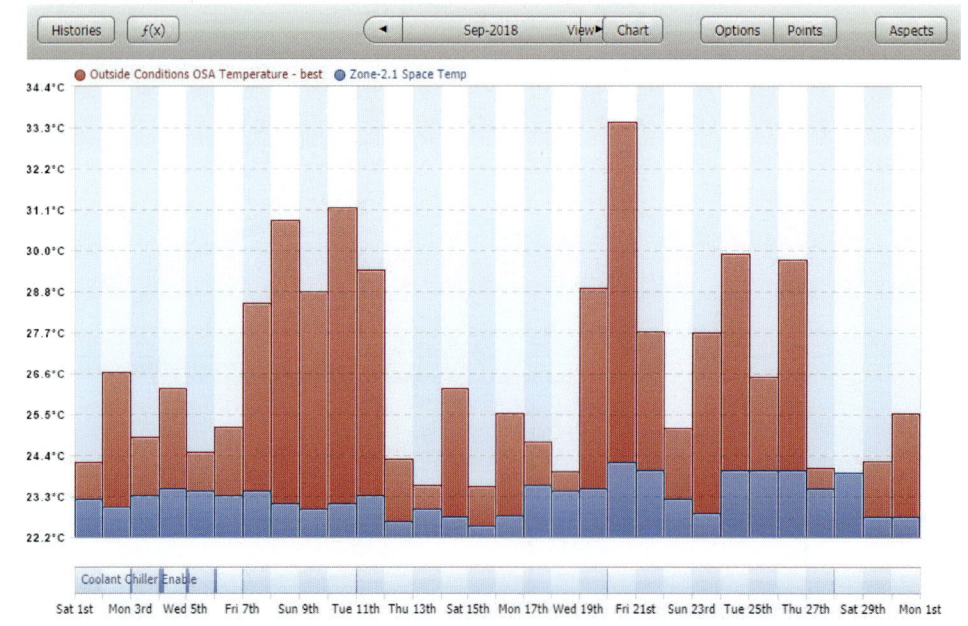

76. Histograms showing daily outdoor (red) and indoor (blue) dry bulb temperatures and chiller use (bottom bars) during the month of September 2018.

CASE STUDY 6 | GLOBAL ECOLOGY RESEARCH CENTRE, PALO ALTO, CALIFORNIA, USA

The Passive Downdraught Evaporative Cooling (PDEC) is marginal to the overall building's cooling strategy both by capacity and location (fig. 77), but it demonstrates the potential of using direct evaporative cooling in this location. The single 10mt high PDEC tower serves the entrance lobby exclusively and although its benefits may be felt in the ground floor open plan lab, when the lab door is closed and the patio doors are open, a lot of the cool air delivered by the low level 1m² opening escapes onto the outdoor (figs 78–79). The tower top termination was designed to enhance the channelling of the prevailing NW winds.

The original specification of the passive downdraught evaporative cooling system comprised of twelve hydraulic nozzles attached to one circular water supply pipe at mains pressure. Currently the system consists of one water supply pipe with five nozzles also at mains pressure. The system is controlled by the Building Management System (BMS) and it operates when the outdoor temperature reaches 29.5°C and the RH is below 36%. The water requirement of such a small system is marginal, considering that it operates only few days a year, when the weather reaches peak conditions during the hottest months of July and August.

OCCUPANTS' PERCEPTION

A survey of the occupants conducted in October 2007 revealed a very high level of satisfaction with the internal environment.

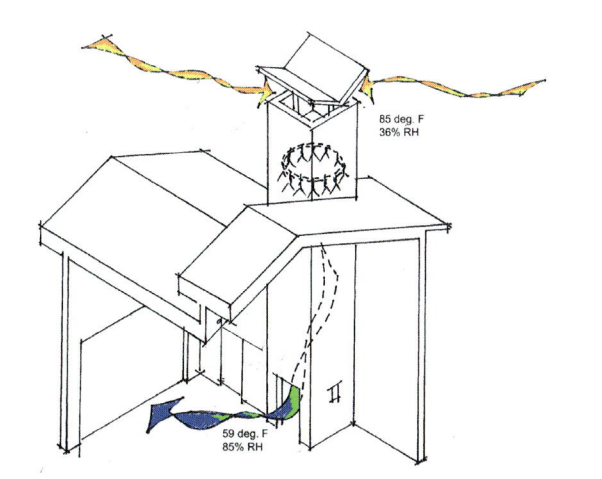

77. PDEC strategy in the lobby area.
78. PDEC Tower, view of interior.
79. PDEC tower outlet into entrance lobby.

CASE STUDY 6 | GLOBAL ECOLOGY RESEARCH CENTRE, PALO ALTO, CALIFORNIA, USA

Although the sample was small (20 subjects) the results show a high level of consistency. The majority of the subjects were located on the open plan first floor offices.

The results from the 2007 survey (fig. 80) show a high level of satisfaction with thermal comfort and air quality. Design and image to visitors were regarded as particularly satisfactory, exceeding a mark of 6 on a scale 1 to 7. However, from detailed analysis of the results and occupants' comments it is apparent that there is a slight overheating problem for a short period in summer (especially on the top floor) but this seems to be not a major reason for concern by the staff. Looking at the data in more detail:

Temperature in summer

71% thought that the temperature overall is comfortable. However, when prompted on whether it is too hot or too cold, 59% perceived it to be between slightly warm and too hot.

Air quality in summer

For the majority of the staff, air was fresh, odourless and generally satisfactory. For 47% of the occupants it was dry, whereas only 12% considered it humid. This can be clearly related to the fact that in the majority of the offices there is radiant cooling rather than direct evaporative cooling.

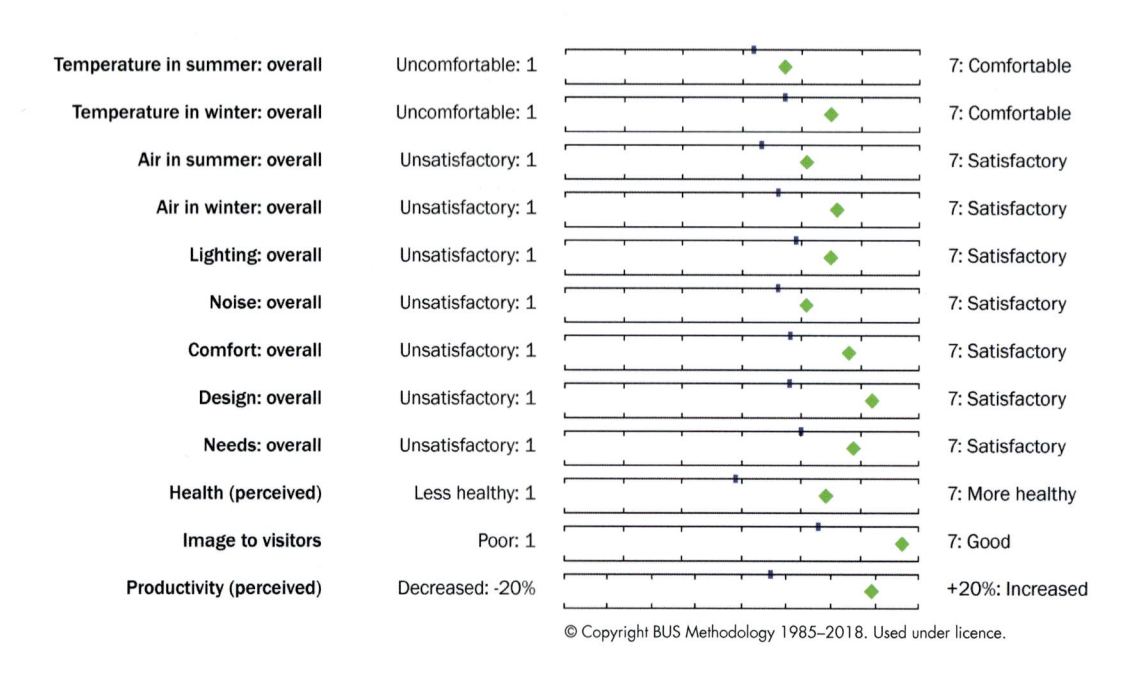

Category	Low	High
Temperature in summer: overall	Uncomfortable: 1	7: Comfortable
Temperature in winter: overall	Uncomfortable: 1	7: Comfortable
Air in summer: overall	Unsatisfactory: 1	7: Satisfactory
Air in winter: overall	Unsatisfactory: 1	7: Satisfactory
Lighting: overall	Unsatisfactory: 1	7: Satisfactory
Noise: overall	Unsatisfactory: 1	7: Satisfactory
Comfort: overall	Unsatisfactory: 1	7: Satisfactory
Design: overall	Unsatisfactory: 1	7: Satisfactory
Needs: overall	Unsatisfactory: 1	7: Satisfactory
Health (perceived)	Less healthy: 1	7: More healthy
Image to visitors	Poor: 1	7: Good
Productivity (perceived)	Decreased: -20%	+20%: Increased

© Copyright BUS Methodology 1985–2018. Used under licence.

80. Summary of post occupancy survey results for Global Ecology Research Centre, Palo Alto, California, ©Usable Buildings Trust.

Control

The majority thought they had little or no control on heating and noise especially. However, 44% thought they had good control of ventilation, with a result that is significantly higher than benchmark. For cooling control 34% thought they have some control, which is no different from benchmark but lower than the scale midpoint.

The occupants' comments on comfort and design conveyed a sense of general satisfaction with building: *"Very comfortable and quiet work space. – They did a great job with the place. Small eco footprint. Plenty of room. – Like the natural lighting and quiet. – The building is very comfortable all year long."*

However, the comments revealed a few problems during the hottest days and stratification of temperatures on the upper floor: *"Great building! Very easy to work in, with only a few minor complaints during late summer (for 3–4 weeks only). – Efficient. Temperature is great except on the very hottest days (38°C) – In the main lab where I work it is always cool to keep the computers working so it's nice down here but upstairs can get bad. – Upstairs needs more air circulation."*

LESSONS LEARNED

POSITIVE ASPECTS OF PROJECT IMPLEMENTATION

Energy saved

The predicted total energy consumption resulting from the combined low energy strategies for the entire building is 97kWh/m^2·yr compared to 194kWh/m^2·yr for a conventional building. This notional performance could not be verified but it is anticipated that the increased use of mechanical ventilation in the open plan offices and the addition of individual coolers in the perimeter south facing offices will have increased the predicted energy consumption.

Economic benefits

Projected energy savings represent an annual cost saving of $9,700/yr based on $0.10/kWh.

Environmental benefits

The projected energy savings represent a reduction in CO_2 emissions (against a conventional building) of approximately 52.60kgCO_2/m^2·yr.

PROBLEMS ENCOUNTERED AND OPPORTUNITIES FOR IMPROVEMENT

Radiant cooling system

The radiant cooling system of the first-floor open plan office area relies on both the ceiling panels and radiant floor slab. However, the overall performance of the system is not as good as reported in 2010 due to higher gains since the original survey (increased occupancy and warming temperatures) and, also, due to the limited heat exchange between the floor and the air space above. To obviate this problem, the building managers have installed an air circulation system that links to an air handling unit in the server room. This helps cool the open office space on the first floor but clearly increases energy consumption from fans and pumps.

PDEC system

Problems were experienced at the beginning of operation with dripping of brass nozzles due to blockage and insufficient water pressure. This was addressed by introducing a high-pressure system in the tower but the noise generated was a cause of annoyance for the occupants and the system was eventually reverted back to low pressure. In this respect the PDEC tower works as a shower tower rather than a misting system. Presumably, the small number of nozzles, the drainage point at the bottom and the fact that the tower is only used for a short period of time in summer contributes to the water dripping not being perceived as a particular problem. However, mains water pressure could still be used if combined with compressed air (pneumatic nozzles) as used in the Madrid Solar Decathlon House.

Control system

The small amount of water used for the PDEC tower is UV treated and particle filtrated. Although the BMS controls the operation of the nozzles through temperature and humidity sensors, the opening of the tower outlet into the lobby is manually controlled. It is typically left open during peak summer conditions and the occupants are encouraged to take the initiative to operate and control the system as needed.

SUMMARY

The cooling strategy of the GERC is currently quite different from the original vision of design team and of the client. The main strategy which mostly relied on natural ventilation with complementary cooling from the radiant ceiling panels and the careful zoning of the areas needing mechanical cooling seems to have been altered. Currently, mechanical ventilation and cooling have been extended to the open plan offices and the south facing cellular offices.

The original outcome of the survey visit in 2007 showed that the radiant ceiling panels performed well and that the combination of convective cooling when possible and radiant cooling during peak summer conditions produced a comfortable work environment where the occupants felt able to control and adjust to their own preferences. After 11 years, the rising summer temperatures, increased overheating episodes and the evident disenchantment with the possibility of engaging with the manual operation of the vents on the part of occupants and building managers have led to the natural ventilation system being abandoned in summer.

According to the building managers: *"Window actuators were deemed unnecessary with the current forced-air ventilation scheme (and too difficult/costly to implement). Although open windows sound like a wonderful idea, in practice they increase ambient noise significantly – due to traffic, construction, and noisy outdoor equipment, – and let in more dust and allergens than a filtered inlet. Automating the windows with the current system would save little if any operating costs. Incidentally, the current system actually provides substantially more fresh air during occupied times than was provided with the original system (which required the windows to be closed as the day warmed up)."*

Although these arguments might seem initially compelling, it must be said that mechanical ventilation is still responsible for a large proportion of energy consumption due to the use of pumps and fans. Moreover, the site is away from heavy traffic routes and vehicular traffic which reduces the impact of noise and outdoor pollution but apparently local generators and a temporary construction project nearby have exacerbated disruption. The reliance on mechanical air filtration is very much dependent on a good maintenance regime which avoids common problems linked to poorer indoor air quality due to recirculation and bacterial contamination. The current set up based on the maximisation of an oversized mechanical system, which essentially replaces the function of the natural ventilation system, is somehow far from the low-carbon best practice vision of the original design. Ironically the contentious point of the maintenance engineers against the natural ventilation system is the operability of the vents, which are considered costly, if fully automated, and unreliable, if left to manual control. However, there are many examples of successful natural ventilation achieved by automated high-level windows and vents, with manually controlled low level windows. Equally, frustration with actuators which fail after just a few seasons is understandable. Actuators must be designed to be robust, with a long service life, if natural ventilation is going to be widely accepted.

LOW CARBON BUILDING DESIGN MUST BE FLEXIBLE AND RESILIENT

While maintenance is an important issue (as is the case with all mechanical systems), it is also a question of design. Not only detailed design, such as sizing of windows and opening areas, but also initial strategic design which could have increased the resilience of the building. For example, greater thermal capacitance of internal linings, in combination with night-time ventilation to reduce the building's cooling demand and provide greater resilience to warming summers.

The performance of the PDEC tower in this building, although marginal, contributes to the overall cooling strategy of the building and complements well the main radiant cooling system. It could be argued that the tower was designed with a symbolic intention additional to that of creating an opportunity for the occupants to relax in a cool space where they can take direct control of their comfort. It was observed, however, that the positioning of the tower outlet relative to the large openable patio doors could create the risk of cool air being spilled outside and not entrained within the indoor space.

Reports from the first three years of building occupation indicated overheating occasionally experienced on the first floor by some for some weeks in the summer when outdoor temperatures reach 38°C. This has clearly changed over the years with warming summers and altered perceptions. This and the reduced use of the natural ventilation system have increased cooling loads and the chilled slabs and ceiling panels don't seem to provide sufficient cooling. Additionally, chilled slabs in a densely occupied space could provide poor contact between the slab and the surroundings due to a large area covered by furniture reducing the surface area available for radiant exchange. The building managers have added additional air circulation from an existing air handling unit in order to top up the cooling when required. Although this appears a bit of a compromise, compared to the original ambition of a nearly passive/hybrid building, the work of the current maintenance team has optimised the radiant/evaporative roof system in order to reduce water consumption and the need for chiller top up.

REFERENCES

- EHDD (2007). *'GERC Design Report'*. EHDD Architects (Unpublished).

CONCLUSIONS

Part 2 has looked at built examples of various forms of natural cooling strategies and systems in the context of their architectural integration within six buildings around the world, with very different climates, microclimates and cultural contexts. Amongst them there are commonalities based on cooling delivery typology and architectural integration.

The Torrent Research Centre, the Phoenix Courthouse and the Malta Stock Exchange buildings are located between latitudes 23°N and 36°N and experience dry to composite climates with a strong potential for passive cooling during peak conditions. These buildings have opted for an architectural integration within a large central space of the building (see Chapter 2.4 – Central Atrium or Concourse). They are either large office or semi-public spaces and the cooling strategies used are flexible and easy to integrate in the existing configuration of the design with little need for additional dedicated spaces.

The GERC and the Zion Visitor Centre are both located at latitude 37°N and experience dry summers with a strong potential for convective cooling (when outside air is lower than that desired inside), passive downdraught evaporative cooling (when external air is higher than that desired inside) and night-sky radiant cooling (to provide chilled water for internal radiant cooling). They have opted for an architectural integration of the passive downdraught evaporative cooling system which uses cool towers, which are fairly widely used in the south west of the United States (see Chapter 2.4 – Central Shaft or Concourse). They are smaller public and office buildings where the cooling strategies are integrated in the mode of an 'add on' such as the PDEC delivery tower or the radiant ceiling panels.

The CSET building is located at latitude 28°N and has a warm and humid summer which required a hybrid strategy combining convective and radiant cooling with de-humidification of the supply air. The airflow path through the building changes with the seasons, achieved through the combination of a ventilation shaft and double skin façade. The architectural integration is seamless as these are not dedicated spaces for the natural cooling strategy alone but they blend together within architectural elements which have multiple functions.

While the sample of buildings is small, a summary of general observations can be made for more widespread application.

1. The survey of the case study buildings has shown that natural cooling can provide significant energy savings, significant life-cycle cost benefits, and high levels of occupant satisfaction. Some of the buildings have demonstrated all of these characteristics, while some have demonstrated small energy savings and a relatively low level of occupant satisfaction.
2. The six buildings studied included buildings only cooled using natural ventilation and passive downdraught evaporative cooling, buildings using both passive downdraught evaporative cooling and active downdraught cooling, and buildings using only active downdraught cooling. Generally, hybrid buildings are more complex and involve a greater management challenge.
3. Misting nozzles have been shown to be generally the most efficient way of delivering direct evaporative cooling (in terms of water use compared to other evaporative media). They provide a flexible, easy to control and low maintenance technique.
4. Cellulose and other porous panels have been used successfully, but tend to be inefficient in water use, and difficult to control.
5. Cooling coils have been shown to be effective in delivering active downdraught cooling (and saving fan energy) in humid conditions.
6. Maintenance of passive cooling systems is regarded as far less onerous than for hybrid or conventional air-conditioning. However, hybrid systems require expertise to run them successfully, as they are more complex and building managers require training.
7. Occupant surveys reveal that case study buildings in India and China were more favourably perceived by their occupants than those in the USA. This suggests that culture as well as climate influences the expectations (and perceptions) of building occupants, although the reasons for these differences are likely to be multi-faceted and require further exploration.
8. There is significant variation across the sample in terms of occupant satisfaction. Although 50% of the sample had internal environments which were perceived as being significantly better than benchmark values, a similar number had results which were

significantly below benchmark. However, the sample of buildings is too small to derive wider generalisations about the 'acceptability' of passive and hybrid cooling solutions.

The review of these six case studies also highlighted further common issues which transcend the building type and location and can be referred back to essentially two aspects: strategic design and building operation.

Strategic design

The study revealed that, with few exceptions, despite the best intentions and efforts to create low-energy, sustainable exemplars employing natural forms of cooling, most of the buildings have found that over time these aspirations are challenged by a number of factors which range from occupants' comfort expectations, rising summer temperatures, poor maintenance, and lack of inbuilt resilience in the design strategy of the building. These are very complex issues which are difficult to generalise given the wide climatic and socio-cultural context of each building. However, taking Bill Bordass' advice to designers of the need for simplicity (Bordass, 2011) as potentially the only safe passage towards buildings that work, a pattern starts to emerge and it is clear that simplicity of strategy and delivery will yield smoother operation as well as greater occupant satisfaction.

This is the case for buildings such as the Torrent Research Centre, the Zion National Park Visitor Centre and the Global Ecology Research Centre where during the initial years of operation both occupants' perception and cooling performance achieved satisfactory levels using the intended strategies. In fact, if the metrics of assessment become purely the building performance in relation to the cooling strategy, then the group of satisfactory examples extends to other cases such as the Phoenix Courthouse, which delivers in the central transitional space internal air temperatures which are 10°C lower than the outdoors. However, the occupants' perception of the space is greatly affected by their comfort expectations, which demand cooler conditions even in transitional areas. There are cases, such as that of the CSET building, where instead, despite a complex environmental strategy, occupant satisfaction is good although the cooling performance and natural ventilation strategy are compromised by poor maintenance and blocking of the intended airflow path. In the case of the Malta Stock Exchange, the cooling performance is compromised by related and unresolved maintenance issues which also coincide with a low occupant satisfaction.

On the issue of resilience of the design strategy, it is fundamental to follow every opportunity to exploit climatic heat sinks through the choice of environmental strategies, spatial layouts and materiality in view of not only past and current climatic conditions but also future ones. Failing to address these aspects is ultimately failing the feasibility of a natural ventilation and natural cooling strategy in the long term. Climate change is no longer something that we need to prevent, but it is something that we need to mitigate and adapt to. We can already observe its detrimental effect and how global warming, extreme weather events and heatwaves can all unfavourably affect building performance and convert natural ventilation and mixed mode strategies into mechanically ventilated strategies in numerous buildings.

The case of the GERC building in California illustrates well this point and the need for added climatic resilience. The conversion of the natural ventilation strategy in the open plan offices to fully mechanical is emblematic of recurrent issues and barriers to natural ventilation (such as noise, pollution and increasing summer temperatures). The interesting finding, however, is that the mechanical ventilation mainly uses fresh air from outside (whenever this is cooler than setpoint) supplementing it with mechanical cooling only during peak conditions. Whilst this proves that outdoor air can keep the temperatures within comfort for most of the time (hence the potential for natural ventilation) it also suggests that the amount of fresh air introduced through natural ventilation via the manually operated vents is insufficient. This can be due to the insufficient opening areas or inadequate control (i.e. occupants not opening wide or often enough to produce the required air change rates) or both.

This case highlights another important issue: that of building resilience. A design strategy which exploits natural heat sinks, such as high capacitance materials internally coupled with night-time ventilation, will stabilise internal temperatures, making the building more resilient to extremes of ambient temperature. Building resilience will be enhanced by avoiding complete dependence on mechanical cooling systems, which, with their high electrical energy consumption and waste heat discharged to the outside, can create a vicious circle, exacerbating the overheating which the cooling system was designed to address *(Schiano-Phan et al., 2015)*.

Within strategic design and architectural integration, another recurrent issue is the importance of providing a clear airflow path and sufficient well-integrated opening areas (both inlet and outlet areas). This is important for all forms of convective cooling, but in the case of downdraught cooling, where the temperature of discharged air is often lower than the outdoor temperature and exhausting at high level may not be possible, a low discharge point is essential or stagnation will occur. A positive example of this design strategy is implemented in the Torrent Research Centre where the perimeter towers have openings at both high and low level depending on the ventilation mode and on the relative temperature between inside and outside air. Also, in the case of cross ventilation, the design of a realistic airflow path is essential as too often over-optimistic strategies, which assume the connection of adjacent spaces to achieve cross ventilation (i.e. between the open plan and perimeters offices of the GERC), result in a reduction of the inlet or outlet opening areas, whenever the perimeter offices remain closed due to privacy or acoustic reasons. Provision of dedicated bulkheads or additional opening to guarantee the delivery of cross ventilation is desirable to avoid failure of the strategy.

Building operation

The other common aspect which has emerged from the review of the case studies is the role that building management, and the presence of an effective maintenance team, can make on site. The presence of a dedicated management and maintenance team 24/7 in the building is clearly very beneficial in any large building. However, it becomes essential in buildings where the introduction of innovative ventilation and cooling strategies can require additional care and attention to prevent and troubleshoot any issues related to operation and control. A positive example of this practice is given by the Torrent Research Centre where to this date the downdraught evaporative cooling system works well and potential issues with the spitting and dripping of the brass nozzles are prevented by routinely cleaning the nozzles and avoiding blockage. Similar issues have been experienced elsewhere but this led to the abandonment of the strategy in some cases (where the maintenance budget and commitment was not provided) or to expensive replacement practices in others.

Although for complex and larger buildings it is certainly important to have a dedicated and trained management and maintenance team, recent advances in technology and smaller building applications have demonstrated that innovative evaporative cooling systems can be easily and effectively integrated in housing provided that the right opening areas and airflow paths are identified. One such example is that of the Nottingham HOUSE at the Solar Decathlon in Madrid (see Chapter 4).

The occupants' perception of comfort in the reviewed buildings, whilst being a subjective factor, does represent an essential criterion for their evaluation of a building. Referring back to *Leaman & Bordass (2006)*, perceived comfort and productivity in buildings are strongly affected by five *'killer variables'*, which are: 1. Personal control; 2. Responsiveness to need; 3. Ventilation type; 4. Workgroups; and 5. Design intent. Through the case studies reviewed in this section it was observed similar correlations between occupants' satisfaction and the opportunity for personal control of the surrounding environment (ranging from the possibility to open windows to being encouraged to activate the passive evaporative cooling system during hot days). This was observed in smaller buildings where smaller workgroups would be identified and thermal zoning adjusted accordingly like in the early performance of the Global Ecology Research Centre. Well-informed occupants and knowledge of the design intent also plays a role in the positive perception of the building where scientists sensitive to climate change themes, working at the GERC expressed the highest satisfaction out of the six buildings sampled. Rapid responsiveness to need of the occupants on the part of the building management and maintenance team were also observed at the Torrent Research Centre, Zion National Park Visitor Centre and Global Ecology research centre where the overall comfort was rated satisfactory and where coincidentally an on-site building manager was present. For the *'ventilation type'* criterion, it was also observed that buildings with shallow plan and simpler ventilation strategies performed better and yielded greater occupant satisfaction.

In conclusion, the review of the case studies has shown that: a) various forms of natural cooling are viable for achieving indoor comfort during hot periods provided the type of cooling is selected according to the highest climatic applicability; b) better performance is achieved when the integration of the cooling strategy into the building is carefully designed from the onset and is part of an holistic approach to design; c) the current and future limitations posed by climate change, variations in occupancy and microclimatic barriers (noise and pollution) can often compromise the long term application of these natural cooling strategies. However, diverting to mechanical cooling is not a solution as it will only exacerbate the need for additional cooling and contribute detrimentally to the urban heat island effect, creating a vicious circle and slowing down the process of mitigation and adaptation to climate change. Therefore understanding the nature of these barriers and improving building performance through strategic design and optimised operation is the way forward towards built environments which are healthier, more sustainable and climatically resilient.

REFERENCES

- Bordass, W. (2011). *'Saving Money, Saving Energy, Saving Carbon'*. Edge-CIBSE President's Debate, Arup Bristol, 8 September 2011.
- Leaman, A. & Bordass, B. (2006). *'Productivity in Buildings: the 'Killer' Variables'*. In 'Creating the productive workplace' Ed. Derek Clemens-Crome. Taylor & Francis.
- Schiano-Phan, R., Weber, F., Santamouris, M. (2015). *'The Mitigative Potential of Urban Environments and Their Microclimates'*. Buildings 2015, 5: 783–801.

IMAGE CREDITS

PREFACE

Fig. 01. Author: Charles-Dominique-Joseph Eisen (1720–1778). (CC)Creative Commons 1.0 Universal (CC0 1.0) - Public Domain Dedication.

PART 1

CHAPTER 1

Fig. 01. Author: Charles K. Wilkinson ca. (1928–1930). (CC) Creative Commons 1.0 Universal (CC0 1.0) - Public Domain Dedication.
Fig. 02. ©Brian Ford.
Fig. 03. From: Fathy, H. (1986). 'Natural Energy and Vernacular Architecture. Principles and Examples with Reference to Hot Arid Climates'. ©University of Chicago Press.
Fig. 04. After: Beazley, E. & Harverson, M. (1986). 'Living with the Desert: Working Buildings of the Iranian Plateau'. ©Mirentxu Ulloa.
Fig. 05. After: Beazley, E. & Harverson, M. (1986). 'Living with the Desert: Working Buildings of the Iranian Plateau'. ©Mirentxu Ulloa.
Fig. 06. After: Cain, A., Afshar, F., Norton, J. & Daraie, M. (1976). 'Traditional Cooling Systems in the Third World'. ©Juan A. Vallejo.
Fig. 07. ©Brian Ford.
Fig. 08. ©Brian Ford.
Fig. 09. After: Ernest, R. & Ford, B. (2012). 'The Role of Multiple Courtyards in the Promotion of Convective Cooling'. ©Juan A. Vallejo.
Fig. 10. After: Ernest, R. & Ford, B. (2012). 'The Role of Multiple Courtyards in the Promotion of Convective Cooling'. ©Juan A. Vallejo.
Fig. 11. ©Brian Ford.
Fig. 12. ©Brian Ford.
Fig. 13. After: Ford, B. & Hewitt, M. (1996) 'Cooling without Air-Conditioning – Lessons from India'. ©Brian Ford.
Fig. 14. ©Brian Ford.
Fig. 15. ©Brian Ford.
Fig. 16. ©Brian Ford.
Fig. 17. From: Lau, B., Ford, B. & Hongru, Z. (2014). 'Chapter 6: Traditional Courtyard Housing in China', in 'Lessons from Vernacular Architecture'. ©Brian Ford.
Fig. 18. From: Lau, B., Ford, B. & Hongru, Z. (2014). 'Chapter 6: Traditional Courtyard Housing in China', in 'Lessons from Vernacular Architecture'. ©Brian Ford.
Fig. 19. ©Brian Ford.
Fig. 20. ©Brian Ford.
Fig. 21. ©Dean Hawkes.
Fig. 22. ©Juan A. Vallejo.
Fig. 23. ©Nader V. Chalfoun.
Fig. 24. ©Bill Cunningham.
Fig. 25. ©Bill Cunningham.
Fig. 26. ©Brian Ford.
Fig. 27. ©Brian Ford.
Fig. 28. ©Brian Ford.
Fig. 29. ©Brian Ford.
Fig. 30. ©Mario Cucinella Architects.
Fig. 31. ©Mario Cucinella Architects.
Fig. 32. ©Peter Cook.
Fig. 33. ©Peake Short.
Fig. 34. ©Peake Short.
Fig. 35. ©Peter Cook.
Fig. 36. ©Short Ford Architects.
Fig. 37. ©Short Ford Architects.

CHAPTER 2

Fig. 38. ©NASA.
Fig. 39. ©Enrique Browne.
Fig. 40. ©Brian Ford.
Fig. 41. ©Mario Cucinella Architects.
Fig. 42. ©Mario Cucinella Architects.
Fig. 43. ©Mario Cucinella Architects.
Fig. 44. ©Juan A. Vallejo.
Fig. 45. ©Rosa Schiano-Phan.
Fig. 46. ©Edward Ng.
Fig. 47. ©Estudio Carme Pinós.
Fig. 48. ©Estudio Carme Pinós.
Fig. 49. ©Bennets Associates.
Fig. 50. ©Peter Cook.
Fig. 51. ©Peter Cook.
Fig. 52. ©Gianni Berengo Gardin.
Fig. 53. ©Gianni Berengo Gardin.
Fig. 54. ©Fondazione Renzo Piano.
Fig. 55. ©Atelier Ten.
Fig. 56. ©Atelier Ten.
Fig. 57. ©Atelier Ten.
Fig. 58. ©Atelier Ten.
Fig. 59. ©Nina Maritz Architects.
Fig. 60. ©Nina Maritz Architects.
Fig. 61. ©Nina Maritz Architects.
Fig. 62. ©Brad Feinknopf.
Fig. 63. ©Architects Pringle Richards Sharratt.
Fig. 64. ©Architects Pringle Richards Sharratt.
Fig. 65. ©Brian Ford.

Fig. 66. ©Brian Ford.
Fig. 67. ©Brian Ford.
Fig. 68. ©Brian Ford.
Fig. 69. ©Brian Ford Associates.
Fig. 70. ©Karan Grover & Associates.
Fig. 71. ©Brian Ford Associates.
Fig. 72. ©Rosa Schiano-Phan.
Fig. 73. ©StudioKahn.
Fig. 74. ©StudioKahn.
Fig. 75. ©Pedro Pegenaute.
Fig. 76. ©Pedro Pegenaute.
Fig. 77. ©Wolfgang Motzafi-Haller.
Fig. 78. ©Evyatar Erell.
Fig. 79. ©David Pearlmutter.
Fig. 80. ©DesignInc.
Fig. 81. ©Scott Francis.
Fig. 82. ©Brian Ford.
Fig. 83. ©Brian Ford Associates.

CHAPTER 3

Fig. 84. ©Juan A. Vallejo.
Fig. 85. ©Juan A. Vallejo.
Fig. 86. ©Juan A. Vallejo.
Fig. 87. ©Juan A. Vallejo.
Fig. 88. ©Juan A. Vallejo.
Fig. 89. ©Juan A. Vallejo.
Fig. 90. ©Juan A. Vallejo.
Fig. 91. From: Salmerón, J. M., Sánchez, F. J., Sánchez, J., Álvarez, S., Molina, J. L. & Salmerón, R. (2012). 'Climatic applicability of downdraught cooling in Europe'. ©Jose Manuel Salmerón.
Fig. 92. From: Salmerón, J. M., Sánchez, F. J., Sánchez, J., Álvarez, S., Molina, J. L. & Salmerón, R. (2012). 'Climatic applicability of downdraught cooling in Europe'. ©Jose Manuel Salmerón.
Fig. 93. From: Xuan, H. & Ford, B. (2012). 'Climatic applicability of downdraught cooling in China'. ©Huang Xuan.
Fig. 94. From: Xuan, H. & Ford, B. (2012). 'Climatic applicability of downdraught cooling in China'. ©Huang Xuan.
Fig. 95. ©Juan A. Vallejo & Pablo Aparicio.
Fig. 96. ©Juan A. Vallejo & Pablo Aparicio.
Fig. 97. ©Juan A. Vallejo & Pablo Aparicio.
Fig. 98. ©Juan A. Vallejo & Pablo Aparicio.

CHAPTER 4

Fig. 99. ©ARUP.
Fig. 100. ©ARUP.
Fig. 101. ©Foster & Partners.
Fig. 102. ©Foster & Partners.
Fig. 103. ©Foster & Partners.
Fig. 104. ©Brian Ford Associates.
Fig. 105. ©Brian Ford & Juan A. Vallejo.
Fig. 106. ©Brian Ford & Juan A. Vallejo.
Fig. 107. ©Brian Ford & Juan A. Vallejo.
Fig. 108. ©Short Ford Architects.
Fig. 109. ©Short Ford Architects.
Fig. 110. ©Short Ford Architects.
Fig. 111. ©Annette Kisling.
Fig. 112. ©Annette Kisling.
Fig. 113. ©Sauerbruch-Hutton.
Fig. 114. ©Short Ford Architects.
Fig. 115. ©Mario Cucinella Architects.
Fig. 116. ©Short Ford Architects.
Fig. 117. ©Edward Cullinan Architects.
Fig. 118. ©Edward Cullinan Architects.
Fig. 119. ©Edward Cullinan Architects.
Fig. 120. ©Bennetts Associates.
Fig. 121. ©Bennetts Associates.
Fig. 122. ©Bennetts Associates.
Fig. 123. ©Bennetts Associates.
Fig. 124. ©Marsh:Grochowski.
Fig. 125. ©Marsh:Grochowski.
Fig. 126. ©Brian Ford & Juan A. Vallejo.
Fig. 127. ©Mingwei Sun.
Fig. 128. ©Tim Griffith.
Fig. 129. ©Tim Griffith.
Fig. 130. ©Rosa Schiano-Phan.
Fig. 131. ©Rosa Schiano-Phan.
Fig. 132. ©Cullinan Studio.
Fig. 133. ©Cullinan Studio.
Fig. 134. ©Cullinan Studio.
Fig. 135. ©Brian Ford Associates.
Fig. 136. ©Brian Ford Associates.
Fig. 137. ©University of Nottingham.
Fig. 138. ©University of Nottingham.
Fig. 139. ©University of Nottingham.
Fig. 140. ©University of Nottingham.
Fig. 141. ©Mario Cucinella Architects.
Fig. 142. ©Mario Cucinella Architects.
Fig. 143. ©Mario Cucinella Architects.
Fig. 144. ©AICIA.

CHAPTER 5

Fig. 145. ©Juan A. Vallejo.
Fig. 146. ©Juan A. Vallejo.
Fig. 147. ©Juan A. Vallejo.
Fig. 148. ©Juan A. Vallejo.
Fig. 149. ©Juan A. Vallejo.
Fig. 150. ©Juan A. Vallejo.
Fig. 151. ©Brian Ford.
Fig. 152. ©Rosa Schiano-Phan.
Fig. 153. ©Kwok & Grondzik.
Fig. 154. ©Kwok & Grondzik.
Fig. 155. ©Rosa Schiano-Phan.
Fig. 156. ©Brian Ford Associates & Mario Cucinella Architects.
Fig. 157. ©Rosa Schiano-Phan.
Fig. 158. ©Juan A. Vallejo.
Fig. 159. ©Juan A. Vallejo.
Fig. 160. ©Brian Ford.
Fig. 161. ©Juan A. Vallejo.
Fig. 162. ©Juan A. Vallejo.

CHAPTER 6

Fig. 163.   ©Architecture Project.
Fig. 164.   ©Mario Cucinella Architects.
Fig. 165.   ©Mario Cucinella Architects.
Fig. 166.   ©Mario Cucinella Architects.
Fig. 167.   ©Mario Cucinella Architects.
Fig. 168.   ©Juan A. Vallejo.
Fig. 169.   Author: Bernard Gagnon. (CC) Creative Commons Attribution-ShareAlike 3.0 Unported (CC BY-SA 3.0).
Fig. 170.   ©Juan A. Vallejo.
Fig. 171.   ©Juan A. Vallejo.
Fig. 172.   ©Juan A. Vallejo.
Fig. 173.   ©Juan A. Vallejo.
Fig. 174.   ©Juan A. Vallejo.
Fig. 175.   ©Juan A. Vallejo.
Fig. 176.   ©Juan A. Vallejo.
Fig. 177.   ©Juan A. Vallejo.
Fig. 178.   ©Juan A. Vallejo.
Fig. 179.   ©Mario Cucinella Architects.
Fig. 180.   ©Mario Cucinella Architects.
Fig. 181.   ©Natural Cooling Ltd.
Fig. 182.   ©Natural Cooling Ltd.
Fig. 183.   ©Natural Cooling Ltd.
Fig. 184.   ©Natural Cooling Ltd.
Fig. 185.   ©Rosa Schiano-Phan & Juan A. Vallejo.
Fig. 186.   ©Juan A. Vallejo.
Fig. 187.   ©Geoff Whittle & Brian Ford.

CHAPTER 7

Fig. 188.   ©Brian Ford.
Fig. 189.   ©Brian Ford.
Fig. 190.   ©naco.
Fig. 191.   ©Juan A. Vallejo.
Fig. 192.   ©WindowMaster.
Fig. 193.   ©DesignInc.
Fig. 194.   ©Short Ford Architects.
Fig. 195.   ©Nina Maritz Architects.
Fig. 196.   ©Brian Ford.
Fig. 197.   ©Brian Ford.
Fig. 198.   ©Ingeniatrics/Frialia.
Fig. 199.   ©Brian Ford.
Fig. 200.   ©University of Nottingham.

PART 2

INTRODUCTION

Fig. 01.   ©Natural Cooling Ltd.

CASE STUDY 1

Fig. 02.   ©Brian Ford.
Fig. 03.   ©Juan A. Vallejo.
Fig. 04.   ©Abhikram Architects.
Fig. 05.   ©Brian Ford Associates.
Fig. 06.   ©Brian Ford Associates.
Fig. 07.   ©Brian Ford Associates.
Fig. 08.   ©Brian Ford.
Fig. 09.   ©Brian Ford.
Fig. 10.   ©Usable Buildings Trust.
Fig. 11.   ©Usable Buildings Trust.
Fig. 12.   ©Brian Ford.
Fig. 13.   ©Brian Ford.

CASE STUDY 2

Fig. 14.   ©Daniele Domenicali.
Fig. 15.   ©Juan A. Vallejo.
Fig. 16.   ©Mario Cucinella Architects.
Fig. 17.   ©Mario Cucinella Architects.
Fig. 18.   ©Mario Cucinella Architects.
Fig. 19.   ©Mario Cucinella Architects.
Fig. 20.   ©Mario Cucinella Architects.
Fig. 21.   ©Mario Cucinella Architects.
Fig. 22.   ©Daniele Domenicali.
Fig. 23.   ©Daniele Domenicali.
Fig. 24.   ©Mario Cucinella Architects.
Fig. 25.   ©Mario Cucinella Architects & Juan A. Vallejo.
Fig. 26.   ©Huang Xuan & Juan A. Vallejo.
Fig. 27.   ©Huang Xuan.
Fig. 28.   ©Huang Xuan.

CASE STUDY 3

Fig. 29.   ©Scott Frances.
Fig. 30.   ©Juan A. Vallejo.
Fig. 31.   ©Richard Meier Architects.
Fig. 32.   ©Scott Frances.
Fig. 33.   ©Rosa Schiano-Phan.
Fig. 34.   © Eric E. Johnson.
Fig. 35.   ©Daniel Griffin.
Fig. 36.   ©Daniel Griffin.
Fig. 37.   ©Rosa Schiano-Phan.
Fig. 38.   ©Rosa Schiano-Phan.
Fig. 39.   ©Usable Buildings Trust.

CASE STUDY 4

Fig. 40.   ©Architecture Project.
Fig. 41.   ©Juan A. Vallejo.
Fig. 42.   ©Architecture Project.
Fig. 43.   ©Brian Ford Associates.
Fig. 44.   ©Architecture Project.
Fig. 45.   ©Geoff Whittle & Brian Ford.
Fig. 46.   ©Brian Ford Associates.
Fig. 47.   ©Brian Ford Associates.
Fig. 48.   ©Usable Buildings Trust.

CASE STUDY 5

Fig. 49.   ©Rosa Schiano-Phan.
Fig. 50.   ©Juan A. Vallejo.
Fig. 51.   ©Rosa Schiano-Phan.
Fig. 52.   ©Rosa Schiano-Phan.

Fig. 53. ©Daniel Griffin.
Fig. 54. ©Daniel Griffin.
Fig. 55. ©Daniel Griffin.
Fig. 56. ©Rosa Schiano-Phan.
Fig. 57. ©Daniel Griffin.
Fig. 58. ©Rosa Schiano-Phan.
Fig. 59. ©Rosa Schiano-Phan.
Fig. 60. ©Usable Buildings Trust.
Fig. 61. ©Rosa Schiano-Phan.
Fig. 62. ©Rosa Schiano-Phan.

CASE STUDY 6

Fig. 63. ©Rosa Schiano-Phan.
Fig. 64. ©Juan A. Vallejo.
Fig. 65. © Microsoft® Bing™ Maps. Microsoft product screen shot reprinted with permission from Microsoft Corporation.
Fig. 66. ©Rosa Schiano-Phan.
Fig. 67. ©Rosa Schiano-Phan.
Fig. 68. ©Rosa Schiano-Phan.
Fig. 69. ©Daniel Griffin.
Fig. 70. ©Daniel Griffin.
Fig. 71. ©Daniel Griffin.
Fig. 72. ©Rosa Schiano-Phan.
Fig. 73. ©Rosa Schiano-Phan.
Fig. 74. ©Daniel Griffin.
Fig. 75. ©Daniel Griffin.
Fig. 76. ©Ari Kornfeld & Joe Berry.
Fig. 77. ©Daniel Griffin.
Fig. 78. ©Rosa Schiano-Phan.
Fig. 79. ©Rosa Schiano-Phan.
Fig. 80. ©Usable Buildings Trust.

INDEX

A

Abhikram Architects 45, 168, 180
Academy of Sciences, San Francisco, USA 77–113
acoustic attenuation 100, 165
acoustic buffering 29, 42, 86–92
actuator 77, 100, 160–166, 170, 200, 205, 229–230, 257, 265
Adalaj step well, Gujarat, India 8–9
air
    air-conditioning 12–13, 18, 26, 93, 103, 140, 182, 215, 220, 223, 264, 266, 269
    airflow strategies 18, 40, 43–47, 124, 146
    air stratification 151, 154, 157
    air tightness 100
    fresh air 6, 21, 28, 32, 59, 90, 135, 147, 149, 169, 196, 200, 234, 242, 257, 259, 265, 271
albedo 117
Arup 77–78, 114, 212, 217, 221, 223, 273
ASHRAE 57, 70, 74, 120, 136, 139, 149–150, 158–159, 172, 174
Atelier Ten 36–37
atrium 37, 41–44, 50–53, 82, 98–100, 133, 147, 208–209, 214–222, 226–227, 230–236
Avenue of Europe, Seville, Spain 17, 22, 51, 55

B

baffles 43, 52, 85–86, 167–169
Bawa, Geoffrey 25, 102–103, 114
Bennetts Associates 34, 42, 79, 98–99, 114
Bennetts, Rab 79
blinds 91, 97, 119, 163–164, 170
BMS 91, 171, 200, 206, 209, 215, 218, 222, 229, 245, 261, 264
Bordass, Bill 77, 94, 270
breezeways 106–107
Brian Ford Associates 44, 104
buffer space 46, 196
building
    laboratory building 20, 182, 185, 252
    office building 17, 33, 35, 42, 49–52, 102, 111, 131, 155, 187, 190, 224, 244, 249, 259
buoyancy 16–18, 40, 43, 54, 96, 100–101, 110, 121–122, 133–134, 147–153, 168–169, 198, 204, 228–229, 235

C

Casa de Pilatos, Seville, Spain 6–7
Casa di Bianco, Cremona, Italy 92–93
cellulose pads 16, 47, 49, 129, 244–245, 250–251
Centre for Mathematical Sciences, Cambridge, UK 96
CFD 46, 88, 101, 106, 112, 138, 154–159, 201, 217, 230, 235
CH2 Office Building, Melbourne, Australia 50–51, 167
chilled slabs 266
CIBSE 31, 34, 55, 57, 114, 119–120, 136, 146–151, 158, 162, 166, 174, 273

CO_2 17, 27, 166, 208, 236, 249, 263
colour 34, 108, 170, 232
comfort
    thermal comfort 7, 12, 16, 25, 32, 44, 58–62, 66, 102–105, 109–110, 130, 139, 146–147, 154–155, 178, 184, 196, 201, 207–209, 219, 222, 233, 246, 262
Consorcio Building, Santiago, Chile 27
controls 44, 77, 91, 97, 150, 160–161, 165–166, 173, 221, 231, 234–235, 244, 264
cooling
    active downdraught cooling 47, 53, 63–64, 133, 176
    convective cooling 7–9, 18, 22, 39–43, 54, 57, 60–61, 71–75, 84–89, 98, 103, 110, 112, 119–120, 124, 150, 156, 166, 168, 180, 183, 205, 224, 229, 235–236, 242, 256, 264, 268, 271
    cooling degree hours 60, 66, 70
    cooling load 27, 81–82, 107–108, 117–119, 124, 129–130, 143, 172, 186, 200–201, 205–206
    evaporative cooling 4–9, 14–17, 38–39, 44–54, 57–74, 108–115, 124–131, 147, 154–159, 167, 171–172, 180–192, 195, 213, 216–217, 222, 224–237, 239–250, 252–262, 268–272
    ground cooling 36–37
    mechanical cooling 21, 36–37, 54, 206, 237, 264, 271, 273
    passive cooling 15, 28, 56–61, 65–66, 74, 77, 81–84, 108, 117, 119, 141, 150, 158, 174, 192, 206, 228, 244–245, 268–269
    pre-cooling 34–37, 58, 98, 195–197, 242–245
courtyards 5–7, 22, 27, 82–83
Crystal Palace, London, UK 14, 22
CSET Building, Ningbo, China 54, 194–209, 269–270

D

dampers 86, 162–164, 229–235
daylight 18–21, 91–93, 107, 112, 118, 150, 182, 207, 214, 247, 254
design tools 30, 141, 144
diffusers 52, 217, 221
dynamic thermal simulation 152–154, 205

E

Earth Centre, Doncaster, UK 36–37
Edward Cullinan Architects 91, 96–97, 104
electrostatic precipitators 165
emissivity 34
empirical tradition 15, 80, 102
energy consumption 27, 37, 95, 100, 107, 112, 177, 187–192, 201, 206, 216, 220, 235, 244, 248–250, 256, 263–265, 271
energy savings 54, 108, 155, 190–192, 199, 205, 208, 216, 235–236, 249, 256, 263, 269
Enrique Browne Arquitectos 27
Estudio Carme Pinos 33
Evapcool 130–131
evaporation 4–7, 16, 32, 32–54, 60, 63, 119, 125–133, 150, 154, 157, 236, 256, 259

278   **INDEX**

evapotranspiration 7, 27
exhaust 32, 40–41, 48–53, 85–90, 96, 100–102, 113, 123–124, 166–168, 182, 185, 197–199, 203–204, 208–209, 218
Experimental Building, Tucson, Arizona 15–18, 122

F

façade
    double skin façade 43, 89–92
    ventilated façade 43, 86
fan 16, 32, 50, 54, 99–100, 108, 133, 184–189, 199, 206, 226–230, 234–236, 242–250, 257, 260, 264–265, 269
Fathy, Hassan 5, 49
Federal Courthouse, Phoenix, Arizona, USA 212–223
Federation Square, Melbourne, Australia 37
Fletcher Building, Leicester, UK 92
fluid dynamics 124, 156
Foster & Partners 82–83, 114

G

glazing 13, 19, 21, 27, 118, 143, 185, 196, 214–215, 218, 256
Global Ecology Research Centre, Palo Alto, California 241, 252–266, 270, 272
GSW Headquarters, Berlin, Germany 43, 89–91, 114
Guthrie, Alistair 77
G-value 118

H

Habitat R&D Centre, Katatura, Namibia 38–39, 168
heat
    heat gains 14, 18, 21, 27, 30–34, 40, 53, 58–62, 89, 91, 94, 108, 117–120, 125–126, 130, 136, 140, 143, 149–153, 172, 183–186, 191, 198, 202, 217–218, 228, 235, 254
    heat sink 32, 34, 54, 108, 140, 197, 200, 255, 260, 270–271
high-performance envelope 108, 256
hot dry climates 32, 108, 171
humidity
    relative humidity 6, 9, 39, 44, 57, 68, 104, 110–113, 126, 133, 185, 200–205, 221, 229, 231, 240
HVAC 217
hybrid strategy 44, 269

I

ice house 4
integrated design 74, 77, 79, 94, 107, 112, 158, 192, 224

J

Joint Research Centre, ISPRA, Italy 141–142

K

Kere, Francis 25
Kuma, Kengo 25
Kuwait School, Gaza, Palestine 151–153

L

latent heat of vaporisation 39
legionella 172
lightwell 11, 52–53, 82–85, 196–205, 209
long wave radiation 3–4
louvres 85, 164, 182, 204
Lowara Office, Vicenza, Italy 35, 259
low carbon 2, 77, 91, 93, 151, 237, 255

M

maintenance 54, 76, 92–95, 112, 143, 161–165, 171–174, 178, 182, 190, 192, 206, 209, 221, 227, 235–236, 247, 251, 265–266, 269–272
maps 56, 60, 65–66, 68, 70–71, 74, 247
Mario Cucinella Architects 28–29, 92–93, 111–114, 141–142, 151, 194, 275
Marsh Growchowski Architects 100
Martin, Sir Leslie 80, 114
mashrabiya 143
misting systems 47, 52, 125
mixed-mode 91, 95–96, 114
monsoon 102–103, 181, 184, 189
monsoon window 103

N

neutral plane 122–123
night sky 3–4, 32, 34–35, 54, 256, 258–259
Nina Maritz Architects 38–39, 168
No1 Moulmein Rise, Singapore 103
noise 21, 28–29, 42, 61, 88–89, 93, 98–102, 107, 166, 178, 232, 239, 247–250, 263–265, 271–273
Nottingham HOUSE, Solar Decathlon Europe 109–110, 117–118, 123, 170, 173, 272
nozzles 16, 17, 47, 51–52, 85–86, 110, 125–128, 161–162, 167–168, 172–174, 182–186, 192, 217–218, 221–222, 227–229, 232–236, 261, 264, 269, 272

O

occupant satisfaction 77, 81, 94, 97, 99, 190, 192, 269–270, 273
Office Building, Catania, Italy 84, 111–113
Optivent 145–149
orientation 25, 30, 43, 104–105, 117, 254–256
outlet 36–37, 40, 51–52, 92, 101–102, 113, 122–124, 167, 199–204, 209, 244, 261, 264, 266, 271
overheating 14, 30, 43, 57, 61, 71–74, 89, 99, 117, 140, 146, 152, 182, 202, 209, 217, 231, 244, 247–250, 254–257, 262, 265–266, 271

P

parasol 106, 142, 152
Passivhaus 58, 200
PDEC 17, 53, 83, 84, 87, 112, 114, 125, 154, 158, 167, 176, 182, 185–192, 205, 210, 212, 215–222, 224, 227, 229, 232, 236–237, 252, 254, 261, 264–267
Peake Short Architects 19
phase change 39, 156
plenum 46, 100, 102
pollution 28–31, 34, 88–89, 93, 107, 165–166, 181, 239, 265, 271, 273
porous ceramic 47–49, 128–133, 154–156
porous media 47, 124–125, 128, 156
post occupancy evaluation 81, 176–177, 200, 202, 207, 233, 244–246, 262
Potterrow Development, Edinburgh, Scotland 80, 98–100
PowerGen/EON HQ, Coventry, UK 34, 79
pressure gradient 121
psychrometric chart 39, 57–59, 110, 126, 133

Q

Queens Building, Leicester, UK 20–22, 94–95, 114, 167

INDEX   279

R

rainwater 49, 216
Rai Praveen Mahal, Orchha, India 8–9
Raphael Vinoly Architects 41
refurbishment 29, 82–94, 222, 226
Renzo Piano Building Workshop 35, 55, 77–78
Richard Meier Architects 52, 212–223
roof ponds 34
rules of thumb 81

S

salsabil 5, 7
Sauerbruch & Hutton 43
screen
    glazed screen 29
    vegetative screen 27
SFC Brewery, Valetta, Malta 19, 34
shading device 14, 30, 91, 141–145, 215
shaft 8, 43–45, 48, 54, 84, 124, 167–168, 196–197, 202, 205, 228
Sheldonian Theatre, Oxford, UK 13
Short Ford Architects 20, 88–89, 92, 95, 167, 180
shower towers 49–50
shutters 4, 29, 92, 102, 140–141, 228
Singapore Management University, Singapore 103–107
sky view factor 26, 34
smoke test 202–205, 235
solar gains 9, 19, 110, 117–119, 144–145, 152, 197, 215, 254
solar geometry 30, 117, 141–142
solar radiation 19, 30, 57–58, 109, 117–118, 142–145, 214, 248, 250
SPACE Performing Arts Centre, Nottingham, UK 100–101
stack height 123, 126, 146
stack ventilation 4, 14, 19, 33, 40–43, 83–86, 91–92, 112, 120–126, 146–147, 183, 185, 202, 204, 206, 218, 224, 242, 244, 251, 256
State Mortgage Bank, Colombo, Sri Lanka 102–103
steady-state 81, 116, 118, 131, 138, 140
step well 8–9
Stock Exchange, Valetta, Malta 44, 53–54, 84, 108, 113, 134, 141, 157, 161, 177, 224–237, 268, 270
street pattern 25, 31, 104–105
sunpath diagram 144
survey 97, 102, 176–178, 187–188, 192, 202, 207, 219–222, 232–233, 245–248, 261–264, 269

T

taktabus 5
temperature
    air temperature 6–9, 16, 29, 32–33, 36–39, 43, 46, 57–62, 71–72, 108–110, 119–133, 139–140, 156, 183, 196, 204
    operative temperature 139, 149–150
    soil temperature 58
    surface temperature 6, 49, 117, 131
    wet bulb temperature 16, 38–39, 53, 62, 64, 125, 129–132, 156, 181, 195, 213, 225, 240
Temple Way House, Bristol, UK 88–89
thermal
    thermal capacitance 6, 9, 18–19, 29, 32, 58, 61, 119, 140, 197, 258, 265
    thermal force 40, 121, 157
    thermal modelling 150, 154–155, 158

thermostats 247, 250
topography 25, 31
Torre Cube, Guadalajara, Mexico 33
Torrent Research Centre, Ahmedabad, India 45, 51, 84, 125, 136, 169, 172, 174, 177, 180, 180–193, 216, 221, 234, 268, 270–272
traffic noise 28, 88–89, 93, 98
transition space 44, 46, 220, 222
Transsolar 79, 114
trickle vents 135
tropics 102–103, 142–143

U

underground labyrinth 36–37
urban heat island 25–26, 213, 223, 273
urban morphology 25, 30, 34, 82–83
U-value 108, 110, 117, 152, 182, 243

V

vegetation 25–27, 242, 255
ventilation
    cross ventilation 40, 103, 169
    downdraught ventilation 38
    mechanical ventilation 91–100, 194, 247, 256, 259, 263–265, 271
    single sided ventilation 40–41, 147
    stack ventilation 4, 14, 19, 33, 40–43, 83–86, 91–92, 112, 120–126, 146–147, 183, 185, 202–206, 218, 224, 242, 244, 251, 256
    updraught ventilation 40, 43, 54, 85–87, 90, 110, 123, 125, 166–167
ventilators 14–15, 32, 134, 160–166, 169–171, 200, 216, 228–229, 235
volumetric heat capacity 60, 120

W

water
    water coils 43–44, 47, 53, 63, 85, 108, 133, 227, 229, 237
    water consumption 126, 131–133, 171–173, 216–222, 245, 258
    water treatment 172–173
weather data 39, 54, 56–60, 65, 74, 94
wet pads 38, 47, 129–130
wind
    wind baffles 169
    wind catcher 5, 50, 168, 203
    wind pressure 31, 51, 113, 148, 150, 156–157, 183
WOHA 103

Z

Zhang's House, Zhouzhuang, China 9–12, 22
Zion National Park Visitor Centre, Utah, USA 47–48, 128, 238–251, 270, 272
Zumthor, Peter 25